INNOVATIVE TEACHING AND LEARNING METHODS IN EDUCATIONAL SYSTEMS

PROCEEDINGS OF THE INTERNATIONAL CONFERENCE ON TEACHER EDUCATION AND PROFESSIONAL DEVELOPMENT (INCOTEPD 2018), OCTOBER 28, 2018, YOGYAKARTA, INDONESIA

Innovative Teaching and Learning Methods in Educational Systems

Editors

Endah Retnowati, Suprapto, Mohammad Adam Jerusalem, Kristian Sugiyarto & Wagiron

Universitas Negeri Yogyakarta, Indonesia

Routledge
Taylor & Francis Group

LONDON AND NEW YORK

Routledge is an imprint of the Taylor & Francis Group, an informa business

© 2020 Taylor & Francis Group, London, UK

Typeset by Integra Software Services Pvt. Ltd., Pondicherry, India

Publisher's Note
The publisher has gone to great lengths to ensure the quality of this reprint but points out that some imperfections in the original copies may be apparent.

Published by: CRC Press/Balkema
Schipholweg 107C, 2316XC Leiden, The Netherlands
e-mail: Pub.NL@taylorandfrancis.com
www.crcpress.com – www.taylorandfrancis.com

First issued in paperback 2021

ISBN 13: 978-1-03-224183-8 (pbk)
ISBN 13: 978-0-367-25792-7 (hbk)

DOI: https://doi.org/10.1201/9780429289897

Table of Contents

Preface   ix

Organizing Committee   xi

Scientific Committee   xiii

Education System

Integrating knowledge and skills-based curriculum in TT-TVET through a blended and
embedded model: An innovative approach at the faculty of technical and vocational
education, Universiti Tun Hussien Onn Malaysia   3
A.R.A. Razzaq, M.H.B. Amiruddin, M.A. Rohiat & N.B. Razali

The effectiveness of academic supervision by school principals   12
S. Darmawanti & H. Usman

The performance of state elementary school supervisors   18
R.S. Maulida & H. Usman

Integrated thematic learning in the 2013 curriculum: Implications for the self-confidence and
academic achievement of primary school students   24
U. Tisngati, C.A. Budiningsih & Sugiman

The effect of technology literacy and learning environment on student motivation in the
educational revolution 4.0   32
A. Saputri Sukirno, H. Kurniawan & H.D. Hermawan

Student perceptions of motivational strategies used by junior high school English teachers in
Kota Yogyakarta   37
G. Ambarini & Ashadi

Study on graduate competence in air-conditioning skills at senior high school with relation to
job market competencies   45
S. Haryadi

Elementary students' performance in mathematical reasoning   52
N. Andrijati, D. Mardapi & H. Retnawati

Evaluation model of the implementation of a quality management system of electrical skills
in vocational high school   59
I.G.B. Mahendra & G. Wiyono

Formal & Informal Education

Strengthening character education through the local wisdom: Indonesian folklore   69
T.A. Rini & P. Mahanani

Forming young citizen characters through youth organizations in Indonesia   77
Wellyana & Marzuki

Revitalization in vocational training centers for improving the quality of human resources   82
I.A. Manalu & R. Asnawi

Learning Models

Theatrical stage of technology and humans in relation to education   91
T. Öztürk

Improving students' critical thinking abilities in probability problems through problem-based learning   97
Rauzah & Kusnandi

Cultural map media as an innovation to overcome cognitive learning difficulties in Social studies at elementary school   104
L. Fatmawati, V.Y. Erviana, D. Hermawati, I. Maryani, M.N. Wangid & A. Mustadi

Learning innovations in citizenship education for strengthening digital and ecological citizenship   113
K.E.R. Marsudi & S. Sunarso

Needs analysis for an electronic module (e-module) in vocational schools   122
S. Oksa & S. Soenarto

Analysis of students' learning readiness in terms of their interest and motivation in achieving students' critical thinking skills   129
R. Putri & A. Ghufron

Study on a test scoring system for vocational secondary schools using Computerized Adaptive Testing (CAT)   135
F.P. Marsyaly & S. Hadi

Implementation of basic graphic design learning skills competence of multimedia in SMK Muhammadiyah Wonosari   141
L.F.A.N.F. Albana & Sujarwo

Professional Teacher

Developing video-based learning resources for music teachers in Singapore   151
A. Bautista, S.L. Chua, J. Wong & C. Tan

Teacher–student communication style and bullying behavior: Sociometry evaluation   159
I. Sholekhah, S. Indartono & D.W. Guntoro

Analysis of students' mistakes in solving algebra word problems using the Newman Procedure   168
R. Keumalasari & Turmudi

Lesson study as an alternative for teacher creativity development in reflecting and improving the quality of learning of the Indonesian language   174
R.W. Eriyanti

Lesson study to improve teacher creativity in solving problems of mathematics learning   179
Y.M. Cholily

Mapping the innovation of Professional Learning Communities (PLC) in primary schools: A review   184
P.S. Cholifah & H.I. Oktaviani

Implementation of snowball drilling learning model on discrete mathematics to improve
student's independence and learning outcomes   190
L. Novamizanti

What lecturers know about their role as an agent of learning: Levels of innovativeness in the
learning process   197
A. Ghufron

Development and quality analysis of a learning media electrical motor installation on the
Android platform, for vocational students   201
B.N. Setyanto & H. Jati

The readiness of vocational secondary schools on forming working characteristics for
industry 4.0   209
H. Mulyani & I.W. Djatmiko

The assessment model for competency certification tests   216
W. Ramadani & D.L.B. Taruno

Contribution of a teacher competency test to identifying teacher performance in vocational
high schools   223
U. Nursusanto & N. Yuniarti

Performance in innovative Teacher Professional Development (TPD) in Indonesia: Does
gender matter?   231
Wuryaningsih, M. Darwin, D.H. Susilastuti & A.C. Pierewan

Needs identification of learning media for people with disabilities   237
Rizalulhaq & R. Asnawi

Implementing gamification to improve students' financial skills in business and management
vocational schools   242
Sukirno, E.M. Sagoro, L.N. Hidayati, Purwanto & D.A.Y Wastari

Author Index   248

Preface of 3rd INCOPTEPD 2018

It is our great pleasure to present the proceedings of the 3rd International Conference on Teacher Education and Professional Development (InCoTEPD) that was held in Yogyakarta (Indonesia) on 20 October 2018. This volume of proceedings provides an opportunity for readers to take note of a selection of refereed papers that were presented during the 3rd InCoTEPD conference.

The 3rd InCoTEPD provides a platform to discuss current issues and challenges related to educational innovation consisting of four sub-themes: innovation in education system at the level early chilhood education, basic education, and higher education; innovation in formal education and non-formal education; innovation of learning models in the field of technology, science, social, culture, economic, sport and health; and innovation in teacher education and professional development.

Exploring the theme, "Educational Innovation: Current Trends", the committee has invited Dr. Ir. Paristiyanti Nurwardani, M.P. (Directorate General of Higher Education, Ministry of Research, Technology and Higher Education) as a keynote speaker, and also Prof. Madya Ts. Dr. Abdul Rasid Bin Abdul Razzaq (Universitas Tun Hussein Onn, Malaysia), A/Professor Sarah K. Howard (University of Wollongong, Australia), (Universitas Negeri Yogyakarta, Indonesia), Associate Professor Dr. Hayriye Tuğba Özturk (Ankara University - Turkey) as invited speakers. More than 90 papers have been submitted to this conference but only 65 of these have been accepted for the presentation after the peer review process and presented at parallel oral sessions related to those four sub-themes of educational innovation.

It has been a great privilege for the Educational Development and Quality Assurance Institution of Universitas Negeri Yogyakarta to serve as host of the 3rd InCoTEPD. With thanks to the contribution of speakers and participants that provided significant and intellectual inspiration to the educational innovation.

Dr. Sunaryo Soenarto, M.Pd..
Conference Chair of the 3rd InCoTEPD
Educational Development and Quality Assurance Institution
Universitas Negeri Yogyakarta

Organizing Committee of 3rd INCOPTEPD 2018

Patron

- Prof. Dr. Sutrisna Wibawa, M.Pd. (Rector of Universitas Negeri Yogyakarta)

Steering Committee

- Prof. Dr. Anik Ghufron, M.Pd. (Universitas Negeri Yogyakarta)
- Prof. Dr. Suwarna, M.Pd. (Universitas Negeri Yogyakarta)

Organizing Committee

Chairman	: Dr. Sunaryo Soenarto, M.Pd.
Secretary	: Dr. Nurhening Yuniarti, M.T.
Treasurer	: Arni Wahyu Budi Lestari, S.Sos.
Coordinator of Speakers and Proceeding	: Endah Retnowati, M.Ed., Ph.D.
Coordinator of Publication and Documentation	: Dr. Sri Andayani, M.Kom.
Coordinator of Parallel Session and Moderator	: Joko Priyana, Ph.D.
Coordinator of Registration	: Rifqi Nur Setyawan, S.Pd.T.
Coordinator of Program	: Dr. Christina Ismaniati
Member of Committee	: Rustam Asnawi, M.T., Ph.D.
	Dr. Samsuri, M.Ag.
	Dr. Kun Setyaning Astuti, M.Pd..
	Joko Sudomo, M.A. .
	Mutaqin, M.T., M.Pd. .
	Dra. Rr. Lis Permanasari, M.Si.
	Drs. Budi Sulistiya
	Agus Riyanto, S.Pd.
	Sudimin, S.Pd.

Scientific Committee of 3rd INCOPTEPD 2018

Scientific Committee

- Dr. Sunaryo Soenarto, M.Pd. (Universitas Negeri Yogyakarta)
- Dr. Marzuki, M.Ag. (Universitas Negeri Yogyakarta, Indonesia)
- Dr. Kasiyan (Universitas Negeri Yogyakarta, Indonesia)
- Joko Priyana, Ph.D. (Universitas Negeri Yogyakarta, Indonesia)
- Dr. Ade Gafar Abdullah (Universitas Pendidikan Indonesia)
- Endah Retnowati, M.Ed., Ph.D. (Universitas Negeri Yogyakarta, Indonesia)
- Ashadi, Ed.D. (Universitas Negeri Yogyakarta, Indonesia)

International Advisory Board

- Prof. Micha De Winter (Utrecht University)
- Prof. Lesley Harbon (Head of School, School of International Studies, Faculty of Arts and Social Sciences, University of Technology Sydney)
- Heidi Layne, Ph.D. (Department of Teacher Education, University of Helsinki, Finland)
- Prof. Dr. Ng Shun Wing (Hong Kong Institute of Education, Hong Kong)
- Prof. Derek Patton (Melbourne Graduate School of Education, Australia)
- Prof. Dorothea Wilhelmina Hancock, Ph.D. (Queensland University of Technology, Australia)
- Prof. Wiel Veugelers (Professor of Education at the University of Humanistic Studies in Utrecht, Member UNESCO Research Group on Global Citizenship Education)
- Prof. Suyanto, Ph.D. (Universitas Negeri Yogyakarta- Indonesia)
- Prof. Dr. Anik Ghufron, M.Pd. (Universitas Negeri Yogyakarta- Indonesia)
- Prof. Dr. Suwarna, M.Pd. (Universitas Negeri Yogyakarta- Indonesia)
- Prof. Sukirno, Ph.D. (Universitas Negeri Yogyakarta- Indonesia)
- Prof. A. K. Projosantosa, Ph.D. (Universitas Negeri Yogyakarta- Indonesia)
- Prof. Suwarsih Madya, Ph.D. (Universitas Negeri Yogyakarta- Indonesia)
- Prof. Darmiyati Zuchdi, Ed.D. (Universitas Negeri Yogyakarta- Indonesia)
- Prof. Dr. Sri Atun (Universitas Negeri Yogyakarta- Indonesia)
- Prof. Sugirin, Ph.D (Universitas Negeri Yogyakarta- Indonesia)
- Prof. Dr. Sunaryo Kartadinata (Universitas Pendidikan Indonesia, Bandung - Indonesia)
- Dr. Alfredo Bautista (National Institute of Education, Singapore)

Education System

Integrating knowledge and skills-based curriculum in TT-TVET through a blended and embedded model: An innovative approach at the faculty of technical and vocational education, Universiti Tun Hussien Onn Malaysia

A.R.A. Razzaq, M.H.B. Amiruddin, M.A. Rohiat & N.B. Razali
Faculty of Technical and Vocational Education, Universiti Tun Hussien Onn Malaysia, Parit Raja, Malaysia

ABSTRACT: Technical and Vocational Education and Training (TVET) plays a very important role in the engine of change to speed up a nation's economic rise and the need for high-quality human resources. The gaps between the skills needed by industry and the skills acquired by the graduates or trainees have always been in question. As such, the quality of TVET teacher and instructor is a critical aspect in closing this gap and minimizing the issue of employability. It is important for TVET teachers to equip themselves with knowledge aspects and skill criteria for realizing the national TVET agenda. The innovation in teaching and learning in TT-TVET should be competency-based or "hands-on" in order to produce a competent teacher and instructors, balance with knowledge-based and skill-based. An innovation approach has been implemented by the Faculty of Technology and Vocational Education (FTVE) at Universiti Tun Hussien Onn Malaysia (UTHM), using a blended and embedded model approach by mapping the academic curriculum accredited by the Malaysian Qualifications Agency (MQA) with Malaysian Skills Certificate (SKM), which is accredited by the Department of Skills Development, Ministry of Human Resource Malaysia. As a result, the approach has successfully produced significant numbers of "ready-made" TT-TVET graduates, each one of which has the academic qualification, the skills certification and also a Vocational Training Officer certificate that will qualify them to be teachers and instructors in the TVET institution.

1 INTRODUCTION

Malaysia has placed crucial emphasis on the importance of technical and vocational education and training (TVET) and its role in equipping young Malaysians for the fourth industrial revolution. In order to achieve Malaysia's vision 2020 of becoming a developed nation, the quality and skills of human resources is crucial to the success of economic transformation. The landscape of TVET in Malaysia has been transforming since it has been widely acknowledged in numerous countries and international union (Yazçayır & Yağcı, 2009) due to technical innovation and globalization (Wilson, 2001), achieving a higher income (ANTA, 2002) and thus decreasing poverty (ILO, 2012).

Due to the importance of TVET, the Malaysian government has established the Technical and Vocational Education and Training (TVET) Empowerment Committee to boost the level and standards of TVET in the country. Thus, as the government is concerned with TVET education, the Empowerment Committee has consulted with the Ministry of Education (MOE) and Department of Skills Division (DSD) on how TVET can help young people secure jobs and entrepreneurship opportunities.

Aligned with these aspirations, the increasing demand of TT-TVET which is not just cognitively good but a highly skilled based in respective fields, appear to be the priority of the whole nation, especially from industries and various public or private institutions. The transformation of TVET automatically leads to major changes in the Teachers Training - Technical and Vocational Education and Training (TT-TVET) program structure. The focus should be on a program structure that has an impact in producing teachers. The gaps between the skills needed by industry and the skills acquired by the graduates or trainees have always been in question among industrial experts and institutions. The innovations in teaching and learning in TT-TVET should have a greater competency-based or "hands-on" focus in order to produce competent teachers and instructors. Most of the countries in the world are facing challenges in this area, especially in producing highly competent and skilled teachers and instructors according to their fields.

Axmann, Rhoades, & Nordstrum (2015) noted that some of the global challenges faced by producing instructors in TT-TVET are as follows:

i. The lack of relevant structures is based on the direct involvement of the industry to recognize the potential of teachers and their career paths to encourage young people.
ii. Limitations of opportunities to carry out more specific skills training to enhance skills competencies in teachers and trainers.
iii. Consistent innovation in the delivery of current pedagogy and technology in teaching.
iv. Teachers and instructors in TT-TVET need to be more active in voice-training to quality training courses that can improve job fulfillment and satisfaction

The TT-TVET teacher training curriculum requires holistic materials in line with the competency-based curriculum characteristics, as suggested by Foyster (1990): i) the curriculum needs to be in line with the job market, ii) the theoretical aspects need to be directly integrated with the skills to achieve the required level of skills, iii) the students should be tutored in the necessary knowledge and skills through content mastery, and iv) learning should be based on individual capabilities.

1.1 Significance of the integration and implementation curriculum blended and embedded model

Previous researchers have shown how important it is to meet the requirements of the industry by having teachers and instructors that are professional experts, both in academia and industrial fields.

Spöttll (2009) stated that a teacher and instructor must like their students, must identify with the school, and must be able to work under psychological pressure. In addition, they may have the role of a social worker, psychologist, mediator, communicator, team worker, expert, and a "knowledge networker".

This is consistent as the teacher and the instructor motivates students to continue their education, and acquire the skills that make them job-ready for a high-technology globalized economy. Through TT-TVET, teachers and instructors have more avenues for continuing education and upgrading their skills. This opens up more opportunities for meaningful and good jobs, and helps to ensure lifelong employability as technology and jobs evolve among their students.

1.2 Challenges in TT-TVET

Challenges faced by TT-TVET teacher training in preparating for the future require a comprehensive curriculum preparation capability to strengthen the profession of TT-TVET teachers. As with many elements of higher education, best practices in pedagogy are researched and documented or shared in other ways between faculty members to the betterment of education. These best practices may include individual course elements, curriculum development, student and faculty interaction and effective presentation of the subject matter. Therefore, it is important that each of these elements be considered for every course, regardless of the delivery format used.

From the First Malaysia Plan to the latest Tenth Malaysia Plan, the TVET sector was divided into education and training. The term 'education' is used by the Ministry of Education, which is responsible for the polytechnics, vocational colleges, technical schools, and community colleges, while the term 'training' is used by ministries such as the Ministry of Human Resources, Ministry of Youth and Sports, Ministry of Rural and Regional Development, and the Ministry of Agriculture and Agro-Based Industry.

There was another accreditation body in addition to DSD in the Malaysian education system. DSD performs accreditation for the skills sector while the Malaysian Qualifications Agency (MQA), a division within Ministry of Education, performed accreditation for the vocational and technical sector as well as the academic sector. The existance of the two bodies with separate standards and process for accreditation resulted in multiple qualification systems.

Axmann et al. (2015) proposed an analytical framework for producing high-quality teacher training in TT-TVET. An effective, efficient and innovative teacher training system based on the country's agenda in the relevant transformation should incorporate four main elements: i) relevant program structure, ii) responsive teacher recruitment, iii) innovation; and iv) highlights teacher quality and communication skills. In the four main elements proposed there are 12 sub elements representing the main element in TT-TVET. Table1 shows the suggested elements.

In addition, in order to create a flexible system of teachers, there must be formal and informal experiential work experience in the industry and the development of training through professional bodies.

Figure 1 shows the appropriate training framework incorporating flexible elements in the TT-TVET teacher training phase.

Table 1.   Main element and sub-element in TT-TVET.

Main element	Sub-element
Relevant program structure	Phase on teacher training
	Industrial training
	Taking part in policy changes
Responsiveness of teacher	Entrepreneurship element
	Gender equality
	Flexible training
Innovation	Core skills
	Pedagogy innovation
	Technology application
Teacher's quality and communication	Knowledge sharing
	Rights as employees/teacher
	Social responsibilities

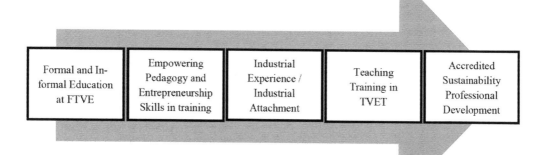

| Formal and Informal Education at FTVE | Empowering Pedagogy and Entrepreneurship Skills in training | Industrial Experience / Industrial Attachment | Teaching Training in TVET | Accredited Sustainability Professional Development |

Figure 1.   Framework of TT-TVET at FTVE, UTHM.

Overall, the unique fragmented TVET landscape, including multiple qualification systems with no central or single agency to coordinate the system, will make each provider act in silo thereby creating limited synchronization and harmonization of TVET sector. Although there was a platform to coordinate the delivery system through the National Skills Coordination Council (MPKK2) (JPK, 2012a), the diverse vision/purpose for offering TVET education will create a wide difference in the policy decision and training delivery.

2 IMPLEMENTATION

The blended and embedded model approach was implemented by merging the two main bodies accrediting the TT-TVET program – the Malaysian Qualifications Agency (MQA) Act 2007 and the National Skills Development Act 2006. This is to ensure the governance of quality assurance of the program structure itself as the combination of COPPA and KAPPK.

2.1 Implementation of TT-TVET and integration of skills at FTVE and UTHM (dual accreditation)

The Faculty of Technical and Vocational Education (FTVE) was launched on 23 June 2011 as a result of rebranding. Prior to that, FTVE was known as the Faculty of Technical Education (FPTeK) and was set up on May 1, 2004. In the early stages, the faculty only offers a program of undergraduate young bachelor's degree in Technical and Vocational Education, which specializes in producing teachers for technical secondary school, vocational secondary school and secondary school (for technical stream subjects). Since 2012, FTVE has developed a technical and vocational education program in line with the development of programs at SMT and SMV. Vocational education (ISMPV) programs designed to meet the needs of teachers at vocational colleges include seven fields of study: Electrical and Electronic, Welding and Fabrication, Building, Creative Multimedia, General Machines, Catering and Air Conditioning and Refrigeration.

FTVE offers an accredited academic program that does not only take into account academic requirements based on MQA qualifications but also the needs of the Malaysian Occupational Safety Standard (NOSS), especially in meeting the needs of stakeholders. Integration between these academic and skills needs through the MQA and NOSS leadership mapping process enables the ISMPV program to be awarded the Malaysian Skills Certificate to Level 3 by the Department of Skills Development (JPK). The implementation of this curriculum will allow each course offered in each field of study with the implementation of teaching and learning as well as the outcome-based education and skills assessment required by the National Occupational Skills Standard (NOSS).

The Malaysian Skills Certificate (SKM) implementation model at FTVE is based on a blended/embedded model that is mapping a number of courses in the Bachelor of Vocational Education (ISMPV) curriculum with NOSS requirements for each SKM. Through this model, the FTVE ISMPV students have the opportunity to gain a bachelor's degree in their respective fields after the completion of their studies with SKM, without the additional period of study. This makes the FTVE program unique as a selling point because it not only benefits graduates but also stakeholders such as vocational colleges, which is indeed a requirement for teaching staff in Vocational College (KV). These 'ready-made' graduates are sure to have the advantage of competing with them in the employment industry.

In general, the implementation of skills training at FTVE is based on the implementation of a single tier skill training program under the Malaysian skills training system, as set out in Section 28 of the National Skills Development Act 2006 [Act 652]. The standard for quality assurance for program accreditation at FPTV is based on the Skill Program Accreditation Practice Code (KAPPK) set by the Ministry of Human Resources.

Hence, in the context of implementing this model in FTVE, it involves nine major elements: curriculum, classroom implementation, equipment and facilities, lecturers and personnel, assignment of lecturers and personnel, assessment, credit score points, recognition, and quality control. Figure 2 shows the SKM implementation model in FTVE.

At FTVE, UTHM has 11 accredited programs approved by the Department of Skills Development (JPK), based on the Code of Conduct Quality Implementation of the Skills Training Program.

i. Multimedia Artist – Authoring [IT-070-2] – Level 2
ii. Multimedia Designer – Authoring [IT-070-3] – Level 3
iii. Shielded Metal Arc Welding Process [MC-024-3: 2012] – Level 3
iv. HVAC Single Phase Air Conditioning Equipment [ME-020-2: 2012] – Level 2
v. HVAC Single Phase Air Conditioning Equipment (Installation, Servicing, Troubleshooting & Repair) [ME-020-3: 2012] – Level 3
vi. Building Constructor [B-010-1] – Level 1
vii. Building Constructor [B-010-2] – Level 2
viii. Single Phase Electrical Installation & Maintenance [EE-320-2-2011] – Level 2
ix. Three Phase Electrical Installation & Maintenance [EE-320-3-2012] – Level 3
x. Machining Operation [MC-050-2: 2012] – Level 2
xi. Machining Operation [MC-050-3: 2012] – Level 3

The implementation of skills training at FTVE is embedded in the ISMPV academic curriculum. Single tier certification blended no exit point refers to the method of conduct for the implementation of a combination of two or more accredited programs to the highest degree of

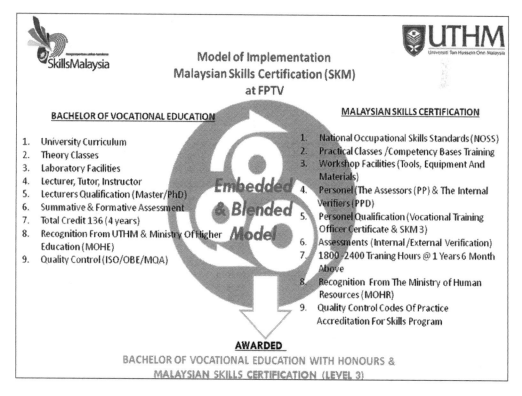

Figure 2. Embedded & blended curriculum module of ISMPV and SKM.

independence in the course of the program. As well as external verification, the award of the skill certificate level only occurs at the highest degree of accredited program. Figure 3 shows 4 years of skills training process conducted in conjunction with academic. Initially, the training process of the Certified Center (PB) must register students online with the Department of Skills Development (JPK) at least 30 days from the date of enrollment at the PB with registration fee. Students are confirmed to be registered with JPK after payment of RM100 (exercise of a single level certification exercise) registration is completed. Generally, the maximum ratio is an Assessor (PP) to 25 students per program.

2.2 *Implementation methods of skill training program at FTVE*

From the aspect of assessment, componence for Level 1 certification only involves assessing the course work completely without undergoing the final exam. For Level 2, certification assessments involve the completion of a 40% work assessment where the assessment fraction involves

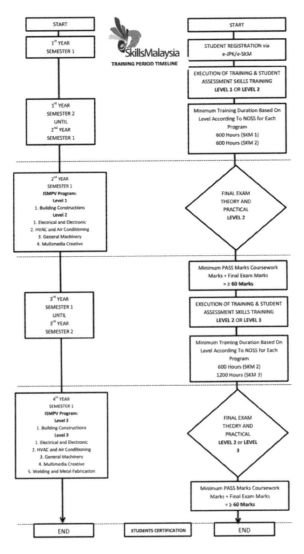

Figure 4. Flow chart of **SKM** embedded/blended learning no exit at single tier approach in ISMPV curriculum design.

15% theory (minimum score 7.5) and 25% practical (minimum score of 12.5). The process of execution and monitoring is fully carried out by the Internal Verification Officer (PPD) and the Assessment Officer (PP) of this process, involving internal verification by the Certified Center Evaluation Panel (PPPB) and the Technical Advisory Committee (TAC). Conversely, aspects of the final examination of the theory and practice are carried out without the presence of External Certified Officers (PPL) appointed from the Department of Skills Development Malaysia.

For the third stage certification process, the assessment component involves course work at 40%, where the assessment involves 15% theory (minimum score 7.5) and 25% practical (minimum score of 12.5). The final exam involves 60% of the assessment score – 20% theory with a minimum score of 15, and 40% practical, with a minimum score of 25. The final examination of theory and practice examinations involves the attendance of External Assessment Officers (PPL). Students who fail in the examination will be allowed to retake within 1 year from the date of final examination results, with a student registration fee of RM 100.00 for any component of the failed assessment. Figure 3 shows a flow chart for the implementation of the training in FPTV.

2.3 *Technical Vocational Education Training (TVET)*

Current teacher training programs do not specify teacher profiles to be produced. However, the transcripts obtained by graduates evidence the subjects and the courses that the graduates have followed to enable them to serve as teachers. Teacher profile production is also important to demonstrate the competence of teachers in knowledge, assessment and the method of teaching that may have changed or not after many years of teaching. Professional development is essential in terms of the mastery of content pedagogy and strengthening training through the recognition of professional bodies and agencies. For example, the Malaysian Construction Industry Development Board (CIDB), the Welding Institute of Malaysia, IBM Malaysia and other agencies recognize the skills of teachers and potential teachers. The Model of Pedagogical Reasoning and Action by Shulman and Wilson (2004) can be adapted in producing teacher profiles to enhance teacher competence in teaching and content knowledge.

Changes in pedagogical delivery must emphasize the process and practice of developing a knowledge base for teaching. Shulman (2013) states pedagogy and action reasoning, requiring teacher transformation in pedagogic knowledge content adapted to meet student abilities. This transformation process requires some of the following combinations: the provision of textual materials including the interpretation process, the ability of teachers to criticize ideas in the form of new analogies or metaphors, and the selection of teaching strategies. Figure 5 shows the criteria in the model proposed by Shulman and Wilson (2004).

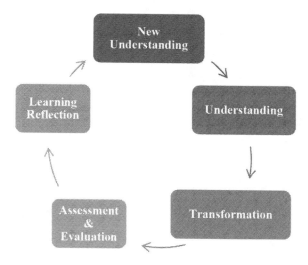

Figure 5. Model of Pedagogical Reasoning and Action (Shulman, 2013).

3   CONCLUSION

Implementation of the Skills Training Program by Single Tier Certification – Blended SKM Exit Point at FTVE, UTHM for pedagogic content has challenges in producing vocational instructors. The lecturer's understanding of this process should be holistically informed so that each course is aligned with outcome-based education (OBE) needs and the process of commissioning students to acquire SKM. As such, graduates are given the opportunity to attend the Prior Achievement Recognition (HEP) program to gain equivalent or higher SKM qualifications by assessing their previous experience. The students are also effectively evaluated to meet academic and skill requirements without having to set aside the quality of FTVE graduates.

The uniqueness of the approach has a implications toward future teachers and instructors of TT-TVET, by having dual certification from MQA and the Malaysian Skill Certification Level 3 with Vocational Training Officer Level 3. Thus, the objective of integrating the knowledge and skills-based curriculum in TT-TVET through a blended and embedded model, produces the "ready-made" graduates of the future complying with the stakeholder's requirement to be the potential lecturers and instructors in public and private institutions. The selling point of this program structure is that students could graduate on time within 3–4 years with the same study fees. A blended and embedded model program provides an attractive career path and personal growth opportunity for participants, and especially FTVE students that intend to be vocational teacher or instructors. Last but not least, the programs also produce skilled workers who are trained and qualified to enhance the competiveness of local industries in the global market.

As a result of this innovation approach, 428 FTVE students have earned SKM (2018 data). Figure 6 shows the proportion of students who completed the courses, by respective field.

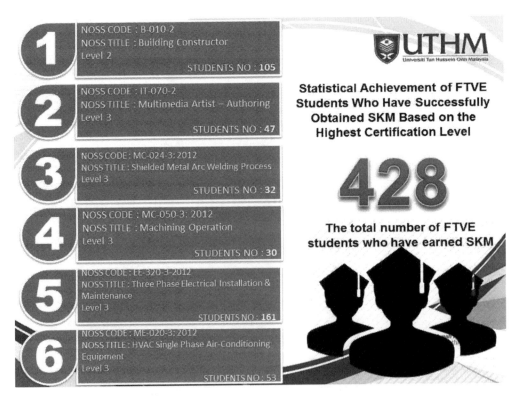

Figure 6.   Statistical achievements of FPTV students who have successfully obtained SKM based on the highest certification level.

Table 2. FTVE lecturers with SKM qualifications according to the NOSS code.

No	Name	Level	NOSS code
1.	Ts. Dr. Lee Ming Foong	2	EE-320-2:2011
2.	En. Mohamad Rafi bin Rahhim	2	EE-320-2:2011
3.	Dr. D'OriaIslamiahRosli	2	IT-070-2
4.	Dr. FazlindabintiAb.Halim	2	IT-070-2
5.	Dr. Azitabte Ali	2	IT-070-2
6.	Pn. ZanariahBinti Ahmad	2	IT-070-2
7.	Pn. Noor HayatiBinti Mustafa	2	IT-070-2
8.	Dr. YusmarwatibintiYusof	2/3	B-010-2 B-010-3
9.	Ts. Muhamad Amin bin Haji Ab Ghani	3	B-010-3
10.	Ts. Dr. Saifullizam bin Puteh	2/3	EE-320-2:2011 EE-021-3:2012
11.	Dr. NormahbintiZakaria	2/3	EE-320-2:2011 EE-021-3:2012
12.	Dr. Tamil Selvan a/l Subramaniam	2/3	EE-320-2:2011 EE-021-3:2012
13.	Pn. MaizathulSahidaBinti Othman	2	EE-320-2:2011

The success of this innovation program is also a reflection of the lecturers in the Department of Vocational Education (JPV). By having this number of man power planning, FTVE successfully comply with the requirement needed by the JPK to appoint these staff to teach SKM program according to their respective field.

As mentioned earlier, in order to qualify to teach the SKM program, the students must at least acquire Vocational Training Officer (VTO) certification and SKM Level 3 certification (minimum requirement) for each respective field. Table 2 shows the lecturers involved in this approach.

REFERENCES

Axmann, M., Rhoades, A. & Nordstrum, L. (2015). *Vocational teachers and trainers in changing world: The imperative of high quality teacher training system.* Geneva: International Labour Office.

Affero, I. & Hassan, R. (2013). Issues and Challenges of Technical and Vocational Education & Training in Malaysia for Knowledge Worker Driven. 10.13140/2.1.4555.2961

Foyster, J. (1990). *Getting to grips with competency-based training and assessment.* Leabrook, S.A: TAFE National Centre for Research and Development.

Huberman, M. (1995). Professional career and professional development: Some intersection. In Guskey, T. & Huberman, M., *Professional development in education: New paradigms and practice,* 77–86. New York: Teacher College Press.

ILO. (2012). *TVET Reform: Design an inclusive skills development program.* Retrieved from www.ilo.org/publns.

Power, L., & Cohen, J. (2005, November). Competency-Based Education and Training Delivery: Status, Analysis and Recommendations. Academy for Educational Development. Dicapai Oktober 26, 2015. http://pdf.usaid.gov/pdf_docs/PNADP013.pdf

Shulman, L.S. (2013). Those who understand: Knowledge growth in teaching. *Journal of Education, 193* (3), 1–11. doi: 10.1177/002205741319300302

Shulman, L.S., & Wilson, S.M. (2004). *The wisdom of practice: Essays on teaching, learning, and learning to teach.* San Francisco, CA: Jossey-Bass.

Spöttll, G. (2009). Teacher education for TVET in Europe and Asia: The comprehensive requirements. *Journal of Technical Education and Training, 1*(1), 1–16.

Yazçayır, N. & Yağcı, E. (2009). Vocational and technical education in EU nations and Turkey. *Procedia-Social and Behavioral Sciences,1*(1), 1038–1042. doi:10.1016/j.sbspro.2009.01.187

Innovative Teaching and Learning Methods in Educational Systems – Retnowati et al. (Eds)
© 2020 Taylor & Francis Group, London, ISBN 978-1-03-224183-8

The effectiveness of academic supervision by school principals

S. Darmawanti & H. Usman
Graduate School, Yogyakarta State University

ABSTRACT: This research aims to reveal the effectiveness of academic supervision conducted by school principals, which consists of the planning, execution, evaluation, and follow-up, and any obstacles and the attempt by school principals to overcome them. This research uses an evaluation method with a descriptive quantitative approach. The research venues were state elementary schools across Yogyakarta, and the sample was 48 teachers selected based on proportional random sampling techniques. The data collection technique was a questionnaire. The data analysis technique was delivered using descriptive qualitative methods with percentages. The outcome of this research highlighted several issues. (1) The planning of academic supervision is very effective. (2) The execution of academic supervision is effective. (3) The evaluation of the execution of academic supervision is effective. (4) The follow-up of academic supervision is effective. (5) The main obstacle experienced by school principals is administrative duties, and hence the academic supervision is constantly delayed. Other obstacles are the principals' reluctance to supervise senior teachers, the lack of administrative support for teachers, and the willingness of teachers to be supervised. (6) The measures put in place to overcome these setbacks include rescheduling the supervision, pre-supervision teacher conferences, staff training and teachers' forums.

1 INTRODUCTION

Today, global education demands a system of education that is capable of developing high-quality human resources. The evolving quality of education is closely in line with efforts to leverage the professional maturity of teachers to anticipate future challenges. Professional maturity is the ability to accept extensive responsibility in educational initiatives, teaching and learning processes, and professional development (Faturahman, 2012). The function of education is a realized and well-planned effort to actualize learning environments and learning processes; to help students actively develop their individual potential with positive aptitudes such as spirituality, religiousness, control, individual well-being, intelligence; and develop the skills needed by the students, society and the nation (RoI, 2003). Therefore, education plays a prominent role in providing professional human resources in every field of work.

School principals have strategic roles in achieving success in the learning process, and this is key to the role of school principal as a supervisor. Supervision helps uncover essential conditions and requisites that guarantees the achievement of the educational purpose (Purwanto, 2012). It is one primary function in educational administration and school principals have an integral role in the supervision of teachers and other staff in their schools. Hence, supervision activities should be conducted regularly in schools as a part of a positive daily routine in order to achieve high-quality learning through a favorable learning process.

Effective supervision is a pivotal factor in developing productivity at work. To make this process effective, partnerships between the school principal and teachers are key. Supervision can become a more effective and satisfying activity for both parties in any setting if there is an explicit contract that can form the basis of the supervisory alliance (Hawkins & Shohet, 2012). Academic supervision by a school principal can cover any assistance needed by the teachers to develop their professional capabilities.

Based on the pre-research observation initiated by state elementary schools across Yogyakarta, it was established that some teachers only teach to deliver their tasks without considering the actual learning process. The teachers had initially made negative assumptions regarding the function of academic supervision by their school principal, and it can be inferred that not all teachers understand the purpose of the academic supervision. The academic supervision activity is yet to be seen as an administrative duty and the implementation of academic supervision was judged as not within their duties. It also revealed that not all school principals structure their academic supervision systematically. Not all school principals deliver proper training to theirteachers in order to elevate the learning quality. Overall, not all school principals delivered the intended academic supervision in response to the needs of teachers in order to improve their competence. The preliminary survey showed that there is little information on the effectiveness of academic supervision by school principals regarding execution, planning, evaluation, and follow-up. The function of academic supervision by school principals has an important role in leveraging the quality of learning in common and education in particular.

In response to the problem identified, this research focuses on the effectiveness of school principals' academic supervision. It includes planning, implementation, evaluation, follow-up, and obstacles in conducting academic supervision and any efforts made to overcome hese issues. Therefore, the outcome of this research is expected to benefit teachers regarding using academic supervision to improve their skills. It will highlight the school principal as the key figure for giving information or advice rather than focussing on the implementation of academic supervision so that the process of learning and teaching can be optimal. And, it will provide government departments of education with a primary reference for the formulation of policies, particularly those in connection with academic supervision by school principals.

2 LITERATURE REVIEW

Academic supervision is a series of activities aimed at assisting teachers to develop their abilities in learning process management, to achieve the purpose of learning (Glickman, 2007; Ambarita, Siburian, & Purba, 2014). It is an internal school examination mechanism adopted by school principals (Ayeni, 2012), and a process of involving teachers in learning-centered dialogue to develop their teaching-learning process and to accelerate students' success (Sullivan & Glanz, 2005).

The primary objective of academic supervision is to observe the implementation of learning and to ensure teachers develop their skills as expected, to increase their conceptual knowledge and teaching skills, and to give motivational support to facilitate improvements in teachers' pedagogical practice and the learning outcome of students (Tyagi, 2010). Furthermore, the purposes of academic supervision are as follows. (1) Assisting teachers in developing their teaching and learning methods. (2) Assisting teachers to integrated the curriculum in their teaching and learning methods. (3) Assisting teachers to train other school staff (Sahertian, 2008). Thus, it can be inferred that the general purpose of academic supervision is to assist teachers in overseeing the goal of education, to guide learning and teaching experiences, to use learning resources and teaching methods, to meet students' learning needs, and to evaluate learning progress and cultivate the culture of work.

3 METHODOLOGY

This was evaluation research with a descriptive quantitative approach. The study was carried out between January and June 2018 in a number of state elementary schools across the Special Region of Yogyakarta. The study sample was 48 teachers at public elementary schools in

Yogyakarta, selected based on proportional random sampling techniques. The data collection method was by questionnaire. The instrument validity was testified using a Pearson correlation. The data analysis technique was descriptive quantitative with percentages.

4 RESULTS

The result of this research identified a series of steps of academic supervision delivered by school principals: planning, implementation, evaluation of implementation, and follow-up. Teachers were the persons involved when school principals delivered supervision at school.

Table 1. The result of academic supervision by school principals.

No.	Academic supervision variable	Result	Percentage	Category
1.	Academic supervision planning	90.04%	X > 90%	Highly effective
2.	Academic supervision implementation	83.31%	76 – 90%	Effective
3.	Evaluation of academic supervision implementation	85.31%	76 – 90%	Effective
4.	Academic supervision follow-up	86.63%	76 – 90%	Effective

The challenges faced by school principals in the implementation of academic supervision include overwhelming administrative duties that delay the supervision, the reluctance of school principals to supervise senior teachers, teachers' keeping incomplete records due to administrative burdens, and teachers' unwillingness to be supervised. The efforts by school principals to address this situation include rescheduling and agreeing a time for academic supervision with teachers, holding a teachers' conference to explain the importance of academic supervision, and regular training and coaching for teachers through an intensive forum at school and cluster level.

5 DISCUSSION

A school supervisor has an important role as a bridge between teachers and education quality, particularly in solving problems (Nirwan, 2017). Teachers may need help from their supervisor with the teaching and learning process. A clear learning process is a factor in effective learning for students, and can influence examination results. Thus, teachers are key to this successful learning.

5.1 *Planning of academic supervision by school principals*

Teachers stated that the level of compliance in academic supervision planning activity by a school principal is 90.04%, which places it in the very effective category. The result of this research shows that school principals were able to plan their academic supervision by determining the goal, target, steps and duration of the supervision. This supervision planning by school principals was oriented to professional coaching for teachers, and was customizable depending on the teacher's needs. Therefore, school principals formulate academic supervision programs based on the needs of the teachers (Makawimbang, 2011), and must prepare any necessary equipment, goals, targets, objectives, methods, techniques and a planned approach in the form of supervising format.

The planning phase is very important as without this the process will not run smoothly. In order for the school principal to succeed in this task, they must have the skills to plan the supervision (RoI, 2011). The benefits of academic supervision planning include (1) as guidance for implementing and observing academic supervision; (2) ensuring all staff members understand the process and need for academic supervision; and (3) guaranteeing funding as well as effectiveness use of the school's material resources (energy, time, cost) (Prasojo & Sudiyono, 2015).

5.2 Implementation of academic supervision by school principals

Teachers stated that the level of compliance by school principals in implementing supervision is 83.31%, which places it in the effective category. This shows that school principals have commendable knowledge in selecting the principles, techniques and approaches to supervision to improve teachers' professional standards. This result is in line with results from Panigrahi, that teachers assume their supervisors have the skills to help them be a more effective teacher (Panigrahi, 2012). To enhance their supervision practices, supervisors should be informed of the supervising model, techniques and principles (RoI, 2011).

5.3 Evaluation of academic supervision by school principals

85.31% teachers stated that the evaluation of academic supervision by school principal was carried out, which placed it in the effective category. The outcome of this research shows that school principals undertook deep evaluation and analysis of the supervision results, which has the potential to improve teachers' professional skills. The evaluation of academic supervision implementation will identify how far the initiated programs are being realized and which agendas should be revised (RoI, 2017). In addition, obstacles to the academic supervision can also be thoroughly identified (Suryani, 2015).

Another concern is the reporting phase of the outcome of supervision, which focuses on the attainment targets. Reporting is a fundamental record and an essential mirror that reveals the quality of supervision and can also identify the comprehensiveness of the supervising methods and how much time the supervisor has spent on implemention (Sullivan & Glanz, 2005). Based on the research data, after the supervision the school principals draft a report to be submitted to the inspector, which is then used as general guidance to deliver the teacher's training.

5.4 The follow-up of academic supervision by school principals

Teachers stated that the level of compliance in follow-up activities after academic supervision is 86.63%, which placed it in the effective category. Thus, school principals had followed-up the academic supervision program by repeating and re-formulating it. Through an effective follow-up mechanism, teachers can be properly assisted to develop their professionalism. A follow-up process is a form of justification, recommendation, and execution of the targeted supervision goals that are conveyed to the head of the education department (RoI, 2017). The steps to implement the follow-up are as follows. (1) Analyse the summary of the scoring outcome. (2) If the goals of academic supervision and learning standards are not met, the supervisors will incur penalties and/or be enrolled in internal training mechanisms. (3) If the goals are not achieved, the academic supervision program will roll over to the next school term. (4) Reformulate the action plan for the next period of academic supervision. (5) Implement the particular action plan in the next school term (Prasojo & Sudiyono, 2015).

5.5 Obstacles faced by school principal in the implementation of academic supervision

Based on the outcome of this research, the obstacles that school principals face during the implementation of academic supervision include (1) the burden of administrative duties causing delays in the supervision schedule; (2) the principal's reluctance to supervise the older/senior teaching members; (3) incomplete record keeping by teachers due to administrative problems; and (4) the unwillingness of the teachers to receive supervision. These results are in line with those of Astuti (2017) that the primary obstacle faced within the implementation of academic supervision is the complexity of the principal's tasks and the impossible schedule.

5.6 The efforts by school principals in overcoming obstacles to academic supervision

Based on the result of this research, the efforts addressed by school principals to overcome obstacles during the supervision include (1) initiating partnerships between teachers and the school principal; (2) explaining the importance of academic supervision via a teacher-principal conference; and (3) initiating regular coaching and training for teachers through forums at school and cluster level. These steps are in line with the basic principle that academic supervision should be implemented systematically and appropriately through a clear plan and on a regular basis (Sahertian, 2008).

6 CONCLUSION

Academic supervision by school principals has so far been effective from different perspectives. (1) The planning of academic supervision is very effective. (2) The execution of academic supervision is effective. (3) The evaluation of the execution of academic supervision is effective. (4) The follow-up of academic supervision is effective. (5) The main obstacle experienced by school principals is administrative duties, and hence the academic supervision is constantly delayed. Other obstacles are the principals' reluctance to supervise senior teachers, a lack of administrative support for teachers, and the unwillingness of teachers to be supervised. (6) The methods to overcome these setbacks include rescheduling the supervision, pre-supervision teacher conferences, training and a teachers' forum.

The result of this study can be used by education departments as a reference to better monitor and guide school principals in academic supervision activities, which are steps to develop teacher competencies, and as self-correction by the school principal in evaluating academic supervision. Effective implementation of academic supervision by the school principal can make a difference to the world of education.

REFERENCES

Ambarita, B., Siburian, P. & Purba, S. (2014). Development of academic supervision model which based on education manajement. *International Journal of Sciences: Basic and Applied Research. 18*(1), 304–314.

Astuti, S. (2017). Supervisi akademik untuk meningkatkan kompetensi guru di laboratorium UKSW [Academic supervision to improve teacher competency in the UKSW laboratory]. *Scholaria: Jurnal Pendidikan dan Kebudayaan. 7*(1), 49–59. doi: 10.24246/j.scholaria.2017.v7.i1.p49-59.

Ayeni, A.J. (2012). Assesment of principals' supervisory roles for quality assurance in secondary school in Ondo State, Nigeria. *World Journal of Education. 2*(1), 62–69. doi: 10.5430/wje.v2n1p62.

Faturahman, P. (2012). *Pengembangan profesional guru* [Teacher professional development]. Bandung, Indonesia: Refika Aditama.

Glickman. D.C, *et al.* (2007). *Supervision of instructional: a development approach.* Needham Heights, MA: Allyn and Bacon.

Hawkins, P. & Shohet, R. (2012). *Supervision in the helping profession.* Milton Keynes, UK: Open University Press.

RoI. (2011). *Supervisi akademik: materi pelatihan penguatan kemampuan kepala sekolah* [Academic supervision: training materials to strengthen the ability of the principal]. Jakarta, Indonesia: Republic of Indonesia.

RoI. (2013). *Undang-Undang Nomor 20 Tahun 2013 tentang Sistem Pendidikan Nasional* [The Law Number 20 of 2013 concerning national education system]. Jakarta, Indonesia: Republic of Indonesia.

RoI. (2017). *Panduan supervisi akademik* [Academic supervision guide]. Jakarta, Indonesia: Republic of Indonesia.

Makawimbang, J.H. (2011). *Supervisi dan peningkatan mutu pendidikan* [Supervision and improvement of education quality]. Bandung, Indonesia: Alfabeta.

Panigrahi, M.R. (2012) *Implemention of intructional supervision in secondary school: Approaches, praspects and problems.* Ethiopia: Haramaya University.

Purwanto, M.N. (2012). *Administrasi dan supervisi pendidikan* [Educational administration and supervision]. Bandung, Indonesia: Remaja Rosdakarya Offset.

Prasojo, L.D. & Sudiyono. (2015). *Supervisi pendidikan* [Supervision of education]. Yogyakarta, Indonesia: Gava Media.

Rahabav, P. (2016). The effectiveness of academic supervion for teacher. *Journal of Education and Practice.* 7(9), 47–55.

Sahertian, P.A. (2008). *Konsep dasar & teknik supervisi pendidikan: dalam rangka pengembangn sumber daya manusia* [Basic concepts & techniques of educational supervision: in the framework of developing human resources]. Jakarta, Indonesia: Rineka Cipta.

Sullivan, S & Glanz, J. (2005). *Supervision that improves teaching strategies & techniques* (2ndedn). Thousand Oaks, California: Corwin Press.

Suryani, C. (2015). Implementasi supervisi pendidikan dalam meningkatkan proses pembelajaran di MAN Sukadamai Kota Banda Aceh [Implementation of educational supervision in improving the learning process in MAN Sukadamai, Banda Aceh]. *Jurnal Ilmiah DIDAKTIKA. 16*(1), 23–42.

Tyagi, R.S. (2010). School-based instructional supervision and the effective profesional development of teachers. *Compare: A Journal of Comparative and International Education. 40*(1), 111–125. doi: 10.1080/0357920902909485.

The performance of state elementary school supervisors

R.S. Maulida & H. Usman
Graduate School, Universitas Negeri Yogyakarta, Indonesia

ABSTRACT: This research was conducted to determine the performance levels of elementary school supervisors in academic supervision, which includes planning, implementing, evaluating, and developing professional supervisors at state elementary schools throughout the Cikupa District, Tangerang Regency, Indonesia. This research used an evaluation method with a descriptive quantitative approach. The sample of this research was 102 teachers at state elementary schools in Cikupa District, determined using proportional random sampling for each school. The data collecting technique was a questionnaire research instrument, which had been validated using product moment correlation and Cronbach's alpha reliability testing. To show the data results, the researchers used descriptive statistics (mean score and standard deviation). The findings of this research are as follows. (1) Supervision planning by school supervisors was at a good performance level of 85.07%. (2) Supervision by school supervisors was at good performance level of 82.47%. (3) Evaluation of supervision by school supervisions was at a good performance level of 81.13%. (4) Developing, guiding and training professional teachers performed by school supervisors was at poor performance level of 55.07%. These research recommendations to elementary school supervisors were designed to maintain the good performance and improve the poor performance in supervision skills, especially in academic and managerial supervision.

1 INTRODUCTION

Education is an important factor in a nation's development and the quality of a nation can be gauged from its education standards. In the Global Talent Competitiveness Index (GTCI) 2018, Indonesia was ranked 77 out of 119 countries in the world. GTCI is a comprehensive report that gives an indicator of how a developing country and city can provide human resources to improve their competitiveness.

Globalisation encourages the education becomes increasingly important because it is the determiner of human resource quality. Development of a good quality education system is a constant task for governments seeking to achieve the goal of national education. The process to improve the quality of education requires supervision and education components that relate to each other in achieving the goal of national education. A school supervisor's duty is to supervise both the academic and managerial staff, fostered through monitoring, assessing, guiding, reporting and follow-up. Nowadays, school supervisors support the improvement of the quality of education that has not yet effectively. Education supervision starts with planning, acting, assessing, developing, guiding, and professional training for teachers. The role of the school supervisor is to help teachers and headteachers understand problems and make decisions (Wiles & Bondi, 1986).

Weak education supervision affects the quality of education, as can be seen from the low national examination results in Indonesia. Supervision by the school supervisor requires improvement because poor education is still an problem in districts such as Tangerang, where the quality of elementary schools decreased. In June 2016, Endang, head of Banten Province Education Quality Assurance, explained that the quality of education in Tangerang Regency was still low because of several factors: in some areas national examination scores that still low, unqualified teachers, and a lot of school-aged children with poor attandance (koranban-ten.com, 29 June 2016). Improving the quality and output of education in Tangerang Regency requires good teamwork, so school supervisors are key, especially in Cikupa District (Ediger, 2002).

The pre-research observations carried out in state elementary schools in Cikupa District, Tangerang, showed that school supervisors do not provide each school with full guidance; in fact, not all school supervisors followed up with schools after carrying out supervision. The school supervisors only provided forms to be filled in by the teachers and principals, which should also be completed by supervisors during their visit. In addition, not all school supervisors conducted supervision in accordance with the established standards and not all completed reports of supervision, which are used as reference works for the professional development of school supervisors.

The results of the pre-research show that the role of school supervisors in carrying out supervision, which includes planning, implementation, evaluation, and professional development, has not been implemented properly. This record is the responsibility of all stakeholders to evaluate the performance of school supervisors at Cikupa District. It is also a very important step in ascertaining the performance of elementary school supervisors in conducting educational supervision of state elementary schools. School supervisors are expected to be able to carry out their duties in supervising their target schools, with established standards, and provide follow-up actions for the school, the principal, and the teachers.

This research aims to determine the performance levels of elementary school supervisors in conducting supervision, which includes planning, implementation, evaluation, and professional development. Therefore, the results of this study are expected to provide benefits for supervisors, principals, and teachers as a reference to improving the quality of education in each school. In addition, the education departments can use the results when making policies to improve quality performance standards of state elementary school supervisors. The supervision performed by school supervisors can build relationships between stakeholders of the education system so that they realize the goals of education (Ekundayo, 2013).

2 LITERATURE REVIEW

Someone's performance at work reflects their ability to do a job at a certain time. The result of performance can be product, behavior, proficiency, and competency. To improve the quality of education, the performance of supervisors became the initial benchmark (Astuti, 2010). The current method of assessment of school supervisors by the education authorities is not sufficient because it is only done through the reports submitted by school supervisors and not using supervisors' performance standard assessments. Furthermore, follow-up action is not systematically undertaken (Niswanto, 2013).

School supervisors are an important subsystem of the whole national education system, and have a key role in improving the quality of education and teacher professionalism in schools (Syamsu, 2018). The role requires certain competencies – knowledge, skill, proficiencies, or capabilities – for an optimal cognitive, affective, and psychomotor performance in accordance with the duties and responsibilities (Priatna, 2016). A school supervisor has the responsibility of making sure all government policies are followed in accordance with the rules (Tok, 2013).

Education supervision is a process of guidance and directs a person or a group to achieve the goal of an educational organization (Daresh, 1989). Education supervision is an action that is made positively, dynamically, and democratically to improve teaching and learning, and all parties that have a role in education (Neagley & Evans, 1980). The educational success of a developed country can be seen from the effectiveness and efficiency of the guidance school supervisor give to teachers (Gulsen, Aysel, & Bahadir, 2014). Professional development support for teachers is derived from "listening and expressing the opinions and needs of individuals" (Leithwood & Sun, 2012). The purpose of education supervision is to develop teaching that helps students achieve and improves teachers' ability to raise the quality of education (Glickman, Gordon, & Ross-Gordon, 2004).

3 METHODOLOGY

This was evaluation research with a descriptive quantitative approach. Researchers used a discrepancy evaluation model to compare the gap between school supervisors' performance and the standards set nationally as the school supervisor assessment criteria. This research was done March to July 2018 at elementary schools in the Cikupa District, Tangerang.

3.1 Population and sample

The population of this research was all elementary school teachers in the Cikupa District, Tangerang. The sample of this research was 102 teachers from that population, determined using proportional random sampling from each school.

3.2 Data collecting technique

The data collecting technique was a questionnaire research instrument. Instrument validity was determined using product moment correlation, and instrument reliability was determined using Cronbach's alpha.

3.3 Data analysis technique

Researchers used descriptive statistics as the data analysis technique. The formula was:

$$P = \frac{f}{N} \times 100\%$$

where P = Percentage, f = score sum of research result, and N = expected score sum.

4 RESULTS

The result of this research showed the steps of supervising taken by the state elementary school supervisors: planning, supervising, evaluating, and developing, guiding, and professional teacher training. Teachers were those involved in supervision by the school supervisor.

Table 1. The results of elementary education supervising.

No.	Variable Supervision	Result	Category	Information
1.	Supervision planning	85.07%	76–90%	Good
2.	Supervision	82.47%	76–90%	Good
3.	Supervision evaluation	81.13%	76–90%	Good
4.	Developing, guiding, and professional teacher training	55.07%	51–60%	Poor

When undertaking education supervision, the school supervisor took necessary preventive actions to solve problems and give guidance to the teachers regarding the learning process. This proved the importance of giving the teachers guidance to provide an effective learning process.

5 DISCUSSION

A school supervisor has an important role in solving problems because supervising is like a bridge between teachers and education quality (Nirwan, 2017). Teachers need help and solutions from school supervisors when facing problems in the teaching-learning process.

A learning clear process is a factor for effective learning by students. Students who receive clear explanations in the learning process are more likely to gain high scores in examinations. Because of that, the teacher is key to successful learning for the students.

5.1 Supervision planning

Supervision planning should be done before educational supervision. In this step, the school supervisor should detail about what they will be doing in the supervision steps. The result of supervision planning variables, according to the teachers, was a 'good' performance of supervision planning at 85.07%. This achievement showed that indicators in supervision planning variables were already fulfilled by the school supervisors when preparing supervision programs. One of the planning forms used by the school supervision program compiles a list of actions or activities that will be done to fulfill the goals of supervision (Department of National Education, 2011).

5.2 Supervising

Supervision needs to be properly organized to avoid any misunderstanding between the school supervisor and the teacher supervised. It also needs a good approach and an appropriate supervision technique. The result of supervising variables, according to the teachers, was a 'good' performance with the implementation level of supervision at 82.47%. Key things to supervise were teachers' performance, application of the curriculum, their learning processes, their availability and their use of school resources. School supervisors had to ensure that the learning process in schools is able to achieve the goals set. The supervision conducted for each teacher was at different personal and professional levels, because the needs and preferences for support for each teacher are different. These different needs require adjustments to the forms of supervision for each teacher (Kalule & Bouchamma, 2014). Moreover, the aim of the supervision is not only to note key details and to report the results of activities, but also to help in correcting and fixing mistakes so that the goals can be achieved.

5.3 Supervising evaluation

Evaluation after supervising is an important task. With evaluation in education supervising, a school supervisor can give feedback to the teachers who need help in the learning process. The school supervisor is an advisor for the teacher, and they have a duty to provide advice or guidance (Kolawole, 2012). The result of supervising evaluation, according to the teachers, was a 'good' performance with the implementation level of supervising evaluation at 81.13%. In accordance with the main tasks of a school supervisor, the results of supervision were reported to the Head of Regency Education Authorities. Evaluation is required to assess gaps that occur between expectations and reality.

5.4 Developing, guiding, and professional teacher training

Developing, guiding, and professional teacher training are follow-up actions after supervision evaluations by the school supervisor. The school supervisor is required to share educational innovations that can help teachers and improve their skills. The result of developing, guiding, and professional teacher training variable, according to the teachers, were a 'poor' performance with the implementation level at 55.07%. School supervisors must take these responsibilities seriously. This can be done through educational activities that can improve teacher skills, such as workshops and seminar (Kotirde & Yunos, 2014). Continuous teacher training becomes a contribution to a successful education process (Zoulikha, 2013).

6 CONCLUSION

According to data analysis from all state elementary schools in the Cikupa District, it can be concluded that school supervisor's performance was good except for developing, guiding, and professional teachers training, which scored in the poor category. The limit of this research was that researchers only studied the steps of supervision by school supervisors. The advantage of this research was that the data were descriptive statistics which made it easier to describe the school supervisor's performance.

Effective school supervision can makes changes in the world of education. The solutions given by the school supervisor were expected to help teachers face problems in the learning process at school.

The results of this research are expected to be used as reference for the Educational Office to improve policies related to the performance of school supervisors in implementing education supervision, so that it follows what is expected and determined by the government. As a result, it should improve the learning process, which leads to the improving the quality of education.

REFERENCES

Astuti, S. (2010). Pengaruh motivasi dan kompetensi pengawas terhadap kinerja pengawas sekolah dasar [Effect of supervisor motivation and competence on the performance of primary school supervisors]. *Jurnal Adminitrasi Pendidikan, 12(2)*.

Daresh, J. (1989). *Supervision as a proactive process.* White Plains, NY: Longman.

Departemen Pendidikan Nasioanl. (2011). *Buku pengawas sekolah* [School supervisor book]. Jakarta: Pusat Pengembangan Tenaga Kependidikan, Badan PSDM dan PMP.

Ediger, M. (2002). The supervisor of the school. *Education, 122*(3), 62.

Ekundayo, H. (2013). Effective supervision of instruction in Nigerian. Secondary School: Issues, challenges and the way forward. *Journal of Education and Practice, 4*(8), 189–197.

Glickman, C.D., Gordon, S.P., & Ross-Gordon, J.M. (2004). *Supervision and instructional leadership: A development approach* (6th ed.). Boston, MA: Allyn and Bacon.

Gulsen, C., Aysel, A., & Bahadir, E.G. (2014). The thoughts of school principals about the effects of educational supervisors on the training of teachers in terms of professions. *Jurnal ScienceDirect, 174* (2015), 103-108, doi: 10.1016/j.sbspro.2015.01.632.

Kalule, Lawrence & Yamin Bouchamma. (2014). Teacher supervision practices and characteristics of in-school supervisors in Uganda. *An article in Educational Assessment Evaluation and Accountability, 26*, 51–72. doi: 10.1007/s11092-013-9181-y.

Kolawole, A.O. (2012). Comparative study of instructional supervisory roles of secondary school principals and inspectors of the ministry of education in Lagos State, Nigeria. *European Scientific Journal, December edition 8*, 37-45.

Kotirde, I.Y., & Yunos, B.M.J. (2014). The processes of supervisions in secondary schools educational system in Nigeria. *Jurnal ScienceDirect, 204*(2015), 259–264, doi: 10.1016/j.sbspro.2015.08.149

Leithwood, Kenneth & Sun, J. (2012). The nature and effects of transformational school leadership: A meta-analytic review of unpublished research. *Journal of Educational Administration Quarterly.* doi: 10.1177/0013161X11436268.

Neagley, L. & Evans, H.D. (1980). *Handbook for effective supervision of instruction.* Engle-wood Cliffs, N. J.: Prentice Hall.

Nirwan, A. (2017). A pattern of empowerment in improving the professionalism of school supervisors. *Jurnal Akuntabilitas Manajemen Pendidikan, 5*(1).

Niswanto. (2013). Manajemen pembinaan pengawas sekolah dasar [Management of elementary school supervisor coaching]. *Jurnal Penelitian Pendidikan, 13*(2).

Priatna, A. (2016). Pengaruh kompetensi terhadap kinerja pengawas sekolah dasar di lingkungan dinas pendidikan kota Bekasi [Effect of competency on the performance of elementary school supervisors in the education office of the city of Bekasi]. *Jurnal Penelitian Pendidikan, XVI*(3).

Syamsu, Y. (2018). Mutu kinerja pengawas sekolah menengah [Quality performance of secondary school supervisors]. *Jurnal Adminitrasi Pendidikan, 25*(1).

Tok, T.N. (2013). Who is an education supervisor? A guide or a nightmare? *International J. Soc. Sci. & Education, 3* (3).

Wiles, J. & Bondi, J. (1986). *Supervision: A guide to prac*tice (2ed.). Columbus: Charles E. Merrill Publishing Company.

Zoulikha, T. (2013). Supervision of primary school teachers an analytical field study. *Jurnal Science Direct, 112*(2014), 17–23. doi: 10.1016/j.sbspro.2014.01.1135.

Innovative Teaching and Learning Methods in Educational Systems – Retnowati et al. (Eds)
© 2020 Taylor & Francis Group, London, ISBN 978-1-03-224183-8

Integrated thematic learning in the 2013 curriculum: Implications for the self-confidence and academic achievement of primary school students

U. Tisngati, C.A. Budiningsih & Sugiman
Yogyakarta State University, Yogyakarta, Indonesia

ABSTRACT: The purpose of this study is to explore the implementation of integrated thematic learning in the applicable curriculum in Indonesia and its implications for the self-confidence and academic achievement of fourth grade students in three primary schools in Pacitan Regency. The research method was descriptive qualitative with a case study approach, carried out in the second semester of the 2017/2018 academic year. Data collection was performed by observation, questionnaire, test, interview, documentation, and triangulation techniques as data validity techniques. Data analysis used the Miles and Huberman model as an interactive activity, including data reduction, a data display, and a conclusion drawing/verification. The results of this study showed, first, that teachers and students have implemented integrated thematic learning in the 'good' criteria, including preliminary activities, core activities, and closing activities. Second, students' self-confidence was in the 'good' criteria with four indicators: paying attention, listening, the courage of students, and problem solving. Third, students' academic achievements vary with the level of understanding of optimal average and they still experienced difficulties or barriers to learning. The results of this research have practical implications for teachers and parents to help students achieve success in the learning process and learning outcomes.

1 INTRODUCTION

Indonesia continues to strive to improve the quality of education, to create high-quality human resources in accordance with the ideals of the nation's founders. This is done through innovations in the educational curriculum that can develop the potential of students in accordance with their characteristics. The curriculum is developed by the government based on psychological, philosophical, social and cultural considerations. Psychologically, the curriculum is used to modify the behavior of students so that they develop according to the stages of mental, physical, and emotional development. Philosophically, curriculum changes can help teachers in the process of finding solutions and the basics in planning effective learning processes to achieve their goals.

The change in curriculum grew from fundamental conceptual changes, followed by structural changes such as the 2013 curriculum (Batmalo, 2016). This curriculum was prepared in line with Article 36 of Law Number 20 of 2003, and contains criteria regarding qualifications for the abilities of graduates that include attitudes, knowledge, and skills (Standards of Graduates' Competencies). This means that graduates are expected to have a spiritual attitude in faith and piety, have noble character, and be intelligent. The curriculum should be able to develop and interest students, acknowledge the diversity of regional and environmental potentials, in accordance with the demands of regional and national development, the demands of the world of work, the development of science, technology and art, religion, the dynamics of global development, and united national and national values.

The 2013 primary school curriculum uses an integrative thematic learning approach from grade 1 to grade 6 and is continuously reviewed based on the results of the implementation evaluations in the field. This can be seen from several revisions at the beginning of the year. Intensive grammar learning is a learning approach that integrates various competencies from various subjects into various themes (Majid, 2014).

The characteristics of thematic learning are teachers acting as facilitators or motivators of learning and not as the only source of learning. Thematic learning requires students to be active in learning according to the characteristics of scientific learning approaches. A scientific approach in learning all subjects includes uncovering information through observation, asking, experimenting, then processing data or information, presenting data or information, continuing by analyzing, reasoning, then concluding, and creating (Majid, 2014). The expectation is that students have optimal learning achievements in terms of attitudes, knowledge, and skills in accordance with 21st century educational characteristics. Thus the tenet of this research is how to implement integrated thematic learning, self-confidence and academic achievement of fourth grade primary school students in Pacitan District through the 2013 curriculum.

2 LITERATURE REVIEW

2.1 Philosophical review of 2013 curriculum development

The education curriculum in Indonesia is based on several philosophical foundations, such as perennialism, essentialism, and constructivism. The application of philosophical flow is undertaken selectively and eclectically to accommodate various interests related to education. Based on its characteristics, the 2013 curriculum is adopted from the flow of perennialism. The curriculum must emphasize the universal theme of human life, so that it can foster rationality and logical thinking. This philosophy in the practice of education can help develop the intellectual and spiritual potential of children. Aristotle and Aquinas (Gutek, 1974) assert that the highest humanitarian activity is ratio, intellectual training, and speculating power. Furthermore, the teacher's role is to be a skilled communicator with a refined rhetoric, by choosing the right words, using appropriate speaking styles, and selecting the right examples and analogies. Teaching must always begin with what children have built as a learning experience and must lead to something new (Gutek, 1974).

The 2013 curriculum also still stipulates the activities of reading, writing, and arithmetic, as well as rational social actions, as basic knowledge and skills that students must possess. It adopts an essentialism philosophy that education is a preparation for civilized citizens. The implication is that standardized tests are still used as an ideal benchmark for assessing students and the eachers responsible for student achievement (Moss & Lee, 2010).

2.2 Integrated thematic learning in the implementation of the 2013 curriculum

According to Randle (2010), integrative thematic learning emphasizes the integration of all disciplines with learning experiences based on the experiences of students and real-world structures, to encourage learning to be more meaningful. According to Majid (2014), the characteristics of thematic learning are as follows: (1) student-centered; (2) direct experience; (3) the separation of subjects is not clear; (4) concepts are presented from various subjects; (5) flexible; (6) include the principle of learning while playing and having fun. Jihad and Haris (2013) argue that thematic learning is integrated learning that uses themes to link several subjects. The combination of several subjects is expected to contribute to students' attitudes, experiences, knowledge and their abilities related to concrete things and meaningful learning.

Integrated thematic learning is applied at the primary school level in Indonesia in the implementation of the 2013 curriculum. It uses a learning approach that involves several subjects to equip students with meaningful experiences. In other words, teaching materials are developed from a context that reflects the real world of students in the hope that they will learn better and meaningfully. Context, conflict, and change are the defining parameters of the integrative learning interdisciplinary studies (Klein, 2005).

An integrative approach among the integration of science, technology, engineering, and mathematics (STEM) subjects has positive effects on students' learning (Becker & Park, 2011). Learning objectives are achieved through the discussion of the material in several themes. The integration of teaching materials with each subject by combining appropriate materials is intended to prevent gaps in the implementation process. For example, the material for first grade students, on the theme "Objects, Animals and Plants Around Me" involves subjects in Indonesian language, mathematics, and civic studies. Fourth grade materials on the theme "My Places of Living" combine Indonesian and science subjects.

2.3 *Student confidence in the integrated thematic learning*

Self-confidence is a personal characteristic in which there is a belief in one's ability to develop as a person and be able to overcome a problem with the best result. Some indicators of confidence according to Lauster (1978) include (1) believe in your own abilities; (2) acting independently in making decisions; (3) having a positive self-concept; (4) and daring to express.

Tisngati and Meifiani (2014) explained that students' self-confidence is an important predictor in future development and learning achievement. Al-Hebaish (2012) found a positive and significant correlation between self-confidence and academic achievement in general. That is, a sense of self-confidence in students is an attitude that is important for improving students' academic achievement in the implementation of integrated thematic learning. This is because students are required to have a critical attitude in responding to social phenomena that occur based on the themes discussed. This is different from the results of Woodman et al. (2010) that discused some benefits of self-doubt, which calls into question the widely accepted positive linear relationship between self-confidence and performance. According to Ghufron and Risnawita (2010), factors that can influence self-confidence include (1) self-concept, (2) self-esteem (see Colquhoun & Bourne, 2011), (3) experience, and (4) education.

2.4 *Students' academic achievement in integrated thematic learning in the 2013 curriculum*

The keys to success in the 2013 curriculum include the leadership of school principals, teacher creativity, student activities, socialization, learning resources and facilities, conducive academic environment, and school participation (Mulyasa, 2014). These variables are related to academic achievement or student achievement. Learning achievement is a measure of the success of student learning activities in mastering the lesson during a specified time. This can be considered as the success of students in carrying out the learning process, which is characterized by an increase in the ability at a certain time and can be expressed in the form of a score. Indicators of academic achievement include changes in students' abilities in aspects of knowledge, attitudes, and skills.

According to Syah (2014), learning achievement is influenced by internal factors (circumstances or physical and spiritual conditions of students), external factors (environmental conditions around students), and the approach to learning factors. In agreement with Hariri and Hattami (2017), the use of technology may produce improvements that significantly affect academic achievements in the learning achievement.

3 METHOD

3.1 *Types of research*

This preliminary study was carried out in the second semester of the 2017/2018 academic year. The research method used is descriptive qualitative with a case study approach. Qualitative research is particularly good at answering 'why', 'what' or 'questions'. The mass of words generated by interviews or observational data must be described and summarized (Lacey & Luff,

2007). The case study is an in-depth study of a social unit such that it produces a well-organized and complete picture of the social unit (Azwar, 2010). This qualitative research has two main objectives: first to describe and reveal and secondly to describe and explain social phenomena in the form of implementing thematic integrated learning in the 2013 curriculum in primary schools.

3.2 Subject and object of the research

Subjects in this study were teachers and 86 fourth grade students in three primary schools in Pacitan Regency. The subject retrieval technique in this study was purposive sampling. The research object studied was integrative thematic learning based on the implemented curriculum in Indonesia, student confidence, and academic achievement.

3.3 Data collection techniques and instruments

Data was obtained by observation, questionnaire, test, interview, documentation, and triangulation/combination techniques. The documentation in this study is in written form, as a plan for implementing theme 8, integrative thematic learning. It uses learning observation for the teacher and students in the introduction, core, and closing activities. Students are given a test at the end of learning on theme 8, Area of Residence, in the form of multiple choice tests, entries, and descriptions. The test is defined as a number of questions that require answers, or questions that must be given a response with the aim of measuring ability level (Mardapi, 2008). Another important technique is interview, which is used to discover things from respondents that are deeper and the number of respondents is small. The interview used is an unstructured interview: a free interview in which the researcher does not use interview guidelines that have been arranged systematically and completely for data collection (Sugiyono, 2011).

3.4 Data validity

The data in this study was tested using the validity test and triangulation. Triangulation techniques are used to obtain data from the same source with different techniques; that is, researchers confirm data from observation techniques, questionnaires, tests, and interviews on the same source. Researchers also confirm data from different sources (teachers and students) using the same data collection techniques, such as observations and interviews. In addition, researchers deepened the reference material to increase confidence in the results.

3.5 Data analysis techniques

The data analysis techniques from the data analysis model by Miles and Huberman (1992), include an interactive activity taking place continuously until complete, so that the data is saturated, including data reduction, data display, and conclusion drawing/verification. The data is reduced with selection techniques, simplification, and focus data obtained from various techniques and sources. After the data was reduced, it is presented in a descriptive form: words, numerical data, tables, pictures. Next the researcher describes and analyzes the data from the interview. At the end of the activity, the researcher conducts an examination of the research data to determine the consistency of the information provided by the subject so that a valid conclusion is obtained.

4 RESULTS

4.1 Integrated thematic learning implementation

The integrative thematic learning process in grade four runs in three stages: preliminary activities, core activities, and closing activities with a scientific approach. Based on observation activities of teachers and students in integrative thematic learning, the data was as follows.

Table 1.   Observation scores for teacher activities.

Description	1	2	3	4
Total score	89	87	91	96
Total indicator	27	29	27	28
Measurement score	3.3	3	3.37	3,42
Criteria *	Good	Good	Good	Good

* Poor (0–1), Fair (1.1–2.5), Good (2.51–3.5), Very Good (3.51–4)

Table 2.   Observation scores for student activities.

Description	1	2	3	4
Total score	90	92	91	94
Total indicator	30	30	30	30
Measurement score	3	3.07	3.03	3.13
Criteria *	GOOD	GOOD	GOOD	GOOD

Based on the recap observation, the data in the table above shows that the activities of teachers and students were in the 'good' criteria.

4.2   *The students' self-confidence*

There are four indicators observed in this process: paying attention, listening, the courage of students, and solving problems (Lauster, 1978), indicating an average criteria of 'good' (see Table 3).

During integrated thematic learning activities, students paid attention to the teacher's explanation. Some students listened to the teacher's direction while others ignored it. For the indicator of courage, students were confident to express their opinions. When the teacher pointed to something, the student was able to answer some questions, confident to convey the results of the work in front of the class, and keen to work on the questions according to the provisions in the book in an orderly manner.

4.3   *The students' academic achievement*

Students' learning achievements varied: some fulfilled the minimum completeness criteria and some did not. Students were still experiencing difficulties following the 2013 curriculum because an integrated thematic approach requires them to be critical, creative, and active in the learning process. Some students did not understand the learning delivered by the teacher. Test results show the level of students' maximum understanding at 30%, followed by optimal understanding (35%), good understanding (22%), and poor understanding (13%). The level of maximum understanding shows that students had understood the learning material learned. Optimal and good understanding show that students were able to understand learning with few obstacles. Then, poor understanding shows that students did not understand the learning material.

Table 3.   Observation score of students self-confidence.

Description	1	2	3	4
Total score	39	34	43	49
Mean	3	2.62	3.33	3,77
Criteria *	Good	Good	Good	Very Good

5 DISCUSSION

In general, integrated thematic learning activities have been carried out well in accordance with the design of the learning implementation and are a condition before the teaching and learning activities. The scientific approach used is relevant to Venville, Rennie, and Wallace (2004), as a learning approach that emphasizes students observing, asking, reasoning, trying and networking in school learning activities (Rusman, 2015).

The role of strategic teachers is related to the success of integrative thematic learning with a scientific approach, namely in pedagogic competencies, such as the development of learning plan tools, the implementation of teaching and learning interactions, the assessment of learning achievement, and the implementation of follow-up on the results of learning achievement. The integrative thematic model gives teachers the convenience of transferring learning experiences based on contextual themes, which is relatively easy to teach and learn (Clarke & Braun, 2013). This can also have a positive effect on students (Becker & Park, 2011).

However, there are difficulties for the teacher during the learning process, such as conditioning students 'activity during the learning process, returning students' enthusiasm before break time and fostering students 'motivation so this can affect students' academic achievement. This condition requires the teacher's pedagogic competence in understanding student characteristics, because intelligence, prior knowledge, learning styles, motivation, and social-cultural factors greatly affect the process and the result of learning (Budiningsih, 2011). It also requires teacher skills to attract students' interest, self-esteem, confidence, and attention.

It can be stated that confidence is likely to be influenced by strategic interest, perhaps consciously processed (Charness, Rustichini, & Van de Ven, 2018). There is also a positive correlation between self-esteem and academic performance (Colquhoun & Bourne, 2011). Self-esteem, goal orientation components and academic achievement are correlated (Rahmani, 2011). There were significant relationships between students' academic achievement and student engagement as well as between their academic achievement and especially the dimensions of cognitive engagement, behavioral engagement and sense of belonging (Gunuc, 2014).

Self-confidence could be predicted by cognitive and beliefs of competence in reasoning abilities (Kleitman & Stankov, 2007). Students must have positive perceptions and attitudes toward learning activities to succeed in academic achievement. Trust and self-confidence operators performed in supervisory control systems (Lee & Moray, 1994). Positive thinking can improve welfare, but it can also be self-defeating. As higher self-confidence enhances motivation this gives anyone with a vested interest in a student's performance an incentive to build up and maintains their self-esteem (Tirole, 2002).

Educators also can improve the students' academic achievements (Rivkin, 2005) and assist in the development of confidence and competency (Perry, 2011) in thematic and integrative learning processes. This can be developed, on a large scale, by identifying people who are in transition, with a need to change and therefore grow their self-confidence and create experiences, for example where people study cases of outstanding leaders, and use team-coaching or peer-coaching (Hollenbeck & Hall, 2004). In addition to teachers, parents factors have an important role to help students succeed in learning, in both material support and attitudes and expectations of student achievement (Yamamoto & Holloway, 2010). The strongest associations are found when the families have high academic expectations for their children, develop and maintain communication with them about school activities, and help them to develop reading habits (Castro et al, 2015).

6 CONCLUSION

The results of this study indicate first that teachers and students have implemented integrated thematic learning in the 'good' criteria in accordance with aspects of learning, which include preliminary activities, core activities and closing activities. The approach used is scientific. Second, the implications of this learning are the students' confidence in the criteria of 'good average'. This is shown from the four indicators: paying attention, listening, the courage of students and solving problems. Third, there is variation in aspects of students' academic achievement, in the category of optimal understanding through integrated thematic learning. The indicator is that students are able to understand learning with few obstacles. These results have theoretical and practical implications for teachers and parents. Teachers are expected to help students develop student confidence and competence through team-coaching and peer-coaching, as well as parenting programs.

REFERENCES

Al-Hebaish, S.M. (2012). The correlation between general self-confidence and academic achievement in the oral presentation course. *Theory and Practice in Language Studies, 2*(1), 60–65. Academy Publisher Manufactured.

Azwar, S. (2010). *Metodepenelitian* [Research method]. Yogyakarta, Indonesia: PustakaPelajar.

Batmalo, J.B. (2016). ImplementasipendekatansaintifikdalampembelajarantematikintegratifpadakelasV SekolahDasarNegeriNirmalaBantul [Implementation of the scientific approach in integrative thematic learning in the fifth grade of NirmalaBantul State Elementary School]. *JurnalPendidikan Guru SekolahDasar.*

Becker, K., & Park, K. (2011). Effects of integrative approaches among science, technology, engineering, and mathematics (STEM) subjects on students' learning: A preliminary meta-analysis. *Journal of STEM Education, 12*(5), 23–38. doi:10.1037/a0019454.

Budiningsih, C.A. (2003). Karakteristiksiswasebagaipijakandalampenelitiandanmetodepembelajaran [Characteristics of students as a foothold in research and learning methods]. *CakrawalaPendidikan (JurnalIlmiahPendidikan), 1*,160–173. doi:10.21831/cp.v1i1.4198.

Castro, M., Expósito-Casas, E., López-Martín, E., Lizasoain, L., Navarro-Asencio, E., & Gaviria, J.L. (2015). Parental involvement on student academic achievement: A meta-analysis. *Educational Research Review. 14*(2015), 33–46. doi:10.1016/j.edurev.2015.01.002.

Charness, G., Rustichini, A., & Van de Ven, J. (2018). Self-confidence and strategic behavior. *Experimental Economics, 21*(1), 72–98. doi:10.1007/s10683-017-9526-3.

Clarke, V., & Braun, V. (2013). Teaching thematic analysis: Over-coming challenges and developing strategies for effective learning. *The Psychologist, 26*(2), 120–123. doi:10.1191/1478088706qp063oa.

Colquhoun, L.K. & Bourne, P.A. (2011). Self-esteem and academic performance of 4th graders in two elementary schools in Kingston and St. Andrew, Jamaica. *Asian Journal of Business Management 4*(1), 36–57. Retrieved from http://mexwellsci.com/print/ajbm/v4-36-57.

Ghufron, M.N. & Risnawita, S.R. (2010). *Teori-teoripsikologi* [Psychology theories]. Yogyakarta, Indonesia: Ar-Ruzz Media Group.

Gunuc, S. (2014). The Relationships between student engagement and their academic achievement. *International Journal on New Trends in Education and Their Implication, 5*(4), 216–231. Retrieved from www.ijonte.org.

Gutek, G.L. (1974). *Philosophical alternativeineducation.* Columbus, OH: A Bell & Howell Company.

Hariri, M.T. & Hattami, A. (2016). Impact of students' use of technology on their learning achievements. *Journal of Taibah University Medical Sciences, 12*(1), 82–85. doi:10.1016/j.jtumed.2016.07.004.

Hollenbeck, G.P. & Hall, D.T. (2004). Self-confidence and leader performance. *Organizational Dynamics, 33*(3), 254–269. doi:10.1016/j.orgdyn.2004.06.003.

Jihad, A. & Haris, A. (2013). *Evaluasipembelajaran* [Learning evaluation]. Yogyakata, Indonesia: Multi Pressindo.

Klein, J.T. (2005). Integrative learning interdisciplinary studies. *Peer Review, 7*(4), 8–10. doi:10.1108/17506200710779521.

Kleitman, S., & Stankov, L. (2007). Self-confidence and metacognitive processes. *Learning and Individual Differences, 17*(2), 161–173. doi:10.1016/j.lindif.2007.03.004.

Lacey, A. and Luff, D. (2007). Qualitative research analysis. The NIHR RDS for the East Midlands/ Yorkshire & the Humber.

Lauster, P. (1978). *The personality test.* London: Pan Books.

Lee, J.D., & Moray, N. (1994). Trust, self-confidence, and operators' adaptation to automation. *International Journal of Human-Computer Studies, 40*(1), 153–184. doi:10.1006/ijhc.1994.1007.

Majid, M. (2014). *Strategipembelajaran* [Learning strategy]. Bandung, Indonesia: PT RemajaRosdakarya.

Mardapi, D. (2008). *Teknikpenyusunaninstrumentesdannontes* [Techniques for preparing test and non-test instruments]. Yogyakarta, Indonesia: MitraCendikia.

Milles, M.B., & Huberman, M. (1992). *Analisis data kualitatif* [Qualitative data analysis]. Jakarta, Indonesia: Universitas Indonesia Press.

Moss & Lee. (2010). A critical analysis of philosophies of education and INTASC standards in teacher preparation. *International Journal of Critical Pedagogy, 3*(2), 36–46.

Mulyasa. (2014). *Pengembangandanimplementasikurikulum 2013* [Development and implementation of the curriculum 2013]. Bandung, Indonesia: PT. RemajaRosdakarya.

Perry, P. (2011). Concept analysis: Confidence/Self-confidence. *Nursing Forum, 46*(4), 218–230. doi:10.1111/j.1744-6198.2011.00230.x.

Rahmani, P. (2011). The relationship between self-esteem, achievement goals and academic achievement among the primary school students. *In Procedia – Social and Behavioral Sciences 29*9(2011), 803–808. doi:10.1016/j.sbspro.2011.11.308.

Randle, I. (2010). *The* measure of success: integrated thematic instruction. *Journal of Educational Strategies Issues and Ideas, 71*(2), 85–87. doi: 10.1080/00098659709599331.

RoI. (2003). *Undang-undangRepublik Indonesia Nomor 20 Tahun 2003 tentangSistemPendidikanNasional* [Law Number 20 of 2013 concerning National Educational System]. Jakarta, Indonesia: Republic of Indonesia.

Rivkin, S.G., Hanushek, E.A., & Kain, J.F. (2005). Teachers, schools, and academic achievement. *Econometrica, 73*(2), 417–458. doi: 10.1111/j.1468-0262.2005.00584.x.

Rusman. (2015). *Pembelajarantematikterpadu: teoripraktikdanpenilaian* [Integrated thematic learning: practice theory and assessment]. Jakarta, Indonesia: Rajawali Pers.

Sugiyono. (2011). *Metodepenelitiankuantitatif, kualitatif, dan R&D* [Quantitative, qualitative research, and R & D]. Bandung, Indonesia: Alfabeta.

Syah, M. (2014). *Psikologipendidika n* [Educational psychology]. Bandung, Indonesia: PT. Remaja Rosdakarya.

Tirole, R.B.J. (2002). Self-confidence and personal motivation. *The Quarterly Journal of Economics, 117*(3), 871–915. doi:10.1162/003355302760193913.

Tisngati, U. & Meifiani, N.I. (2014). *Studiterhadappolaasuhorangtua, kecemasandankepercayaandiri.* Yogyakarta, Indonesia: NuhaMedika.

Venville, G., Rennie, L., & Wallace, J. (2005). Student understanding and application of science concepts in the context of an integrated curriculum setting. *International Journal of Science and Mathematics Education, 1*(4), 449–475. doi: 10.1007/s10763-005-2838-3.

Yamamoto, Y., & Holloway, S.D. (2010). Parental expectations and children's academic performance in sociocultural context. *Educational Psychology Review, 22*(3), 189–214. doi:10.1007/s10648-010-9121-z.

Innovative Teaching and Learning Methods in Educational Systems – Retnowati et al. (Eds)
© 2020 Taylor & Francis Group, London, ISBN 978-1-03-224183-8

The effect of technology literacy and learning environment on student motivation in the educational revolution 4.0

A. Saputri Sukirno & H. Kurniawan
Universitas Negeri Yogyakarta, Indonesia

H.D. Hermawan
The University of Hong Kong, Hong Kong, P.R. China

ABSTRACT: Globalization in the field of education has led to the educational revolution 4.0, where technology literacy and the learning environment become important components. However, not all schools have complete facilities and suitable infrastructure for this era. This study aims to determine the effect of technology literacy and the learning environment on students' learning motivation in the educational revolution 4.0. This study used quantitative research that shows a causal relationship with a population of students in senior high school. The sample of this study was 120 students of senior high school. The Theory of Reasoned Action (TAR) supports that certain student behavior, in this case student motivation, arises due to external stimuli. This study found three hypotheses that were accepted. This study concludes that there is a significant influence of technology literacy and the learning environment simultaneously on students' learning motivation. This research contributes to educational institutions in decision-making and empirical theory.

1 INTRODUCTION

The educational revolution 4.0 has become a key conversation on the readiness of educational institutions in the face of rapid scientific development. In this era, students must be able to compete with the changing times. They have to develop their learning motivation to raise their achievements within the educational revolution 4.0. Therefore, it is necessary to conduct research to analyze school preparedness in facing the educational revolution 4.0. Demand is growing for a revolution in education (Ciolacu, Tehrani, Beer, & Popp, 2017) and schools need to follow this change.

To face the challenges of the era, senior high schools must be ready for education 4.0. But in reality, the results are still far from the target so adequate preparation is needed to improve the quality of education. This study aims to analyze the readiness of schools in facing education 4.0. Technology literacy and the learning environment are expected to be the significant factors influencing student's learning motivation. The Theory of Reasoned Action (TAR) supports that students behavior, in this case, student motivation, arises due to external stimuli (Trafimow, 2009). The importance of technology literacy lies in that it transcends the conventional technological skill-gaining process by offering a better understanding of digital surroundings, which in turn will equip people with intuitive adaptability to new contexts and the ability to coordinate creatively with others.

Data was collected through questionnaires to 120 students of senior high school. The data was presented in descriptive statistics for independent and dependent variables of each aspect studied based on students' perceptions of school readiness for education 4.0.

Senior high schools have a crucial role to play in shaping the societal transitions necessary and prepare thinkers, for the fourth industrial revolution (Gealson, 2018). The use of technology in learning has brought welcome change and supports active learning.

Students can use technology to find learning resources, and upload tasks assigned by the teacher. But, in reality, not all schools have complete facilities and appropriate infrastructure. This will have an impact on students' motivation in learning. Students play an important role in assessing school readiness for facing the industrial revolution in the field of education. Therefore, research needs to be done to determine the readiness of schools based according to the perceptions of students. The purpose of this study was to determine the effect of technology literacy and the learning environment on student learning motivation in educational revolution 4.0.

2 STUDENT MOTIVATION

Motivation is a theoretical construct used to explain the initiation, direction, intensity, persistence, and quality of behavior, especially goal-directed behavior (Maehr & Meyer, 1997). Motivation is a construct that helps to explain why individuals choose to approach or avoid a task, and, once engaged, whether they put in the effort and persist or simply give up (Kumar, Zusho, & Bondie, 2018). Motivation is an important but invisible factor that can affect students' willingness to learn and rate of learning in either positive or negative ways (Mclaughlin, 2018). In the classroom context, student motivation is used to explain the degree to which students invest attention and effort in various pursuits, which may or may not be the pursuits desired by the teacher. Student motivation is rooted in students' subjective experiences, especially those connected to their willingness to engage in learning activities and students' reasons for doing so (Brophy, 2010).

3 TECHNOLOGY LITERACY

Today the concept of literacy is expanding to include the media and electronic text, in addition to alphabetic and number systems (Musingafi et al., 2012). Technology moves the learning experience in a better direction (Evans, 2001). This suggests the importance of technology in the educational environment. Technology literacy in senior high schools should include knowing the basics of hardware, software, ethics, and etiquette and how to apply these to learning (Ejikeme & Okpala, 2017). Dynamic and stimulating literate environments at home, in the classroom, in the workplace, and in the community are essential to literacy acquisition, development and life-long use (Krolak, 2006).

Use of information technology is a sign of school readiness for education 4.0. Information technology should be coordinated and synchronized to streamline, improve, and facilitate human work (Bubel & Cichoñ, 2017). Information technology provides potential opportunities for obtaining competitive advantage and integration of digital competence in education is highly important for future knowledge acquisition (Gran, 2016). Technology contributes to increased learning outcomes in higher education, and students perceive that technology contributes to flexibility in their studies (Tømte & Olsen, 2013). Technology such as media learning, social media, and internet networks could have several motivational benefits for the students (Jones & Issroff, 2016).

4 LEARNING ENVIRONMENT

Learning environment can be defined as the place where learning occurs (Arnold, 2019). Within a learning environment, students can make exchanges, participate in activities, make social interactions, and communicate through various channels to share work, test knowledge, inquire, research, collaborate, express themselves, and show understanding. Therefore learning environment influences student motivation. Students become active and enjoy learning when school offers a conducive learning environment. Thus it is important to know the effects of the learning environment on student motivation in senior high school.

More challenging learning environments lead to enhanced transfer and potentially better-prepared graduates (Bjork & Linn, 2006). At the same time, a safe environment for students is to participate in learning processes without fear of being wrong or appearing to lack necessary knowledge (Hawe & Dixon, 2017). A learning environment gives students the main responsibility for their own learning by considering them to be active agents (Bruno & Dell'Aversana, 2017). An adequate learning environment can motivate students to study harder. The learning environment provided by the education institution illustrates school readiness in the face of the education revolution.

5 METHOD

This study uses a quantitative approach with correlation research to determine whether there is an influence between variables of technology literacy and the learning environment on student learning motivation in the educational revolution 4.0. The design of this study is the relationship between the variables technology literacy, learning environment, and student motivation. Therefore this research is correlational – namely, research that describes the relationship of the three variables.

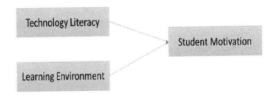

Figure 1. Hypothesized conceptual model.

The sample in this study were 120 high school students, determined using proportional random sampling. Data collection was done using a questionnaire. The data analysis techniques used descriptive statistical analysis and inferential analysis techniques to examine the research variables.

6 RESULT

6.1 *Primary quantitative analysis*

Hypothesis one predicted that technology literacy influences students' learning motivation in the educational revolution 4.0. This hypothesis was supported. The significance level used was 0.05 and the significance value was 0.030 so hypotheses one is accepted. There is a significant influence on technology literacy on student motivation.

Table 1. Descriptive statistics for independent variables.

Coefficients[a]					
	Unstandardized coefficients		Standardized coefficients		
Model	B	Std. Error	Beta	t	Sig.
(Constant)	10.723	5.099		2.103	0.038
Technology literacy	0.189	0.086	0.186	2.199	0.030
Learning environment	0.472	0.083	0.483	5.699	0.000

Table 2. Descriptive statistics for independent and dependent variables.

ANOVA[a]

Model		Sum of squares	df	Mean square	F	Sig.
1	Regression	1355.002	2	677.501	32.083	.000[b]
	Residual	2470.698	117	21.117		
	Total	3825.7	119			

Hypothesis two stated that learning environment influences students' learning motivation in educational revolution 4.0. This hypothesis was supported. The significance level used was 0.05 and the significance value was 0.000 so hypotheses two is accepted. There is a significant influence of learning environment on student motivation.

Hypothesis three predicted that there is a significant influence of technology literacy and learning environment simultaneously on students' learning motivation in the educational revolution 4.0. This hypothesis was supported. The significance level used was 0.05. The obtained F was 32.083. The significance value was 0.000 so hypothesis three is accepted. There is a significant influence on technology literacy and the learning environment on student motivation.

7 DISCUSSION

This study aims to determine the effect of technology literacy and the learning environment on students' learning motivation in the educational revolution 4.0. Today, senior high school, in Indonesia already use technology for distance learning. Through implementing technology, the students can download materials and tasks that teachers have uploaded for each course, then upload their completed assignments. Information technology was originally used only in the subjects of ICT (Information and Communication Technology) but later developed for all subjects. Use of information technology is a sign that schools keep up with the demands of the times. The use of technology by students includes learning media, social media, and internet networks.

The Theory of Reasoned Action (TAR) support that students behavior, in this case, student motivation, arises due to external stimuli (Trafimow, 2009). The external stimuli referred to are technology literacy and the learning environment. The importance of technology literacy is that it transcends the conventional technological skill-gaining process by offering a better understanding of digital surrounding, which in turn will equip people with an intuitive adaptability to new contexts and the ability to coordinate creatively with others. Similar research proves that learning environment has an impact on student behavior, in this case, student motivation (Van Der Kleij, Eggen, Timmers, & Veldkamp, 2012).

8 CONCLUSION

This study concludes that technology literacy and the learning environment influence students' learning motivation in the educational revolution 4.0. There is a significant influence of technology literacy and the learning environment, simultaneously, on students' learning motivation in educational revolution 4.0. This research contributes to educational institutions in decision-making and empirical theory. This research has been informative for science education literature by giving empirical evidence that student motivation to learn is affected by technology literacy and the learning environment.

REFERENCES

Arnold, B.A. (2019). The seven traits of a learning environment: a framework for evaluating mobile learning engagement. *International Journal of E-Education, e-Business, e-Management and e-Learning*, 9(1), 54–60. doi: 10.17706/ijeeee.2019.9.1.54-60.

Bjork, R., & Linn, M. (2006). The science of learning and the learning of science. *APS Observer, 19*(3), 6–7. Retrieved from http://bjorklab.psych.ucla.edu/pubs/RBjork_Linn_2006.pdf.

Brophy, J. (2010). *Motivating Students to Learn*. New York, NY: Routledge 270 Madison Avenue.

Bruno, A., & Dell'Aversana, G. (2017). Reflective practicum in higher education: the influence of the learning environment on the quality of learning. *Assessment & Evaluation in Higher Education, 2938* (July), 1–14. doi: 10.1080/02602938.2017.1344823.

Bubel, D., & Cichoň, S. (2017). Role of information in the process of effective management of the university. *International Journal of Innovation and Learning, 21*(1), 114–125. doi: 10.1504/IJIL.2017.080756.

Ciolacu, M., Tehrani, A.F., Beer, R., & Popp, H. (2017). Education 4.0 – fostering student's performance with machine learning methods. doi: 10.1109/SIITME.2017.8259941.

Ejikeme, A.N., & Okpala, H.N. (2017). Promoting children's learning through technology literacy: challenges to school librarians in the 21st century. *Education and Information Technologies, 22*(3), 1163–1177. doi: 10.1007/s10639-016-9481-1.

Evans, J.R. (2001). The emerging role of the internet in marketing education: from traditional teaching to technology based education. *Marketing Education Review, 11*(3), 1–14. doi: 10.1080/10528008.2001.11488753.

Gealson, N.W. (2018). *Higher Education in the Era of the Fourth Industrial Revolution*. doi: 10.1007/978-981-13-0194-0.

Gran, L. (2016). Sammenhengen mellom dannelse og kompetanse – elevens metakognitive læringsprosess [The connection between education and competence – the students' metacognitive learning process], (15), 1–26. doi: 10.7146/lom.v9i15.23367.

Hawe, E., & Dixon, H. (2017). Assessment for learning: a catalyst for student self-regulation. *Assessmenttpdel and Evaluation in Higher Education, 42*(8), 1181–1192. doi: 10.1080/02602938.2016.1236360.

Jones, A., & Issroff, K. (2016). Motivation and mobile devices: exploring the role of appropriation and coping strategies motivation and mobile devices: exploring the role of appropriation and coping strategies, *7769*(February 2017). doi: 10.1080/09687760701673675.

Krolak, L. (2006). The role of libraries in the creation of literate environments literate environments. *International Journal Of Adult And Lifelong Education, 4*(1/4), 5.

Kumar, R., Zusho, A., & Bondie, R. (2018). Weaving cultural relevance and achievement motivation into inclusive classroom cultures. *Educational Psychologist, 53*(2), 78–96. doi: 10.1080/00461520.2018.1432361.

Maehr, M.L., & Meyer, H.A. (1997). Understanding motivation and schooling: where we've been, where we are, and where we need to go. *Educational Psychology Review Expertise in Counseling Psychology, 9*(4), 371–409. doi: 10.1023/A:1024750807365.

Mclaughlin, M.J. (2018). Tracking dynamic changes in student motivation in the english discussion classroom. *New Directions in Teaching and Learning English Discussion, 6*, 201–208.

Musingafi, M.C.C., Chiwanza, K., Coordinator, P., Studies, D., Resolution, C., Campus, M.R., & Science, I. (2012). The role of public libraries in promoting literacy in Zimbabwe, *2*(7), 52–61.

Tømte, C., & Olsen, D.S. (2013). *Kvalitativ undersøkelse om hvordan IKT påvirker læring i høyere utdanning* [Qualitative study on how ICT affects learning in higher education]. Oslo: NIFU.

Trafimow, D. (2009). The theory of reasoned action: a case study of falsification in psychology. *Theory & Psychology, 19*(4), 501–518. doi: 10.1177/0959354309336319.

Van Der Kleij, F.M., Eggen, T.J.H.M., Timmers, C.F., & Veldkamp, B.P. (2012). Effects of feedback in a computer-based assessment for learning. *Computers and Education, 58*(1), 263–272. doi: 10.1016/j.compedu.2011.07.020.

Innovative Teaching and Learning Methods in Educational Systems – Retnowati et al. (Eds)
© 2020 Taylor & Francis Group, London, ISBN 978-1-03-224183-8

Student perceptions of motivational strategies used by junior high school English teachers in Kota Yogyakarta

G. Ambarini & Ashadi
Universitas Negeri Yogyakarta, Yogyakarta, Indonesia

ABSTRACT: The study aims at finding the student perceptions of motivational strategies used by junior high school teachers in Kota Yogyakarta. The research sample was junior high school students from seven schools in Kota Yogyakarta. In groups of motivated A grade students at junior high schools, the results of the data analysis established that the students had very positive perceptions of the use of motivational strategies by English teachers, meaning that the teachers often applied motivational strategies while teaching English. There was a significant difference in the use of motivational strategies in general ($p = 0.000$) with creating the basic motivational conditions (CBMC) used more often than other strategies. There was also a significant difference in the students' perceptions of the use of motivational strategies based on teaching experience ($p = 0.000$): those who had teaching experience of more than 20 years applied motivational strategies more often than those with less than 20 years of experience. This result can give insights to teachers on the importance of using motivational strategies with students, that in turn will positively affect learning achievement, and help them reflect on their teaching practices.

1 INTRODUCTION

1.1 *Background of the study*

Motivation is one of the key factors in Second Language (L2) learning. Highly motivated students are likely to be better L2 learners than those with low motivation (Oxford, Oh, Ito, & Sumrall, 1993; Bernaus, Wilson, & Gardner, 2009; Chiang, Yang, Huang, & Liou, 2014). Motivation may come from inside or outside of learners, and factors surrounding them can be important determinants of the quality of their motivation. Motivational strategies are the efforts made by people around learners to enhance the learners' motivation. Teachers can make an important contributes to student motivation in learning because they have direct contact with the students while they are learning at school.

There have been studies focusing on the relationship between student motivation and achievement (Bernaus & Gardner, 2008; Wang, 2008; Li & Pan, 2009; Kurum, 2011). Additionally, both variables are positively connected, whereas a student with high motivation achieves more than those with low motivation, and vice versa. Many studies on the issue of motivational strategies and their use in language instruction have been conducted in various countries (Cheng & Dörnyei, 2007; Bernaus & Gardner, 2008; Guilloteaux & Dörnyei, 2008; Vibulphol, 2016). The findings of the studies that focus on the use of motivational strategies in different settings suggest that some strategies may be universal while others are culturally dependent. Thus, there is a need to investigate the use of motivational strategies in Indonesia to get detailed information in this specific context.

As can be seen from various previous studies, there is a positive relationship between motivational strategies and student motivation. Therefore, the sample was controlled by only involving schools that had students with high motivation, to determine the strategies that are likely to work best when implemented in Indonesia. This study investigated the use of motivational strategies by English teachers from the students' perspective as the students are the ones directly influenced by the methods. Looking at this issue from the students' perspective should determine which strategies that are well-received by the students.

1.2 Formulation of the problem

There are three sub-questions that the study tries to answer:

1. What are the students' perceptions of the use of motivational strategies by English junior high school teachers in Kota Yogyakarta?
2. Is there any significant difference in the students' perceptions of the use of motivational strategies by junior high school English teachers in Kota Yogyakarta?
3. Is there any significant difference in students' perceptions of the use of motivational strategies by junior high school English teachers in Kota Yogyakarta based on teaching experience?

The first research question is answered with a descriptive analysis of the data with the assumption that the use of motivational strategies is high. The hypotheses for the second and third research questions are:

H1: there is a significant difference in the use of motivational strategies.

H2: there is a significant difference in the use of motivational strategies based on teaching experience.

2 LITERATURE REVIEW

2.1 Motivation in L2 learning

Murphy and Alexander (2000) defined motivation as the term used to describe the force to make someone do something that will influence the quality and persistence of their behavior. Regarding motivation in L2 learning, according to Dörnyei (2001) a classroom is an intricate setting where students spend a large part of their lives and where so much is going on in their lives. Thus, for practical purposes, we need a detailed and eclectic construct that can represent multiple perspectives. Ortega (2013: 168) proposed that motivation is usually understood to refer to "the desire to initiate L2 learning and the effort employed to sustain it." On the other hand, Kimura, Nakata, & Okumura (2001) stated that motivation is the force that drives learners to use the available learning resources optimally to achieve their learning objectives. Therefore, it can be concluded that motivation in L2 learning helps learners generate and maintain learning and is an essential ingredient in acquiring a foreign language.

2.2 Motivational strategies

Hennessey emphasized that no one is intrinsically motivated under any circumstance (Wery, & Thomson, 2013). Wery and Thomson (2013) further explained that extrinsic motivators can be essential to engage reluctant learners. Motivational strategies have been long studied by some researchers; Dörnyei and Csizer (1998) proposed suggestions for sustaining and increasing the motivation of foreign language learners. In his more recent work, Dörnyei (2011:103) stated that the purpose of motivational strategies is to "consciously generate and enhance student motivation, as well as maintain ongoing motivated behavior and protect it from distracting and/or competing for action tendencies." The strategies include four motivational dimensions:

1) CBMC, involving setting the scene for the effective use of motivational strategies.
2) Generating initial motivation (GIM), corresponding roughly to the pre-action phase in the model.

3) Maintaining and protecting motivation (MPM), corresponding to the action phase.
4) Encouraging positive retrospective self-evaluation (EPSE), corresponding to the post-action phase.

2.3   Relevant studies

There have been several studies conducted concerning the use of motivational strategies in classrooms. Cheng and Dörnyei (2007) conducted a similar study entitled "The Use of Motivational Strategies in Language Instruction: the Case of EFL Teaching in Taiwan". The findings of the research showed that some motivational strategies are transferable across diverse cultural and ethno linguistic contexts. The following year, Guilloteaux and Dörnyei (2008) conducted a similar study, called "Motivating Language Learners: A Classroom-Oriented Investigation of the Effects of Motivational Strategies on Student Motivation". The findings contribute to the theory that the motivational strategies used by language teachers do positively affect the students' motivation. A study by Gardner and Bernaus (2008), entitled "Teacher Motivation Strategies, Student Perceptions, Student Motivation, and English Achievement" looked at 31 EFL teachers and their students. The findings of the study indicated that teachers and students agreed on the relative frequency of some strategies, but not on the frequency of other strategies.

Here we investigated the relationship among the following variables: teachers' preference in the use of motivational strategies, students' perception of the use of motivational strategies, and students' motivation. As there have not been many studies conducted to inspect those variables in Indonesia, interesting findings are expected to emerge from this study.

3   RESARCH METHOD

3.1   Participants

The sample was obtained through simple random sampling. There were a total of seven schools included in this study. The selection process of the sample was made through a lottery method.

3.2   Data collection procedures

The students were administered the survey twice, with permission from the English teacher. The students were given information on how to complete the questionnaire; they were assured that the information would be kept confidential and anonymous and it would not affect their grade in any way.

Table 1.   The number of participants from each school.

School	Number of students
SMP A	32
SMP B	31
SMP C	30
SMP D	24
SMP E	28
SMP F	18
SMP G	32
Total	195

3.3 Instruments

There were four instruments used in this study: two types of questionnaire, observation checklist, and interview guideline. The first questionnaire was aimed at assessing student motivation using an Attitude and Motivation Test Battery (AMTB) developed by Gardner (2001) with few modifications. The second questionnaire was used to find out the motivational strategies used by the English teacher and consisted of a list of strategies taken from Dörnyei (2001). The observation checklist also consisted of the same list to see the observable strategies applied in class. The interview guidelines were used as guidance while interviewing the students and teachers during the preliminary study and the main study.

3.4 Data collection techniques

The data was analyzed using quantitative descriptive and inferential statistics. The quantitative descriptive method was used to classify students' perceptions of the use of motivational strategies by their English teachers into qualitative categories. Inferential statistic using the Kruskal-Wallis Test were used to investigate the differences in students' perceptions of teachers' motivational strategies based on teaching experience.

4 RESULT AND DISCUSSION

4.1 Result

Before conducting the main study, the researcher made a preliminary observation to ascertain a general view of the use of motivational strategies in class. While conducting the observation, the researcher also distributed an AMTB adapted from Gardner (1985) to gauge the students' levels of motivation, as in the selection of the sample there was a pre-requirement for the students to have a high level of motivation, with an assumption that the use of motivational strategies was also high.

The items used to measure motivation are classified into three scales: attitudes toward learning the language, motivation intensity, and desire to learn the language (Gardner, 2001).

The categorization of each aspect was determined using a formula proposed by Widoyoko (2009: 238) and is shown in Table 2.

Each aspect had a different number of items and different scales. Therefore, each aspect's categorization was calculated separately. Calculating the category scores for each aspect gave us the student motivation results, as described in Table 3.

From Table 3, we can see that the students had high scores in all aspects, except for the Desire to Learn English in SMP 13 Yogyakarta. The researcher decided to include schools with scores of sufficient or above, resulting in SMP 13 being included in this study. In addition, if we look at the total scores in Table 4, we can see that student motivation was high in every school.

Table 2. Qualitative category of the product (Widoyoko, 2009: 238).

No.	Scores interval (i)	Qualitative category
1.	$X > (\underline{X_i} + 1.8\ sbi)$	Very high
2.	$(\underline{X_i} + 0.6\ sbi) < X \leq (\underline{X_i} + 1.8\ sbi)$	High
3.	$(\underline{X_i} + 0.6\ sbi) < X \leq (\underline{X_i} + 1.8\ sbi)$	Sufficient
4.	$(\underline{X_i} - 1.8\ sbi) < X \leq (\underline{X_i} - 0.6\ sbi)$	Insufficient
5.	$X \leq (\underline{X_i} - 1.8\ sbi)$	Very Low

Table 3. The students' score of each motivational aspect.

School	ALL		MI		DLE	
	Score	Category	Score	Category	Score	Category
SMP A	25.19	High	22.91	High	22.47	High
SMP B	23.23	High	23.77	High	21.71	Sufficient
SMP C	25.33	High	24.50	High	23.20	High
SMP D	24.92	High	24.21	High	22.08	High
SMP E	25.14	High	25.46	High	23.57	High
SMP F	26.94	Very high	25.72	High	25.28	High
SMP G	25.14	High	25.46	High	23.57	High

Table 4. The students' total score of the motivational aspects.

School	Total score	Category
SMP A	70.57	High
SMP B	68.71	High
SMP C	73.03	High
SMP D	71.21	High
SMP E	74.17	High
SMP F	77.94	High
SMP G	74.17	High

Here, we can see that, overall, student motivation was high in every school, meaning that the main research in those schools could proceed.

To answer the first research question of whether the use of motivation strategies was high, we analyzed the data descriptively from the results of the questionnaire. To categorize the data qualitatively for the motivational strategies, we also used the formula by Widoyoko (2009: 238). As the number of items for each type of strategy was different, we calculated the categorization separately. The results are shown in Table 5.

From Table 5, we can see that almost all of the students in each school had high perceptions of the use of each motivational strategy. Some of the scores were very high, and some were sufficient. Knowing that the use of motivational strategies was high in each school, we calculated the data again to ascertain whether the students' perceptions of the use of each strategy was high in all schools, by looking at the mean score. The result of the calculation can be seen in Table 6.

When we calculated the data collectively, we found that all of the strategies were highly used in the sample schools. The results respond to the hypothesis that the use of motivational strategies was high and this influenced the students' high motivation.

Next, we analyzed whether there was any significant difference in the use of each strategy to see which one contributed the most to the students' high motivation. Because the number of items in the questionnaire for each strategy was different, we made the score equal by using the percentage of the scores to get the descriptive data before comparing them to see whether the difference was significant or not.

From the descriptive data, we found that the strategy with the highest score was CBMC at 75.00 and the lowest score was GIM at 69.57. Thus, the CBMC set of strategies was used the most often and GIM was used least.

To determine whether there was a significant difference in the use of the motivational strategies, we conducted non-parametric analysis with a Kruskall-Wallis test, as

Table 5. The score of motivational strategies in each school.

School	CEBC		GIM		MPM		EPSE	
	Score	Category	Score	Category	Score	Category	Score	Category
SMP A	11.90	High	10.43	High	19.03	High	7.06	High
SMP B	10.96	High	9.22	High	18.29	Sufficient	6.41	High
SMP C	13.53	Very high	13.30	Very high	24.06	Very high	9.06	Very high
SMP D	13.20	Very high	12.95	High	24.75	Very high	8.67	Very high
SMP E	10.07	High	8.85	Sufficient	16.25	Sufficient	6.35	High
SMP F	10.78	High	8.88	Sufficient	16.22	Sufficient	5.77	Sufficient
SMP G	13.31	Very high	10.78	High	21.37	High	7.28	High

Table 6. The collective score of motivational strategies.

CBMC		GIM		MPM		EPSE	
Score	Category	Score	Category	Score	Category	Score	Category
12.03	High	10.68	High	20.11	High	7.28	High

Table 7. Descriptive data of the use of motivational strategies.

	Mean	Standard error
CBMC	75.00	1.34
GIM	66.76	1.47
MPM	69.37	1.50
EPSE	72.82	1.72

the data was not normally distributed ($p < 0.05$). We tried to transform the data to make it normally distributed but unfortunately this was unsuccessful. Therefore, we used non-parametric analysis with a Kruskall-Wallis test to determine whether there was any difference in the motivational strategies used. The result of the Kruskall-Wallis test are shown in Table 8.

From Table 8, we can see p-value of 0.000 (< 0.05), which means that there is a significant difference in the use of motivational strategies by junior high school English teachers in Indonesia, with CBMC strategies used the most, followed by EPSE, MPM, and GIM.

Next, we tried to answer the third research question to determine whether there was any significant difference in the use of motivational strategies in general based on teaching experience. The result of the test is presented in Table 9.

Based on the result of the test, we acquired a p-value of 0.000 (< 0.05). This means there was a significant difference in the use of motivational strategies in general based on teaching

Table 8. The result of Kruskall-Wallis test on the use of motivational strategies.

	Score
Kruskal-Wallis H	19.979
df	3
Asymp. Sig.	.000

42

Table 9. The result of Kruskall-Wallis test on the collective data of the use of motivational strategies based on teaching experience.

	Motivational strategies
Kruskal-Wallis H	35.820
df	1
Asymp. Sig.	.000

experience, and the teachers who had more than 20 years of teaching experience applied motivational strategies more often than those with less than 20 years of teaching experience.

4.2 *Discussion*

There was a prerequisite for the students to be able to take part in this study: they were highly motivated in L2 learning. This led to the hypothesis that the motivational strategies used by the English teachers were also high. To test the hypothesis, the data was analyzed descriptively; the category of the score was high, and this showed that, overall, the teachers often applied motivational strategies in teaching English. However, to find out whether the difference in the use of motivational strategies was significant or not, a Kruskall-Wallis test was conducted because the data was not normally distributed even after being transformed. From the result of the test, we obtained a p-value of 0.000 (< 0.05), indicating that there was a significant difference in the students' perceptions of the use of motivational strategies, with the CBMC strategies used more often than the other three sets. When interviewed about the role of an English teacher as an agent to motivate students, most students agreed that teachers played an important role in motivating the students. However, some stated that other factors also had equally important roles, including parent involvement and peer pressure.

The results of this study show that there was a positive relationship between the use of motivational strategies and student motivation in learning English. However, there needs to be further research on how other factors such as parent involvement and peer pressure contribute to student motivation, to determine what extent motivational strategies influence student motivation. Regarding the students' perceptions of the use motivational strategies by English teachers based on teaching experience, we discovered that those who had more than 20 years of teaching experience applied more motivational strategies than those who had less than 20 years of experience. The difference was also significant at p = 0.001. This result is understandable, considering that teachers who had been teaching for a longer time had more experience in dealing with students.

5 CONCLUSIONS

The researcher determined several conclusions from the data analysis. First, in groups of motivated A grade students at junior high schools, the results of the data analysis established that the students had high perceptions of the use of motivational strategies by English teachers, meaning that the teachers often applied motivational strategies while teaching English. There was a significant difference in the use of motivational strategies in general (p = 0.000) with CBMC used more often than other strategies. There was also a significant difference in the students' perceptions of the use of motivational strategies based on teaching experience (p = 0.000), as teachers who had more than 20 years of experience applied motivational strategies more often than those with less than 20 years of experience. This result can give insights to teachers on the importance of using motivational strategies to motivate students, that in turn will positively affect learning achievement, and help them reflect on their teaching practices.

REFERENCES

Al-Mahrooqi, R., Abrar-Ul-Hassan S., & Cofie, C. (2012). Analyzing the use of motivational strategies by EFL teachers in Oman. *Malaysian Journal of EFT Research, 8*(1), 36–6.

Asante, C., Al-Mahrooqi, R., & Abrar-Ul-Hassan, S. (2012). The effects of three teacher variables on the use of motivational strategies in EFL instruction in Oman. *TESOL Arabia Perspectives, 19*(1), 12–22.

Bernaus, M., Wilson, A., & Gardner, R. (2009). Teachers' motivation, classroom strategy use, students' motivation and second language achievement. *PortaLinguarum, (12)*12, 25–36.

Bernaus, M., & Gardner, R.C. (2008). Teacher motivation strategies, student perceptions, student motivation, and English achievement. *The Modern Language Journal, 92*(3), 387–401. Doi: 10.1111/j.1540-4781.2008.00753.x.

Cheng, H.F. & Dörnyei, Z. (2007). The use of motivational strategies in language instruction: the case of EFL teaching in Taiwan. *Innovation in language learning and teaching, 1*(1), 153–174. doi: 10.2167/illt048.0.

Chiang, T.H.C., Yang, S.J.H., Huang, C.S.J., & Liou, H.H. (2014). Student motivation and achievement in learning English as a second language using Second Life. *Knowledge Management & E-Learning, 6*(1), 1–17.

Dörnyei, Z. & Ushioda, E. (2011). *Teaching and researching motivation: second edition.* Edinburgh: Pearson Education Limited.

Dörnyei, Z. (2001). *Motivational strategies in the language classroom.* Cambridge: Cambridge University Press.

Guilloteaux, M.J. & Dörnyei, Z. (2008). Motivating language learners: a classroom-oriented investigation of the effects of motivational strategies on student motivation. *TESOL Quarterly, 42*(1), 55–77. doi: 10.2307/40264425.

Kimura, Y., Nakata, Y., & Okumura, T. (2001). Language learning motivation of EFL learners in Japan – A cross-sectional analysis of various learning milieus. *JALTJournal, 23*(1), 47–65.

Kurum, Y.E. (2011). The effect of motivational factors on the foreign language success of students at the Turkish military academy. *Novitas-ROYAL (Research on Youth and Language), 5*(2), 299–307.

Li, P. & Pan, G. The relationship between motivation and achievement – A survey of the study motivation of English majors in Qingdao Agricultural University. *English Language Teaching, 2*(1), 123–128. doi: 10.5539/elt.v2n1p123.

Murphy, P.K. & Alexander, P.A. (2000). A motivated exploration of motivation terminology. *Contemporary Educational Psychology, 25*, 3–53. doi: 10.1006/ceps.1999.1019.

Ortega, L. (2013). *Understanding second language acquisition.* New York, NY: Routledge.

Oxford, R., Oh, Y.P., Ito, S., & Sumrall, M. (1993). Japanese by Satellite: Effects of motivation, language learning styles and strategies, gender, course level, and previous language learning experience on Japanese Language Achievement. *Forein Language Annals, 26*(3), 359–371. doi: 10.1111/j.19449720.1993.tb02292.x.

Ruesch, A., Bown, J., & Dewey, D.P. (2012). Student and teacher perceptions of the use of motivational strategies in the foreign language classroom. *Innovation in Language Learning and Teaching, 6*(1), 15–27. doi:10.1080/17501229.2011.562510.

Vibulphol, J. (2016). Students' motivation and learning and teachers' motivational strategies in English classrooms in Thailand. *English Language Teaching, 9*(4), 64–75. doi: 10.5539/elt.v9n4p64.

Wang, F. (2008). Motivation and English achievement: An exploratory and confirmatory factor analysis of a new measure for Chinese students of English learning. *North American Journal of Psychology, 10*(3), 633–646.

Wery, J. & Thomson, M.M. (2013). Motivational strategies to enhance effective learning in teaching struggling students. *Support for Learning, 28*(3), 103–108. doi: 10.1111/1467-9604.12027.

Widoyoko, E. P. (2009). *Evaluasi program pembelajaran* [Evaluation of learning programs]. Yogyakarta, Indonesia: PustakaBelajar.

Study on graduate competence in air-conditioning skills at senior high school with relation to job market competencies

S. Haryadi
Pendidikan Teknik Elektro, Pascasarjana Universitas Negeri, Yogyakarta, Indonesia

ABSTRACT: This study aims to determine the competencies developed in the cooling and air-conditioning engineering competency at a vocational high school and their relevance to the needs of the labor market. This learning is descriptive research. The data analysis technique used is descriptive analysis statistics. The results of the study are (1) the relevance of the cooling and air-conditioning engineering competency vocational curriculum for the needs of the labor market in Yogyakarta and its surroundings, (2) competencies needed by the air conditioning and air-conditioning engineering industry in the labor market curriculum and electricity, and (3) competencies needed by the industry in vocational schools for the job market. This study concludes that the curriculum in the relevant category.

1 INTRODUCTION

Education is one of the basic needs in life and has a role in shaping the attitudes and behavior of students. The attitude and good behavior of each student is the main goal of education. This is in accordance with the functions and objectives of Indonesia's national education, such as those contained within the Republik of Indonesia's Law, Number 20 Year 2003 concerning National Education System. Article 3 clearly states that the function of national education is to develop capabilities and shape a dignified national character and civilization (RoI, 2003). The goal is to develop the potential of students to become human beings who believe in and are devoted to God Almighty, have noble character, are healthy, knowledgeable, capable, creative, independent, and who become citizens of democratic and responsible countries.

The behavior of educated individuals who still commit immoral and criminal acts, such as corruption and brawls, is enough to indicate that national education at various levels is considered less successful in forming moral and mindful students with good character than required in the National Education System laws.

This is caused by several core problems. First, the direction of education has lost its objectivity. The school and its environment are no longer where students learn to do something based on moral and moral values, or where they are corrected for their actions, wrong or right, good or bad. Second, the process of maturation does not take place in the school environment. Third, the scope of education process in schools is not wide enough to students and even to teachers. Fourth, the weight of the curriculum is almost entirely oriented to the development of the realm of cognition. So, the development of the affective and psychomotor domains receives almost no attention. Sixth, students are often faced with contradictory values. Seventh, students experience difficulties in finding good examples in their environment.

The Human Development Report issued by the United Nations Development Program (UNDP) reported that Indonesia ranks 113th out of 187 countries on the Human Development Index (HDI) in 2016. The data places Indonesia still below the international average for improvement human resources (HR). The relatively low quality of human resources (HR) must be improved, which requires input from various parties starting from the current government, education, and society in Indonesia. One of the most influential groups in the education sector – who play a role in the development of human resources – is the people who are educated and developed in social, cultural, political, economic matters.

"National development in the field of education is an effort to educate the nation's life and improve the quality of Indonesian people in realizing an advanced, just and prosperous society and enables their citizens to develop themselves, both in terms of physical and spiritual aspects based on Pancasila and the 1945 Constitution." It has been clearly stated that the nation will face the advancement in all life aspects, the industrial sector is not an exception. The Head of Construction Training in Ministry of Public Works, Hediyanto Husaini stated that Indonesia had 6.9 million workers in the construction sector; around 100,000 are experts and 300,000 are skilled workers, the rest are ordinary workers (Wiyanti, 2014). However, the data presented is still lacking information about certified experts.

Reliable and quality human resources (HR) can be achieved through the role of government via schools. Vocational high schools (SMK) shape the character of students for facing current developments in the rapidly changing labor market. Vocational high schools become producers of graduates who are ready to work and have skills in their fields.

Data from the 2017 National Central Statistics Agency (BPS, 2017) show that the number of people in open unemployment in Indonesia is high at all levels of education: junior high school 5.36%, high school (SMA) 7.03%, vocational high school (SMK) 9.27%, Diploma III (D3) 6.35%, and university 4.98%. The data show that vocational high schools have the most unemployed graduates. BPS, from 2013 to 2015, reported an open unemployment rate increase: 2013 = 3.24%, 2014 = 3.33%, and 2015 = 4.07%. This figure shows that the problem of unemployment, which is still high, is a consideration of the government to evaluate education in Indonesia, especially at SMK.

Preparations must be made to develop maximum vocational learning activities and provide hard skills and soft skills. Suryanto (2013) said that the ratio of soft skills and hard skills needed in the business world and industry is inversely proportional to the development in the education system. Furthermore, Neff & Citrin (2013) stated that people's success in business/industry is 80% determined by their mindset and 20% determined by technical skills (Sailah, 2006). But soft skills only account for around 10% of the current curriculum. The soft skills that are desired by industry have not been considered in vocational schools. Vocational education institutions must immediately evaluate and design education systems that also pay attention to the soft skills needed by the industry. Graduates produced by vocational schools must also have personal characteristics that are useful at work, and of higher quality, so that it is easier to penetrate the job market.

The development of capabilities in accordance with the competence of cooling and air-conditioning techniques in vocational schools is carried out by taking into account national education standards, and especially educational content standards. Educational content standards contain criteria about the scope of the subject matter and the level of competency of expertise. Subjects are one component that exists in educational content standards. The subject contains teaching that is delivered to achieve a goal directly with students. Vocational school prepares ready-made graduates with its own characteristics, namely the emphasis on the psychomotor domain. Increasing the psychomotor domain is supported by learning related to practice. In the competence of 12th grade cooling and air-conditioning techniques, such as maintaining commercial refrigeration compressors, assembling piping units for commercial refrigeration units, practical learning teaches cooperation, skill, honesty, thoroughness, communication skills and so on.

The above description can be used to carry out research to answer the existing problems about the relevance of the competence of cooling and air engineering skills for the job market. The current state of expertise competence is still in line with the labor market. Competencies in air-conditioning skills are chosen because there are many opportunities for graduates to find work. There has been a large increase in buildings in Yogyakarta. This addition of space is directly proportional to the need for air conditioning (AC) in each room and directly proportional to the needs of the workforce. It can be assumed that AC needs will increase every year, therefore also increasing the need for human resources (HR) for this workforce. The research conducted is expected to show the relevance of the competence of cooling and air engineering skills offered by the vocational schools, especially in air conditioning engineering competency program in vocational school.

1.1 *Identification problems*

In the background above there are several problems identified to avoid obscurity or inaccuracies in research. Identification of the problems described above are relevant to the fact that there is a 5.36% open unemployment rate at the junior high school (SMP) level, 7.03% at senior high school (SMA), 9.27% at vocational high school (SMK), 6.35% at Diploma III (D3), and 4.98% at university level. The data also showed that vocational high schools have the most unemployed graduates. The Central Statistics (BPS) showed that the unemployment rate increased from 2013 to 2015 (2013 = 3.24%, 2014 = 3.33%, 2015 = 4.07%). This shows that the problem of unemployment, which is still high, is a consideration of the government to evaluate education in Indonesia, especially at SMK.

Vocational high school (SMK) is a formal vocational education institution established by the government as an effort to prepare the nation's children to compete in the business world or industry (DU/DI) as stated in the Republic of Indonesia's Law, No. 20 of 2003. Based on the objectives of vocational secondary education, this means that SMK graduates must have competencies in accordance with the Indonesian National Work Competency Standards (SKKNI) so that SMK graduates can be easily absorbed into the world of work (RoI, 2013)

Instructional materials are knowledge, skills and attitudes that must be mastered by students in order to meet established competency standards. Subject matter occupies a very important position in the entire curriculum, which must be prepared to meet the implementation of learning goals. The target must be in accordance with the Competency and Basic Competency Standards (SKKD) that should be achieved by students. That is, the material determined for learning activities should truly support the achievement of competency standards and basic competencies, as well as achieving indicators.

Subjects are one component that exists in educational content standards. The subject contains teaching delivered to achieve a goal directly with students. Vocational school prepares ready-made graduates with its own characteristics, namely the emphasis on the psychomotor domain. Increasing the psychomotor domain is supported by learning related to practice. In the competence of 12th grade cooling and air conditioning techniques, such as maintaining commercial refrigeration compressors and assembling piping units for commercial refrigeration units, practical learning teaches cooperation, skill, honesty, thoroughness, communication skills, and so on.

1.2 *Limitation of problems*

This research will focus on the competence of expertise in refrigeration and air conditioning techniques related to the following:

1.1.1 Problems are limited to the scope of the competence of air conditioning and air conditioning engineering skills in vocational schools
1.1.2 Implementation of the air engineering skills competency program in vocational schools
1.1.3 Suitability for work that is very much needed by the air conditioning service industry
1.1.4 The study was conducted in the area of the labor market around air conditioning (AC) companies in Yogyakarta

1.3 *Problem formulation*

In the identification of the problems mentioned above, a number of issues were formulated as follows:

1.1.5 What is the level of appropriateness of the program provided in air engineering skills competence in vocational schools?

1.1.6 How can the implementation of the air engineering skills competency program in vocational schools be maintained?

1.1.7 What is the suitability for work needed by the air conditioning service industry?

1.1.8 What is the level of relevance between the competencies of graduates of the air and mechanical engineering vocational course and competency in the field of air conditioning services?

2 LITERATURE REVIEW

2.1 *Vocational education*

Vocational secondary education prioritizes the development of the students to carry out certain types of work. The role of education in improving human resources contained in Law No. 20 of 2003 concerning the National Education System states that "National development in the field of education is an effort to educate the life of the nation and improve the quality of Indonesian people in realizing an advanced, just and prosperous society and enabling its citizens to develop themselves, both with regard to physical and spiritual aspects based on Pancasila and the 1945 Constitution" (RoI, 2003). Rauner and Maclean (2008) stated that "vocational education" denotes the appropriation of the entire range of skills that are acquired within and for the employment system. Of all the above definitions, vocational education is work-oriented education, where graduates of vocational education institutions are expected to be able to compete for jobs and carry out tasks according to their field of work. Vocational high school (SMK) is a form of formal education unit that organizes vocational education at the secondary education level as a continuation of junior high school, MTs, or other equivalent forms. Schools with vocational-type education can be called vocational high schools (SMK) or vocational madrasah aliyah (MAK), or other equivalent forms.

Changing times have been followed by technological and other developments. Changes include the addition of multi-story buildings, apartments, malls, and hotels. Jobs appear in new sectors and this includes opportunities for vocational school children. This challenge will be met by the vocational school in the field of expertise, namely the technology and engineering expertise will be supplied by the electricity engineering expertise study program, consisting of two expertise competencies: installation techniques for electric power utilization, and cooling and air conditioning techniques. Cooling and air conditioning techniques, which are very closely related to air conditioning (AC) are used to repair or service air conditioners. There are about 10 AC companies in Indonesia and workers are usually tied to one brand, such as LG, Samsung, Panasonic, etc. Vocational schools must see the opportunity for change and also the effects of the industrial revolution 4.0.

2.2 *Graduates competency standards (SKL)*

The competencies taught in vocational schools are standardized and regulated by the government, listed in the SKL, and include the criteria for the qualifications that students must possess to complete certain education units (RoI, 2013), as shown in Table 1. Susanto (2014) stated that the preparation of SKL refers to the Indonesian National Work Competency Standards (SKKNI). SKL is the main reference in the development of core competencies (KI) and basic competencies (KD).

2.3 *Competence in air conditioning and refrigeration engineering skills (TPTU)*

TPTU is one of the competencies of the electricity expertise program, which aims to equip graduates with attitudes, knowledge and skills to be competent in carrying out the work of installing, operating and caring cooling systems in restaurant, office, supermarket, and so on. Beside that the competency program aims to prepare students to be able to control temperature, humidity as well as the air quality of various buildings. The air conditioning techniques are very closely related to

Table 1. Competency standards for SMK/MAK graduates.

Ability	Qualification dimensions
Attitude	Having a behavior that reflects the attitude of the people of faith, noble, knowledgeable, confident, and responsible for interacting effectively with the social and natural environment and in placing themselves as a reflection of the nation in world relations
Knowledge	Having factual, conceptual, procedural, and metacognitive knowledge in science, technology, art and culture, with insights on humanity, nationality, state and civilization related to the causes and effects of phenomena and events
Skills	Having the ability to think and act effectively and creatively in the abstract and concrete realm as the development of what is learned independently in school

air conditioners (AC), for AC repairs or servicing. There are about 10 AC companies in Indonesia and workers are usually tied to one brand, such as LG, Samsung, Panasonic, etc.

2.4   Service industry

The refrigeration and air conditioning sector is a component of the electro technology industry. The refrigeration and air conditioning sector includes the installation, maintenance and repair of both domestic and commercial refrigeration and air conditioning. The refrigeration and air conditioning workforce is employed in industries such as construction services and repairs and maintenance. Vocational education and training (VET) is required for refrigeration and air conditioning sector-related occupations such as air conditioning and refrigeration mechanics and electrical or telecommunications trades assistant.

2.5   Evaluation model

Evaluation of the discrepancy model is used to determine the suitability of a program where evaluators can compare between supposed and expected to occur/criteria (standards, competency needs) and what actually happened or was implemented (performance, graduate competencies) to determine whether there are gaps or discrepancies. Provus said that the evaluation of the discrepancy model emphasizes the understanding of the program system, where evaluation is used as a tool for program development as well as an assessment of the program (Provus, 1971). In addition, the evaluation was divided into five stages, where each stage had program implementation indicators and standards as program criteria, as shown in Table 2.

Table 2. Discrepancy model.

Stage	Implementation	Conformity standards
1	Program planning: A. Input dimensions B. Process dimensions C. Output dimension	Compared to planning criteria (standard)
2	Program implementation	Compared to planning criteria (standard) Compared to program planning A. Input dimensions B. Process dimensions
3	Temporary products	Compared to program planning A. Process dimensions B. Output dimension.
4	The final product	Compared to program planning A. Output dimension.
5	Implementation cost program	Compared to implementation costs Other programs with the same product. Program

The focus of the study was the competence of graduates, as the final product of vocational learning programs, while the criteria used were competency needs in the service industry and IQF. Therefore, the stages carried out were the design of making instruments and setting standards and gap criteria by looking for related theories, obtaining data by distributing questionnaires, comparing questionnaires with standards to get results and discussing gaps, determining conclusions by comparing the results of the gap with the criteria, and providing suggestions regarding the conclusions obtained.

3 RESEARCH METHODS

The type of research used is the evaluation of the discrepancy model by identifying the gap between the standard and the reality of the matter under study (Issac & Michael, 1981). Provus said that this was used as a program evaluation, where the achievement of educational outcomes was an output of the interaction of educational inputs during the education process (Provus, 1969). Provus defines evaluation as a tool to make judgments about the weaknesses and strengths of an object based on standards and performance. This model is also considered to use a formative approach and is oriented towards system analysis, while the achievement is focused on what actually happened. The discrepancy evaluation model (Provus, 1971) is a program evaluation model that emphasizes the importance of understanding the system before evaluation. This model is a problem-solving procedure to identify weaknesses (including in the selection of standards) and to take corrective actions. The evaluation process is the way of facilitating the comparison of program achievements with standards, while at the same time identifying standards to be used for future comparison. The four stages for the model are Definition stage, Installation stage, Process stage, Product stage, and Optional stage cost benefit.

4 RESULTS AND DISCUSSION

The program provided in the air engineering skills competence in vocational schools is appropriate for the times. The implementation of the air engineering skills competency program in vocational schools is directly proportional to the core competencies and basic competencies in the labor market, with the level of work needed by the service industry in the field of air conditioners, even with different AC brands. The relevance between the competencies of graduates of the air and mechanical engineering vocational course and the field of air conditioning is very high. So vocational graduates already have the skills to join the business and industrial world (DUDI).

5 CONCLUSION

The results of this discussion indicate that the relevance of the competence of air-conditioning engineering expertise to the job market, especially in the Yogyakarta area, generally falls into the relevant category.

REFERENCES

Arikunto, S. (2014). *Evaluasi program pendidikan* [Evaluation of educational programs]. Jakarta, Indonesia: Bumi Aksara.
Beer, A., & Brooks, C. (2011). Information quality in personality judgment: The value of personal disclosure. *Journal of Research in Personality, 45*(2), 175–185.
Bennett, T.M. (2006). Defining the importance of employability skills in career/technical education. *Dissertation (unpublished)*. Auburn, Alabama: The Graduate Faculty of Auburn University.
Guthrie, H. (2009). *Competence and competency-based training: What the literature says*. Adelaide: National Centre for Vocational Education Research (NCVER) Australia.

Jatmoko, D. (2013). Relevansi kurikulum SMK kompetensi keahlian teknik kendaraan ringan terhadap kebutuhan dunia industri di Kabupaten Sleman [The relevance of the junior high school curriculum in light vehicle engineering expertise to the needs of the industrial world in Sleman Regency]. *Jurnal Pendidikan Vokasi,3*(1), 1–13.

Madar, A.R. & Buntat, Y. (2011). Elements of employability skills among students from community colleges Malaysia. *Journal of Technical, Vocational & Engineering Education*, 4 December 2011, 1–11.

Mariah, S., & Sugandi, M. (2010). *Kesenjangan soft skills lulusan SMK dengankebutuhantenagakerja di industri* [Gaps in soft skills of vocational school graduates with the needs of workers in industry]. Retrieved from http://repositori.kemdikbud.go.id/240/.

Pavlova, M. (2009). Technology education as an effective way of providing vocational education within secondary schooling. In: M. Pavlova (Ed.), *Technology and vocational education for sustainable development, Technical and vocational education and training: Issues, concerns and prospects* (pp. 5–26). Dordrecht: Springer.

Provus, M.M. (1969). *The discrepancy evaluation model: An approach to local program improvement and development*. Washington, D.C.: Pittsburgh Public School, PA.

Rees, C., Forbes, P & Kubler, B. (2007). *Student employability profiles: A guide for higher education practitioners*. United Kingdom: The Higher Education Academy.

RoI. (2003). *Undang-undang Nomor 20 Tahun 2003 tentangSistemPendidikan*Nasional [Law Number 20 Year 2003 concerning National Education System]. Jakarta, Indonesia: Ministry of Education, Republic of Indonesia.

RoI. (2013). *Peraturan Pemerintah Nomor 32 Tahun 2013 Tentang Perubahan Standar Nasional Pendidikan* [Government Regulation Number 32 of 2013 concerning Change of National Education Standard]. Jakarta, Indonesia: Republic of Indonesia.

RoI. (2015). *Kerangkakualifikasinasional Indonesia* [Indonesian national qualifications framework], Jakarta, Indonesia: Ministry of research, technology and higher education, Republic of Indonesia.

RoI, Badan Pusat Statistik. (2015). *Direktoriperusahaankonstruksi DIY* 2015 [Directory of construction establishment of Daerah Istimewa Yogyakarta 2015] (BPS Publication No. 34530.1601). Retrieved from https://yogyakarta.bps.go.id.

RoI, Badan Pusat Statistik. (2017). *Statistik Indonesia 2017* [Indonesian Statistics 2017] (BPS Publication No. 03220.1709). Retrieved from https://yogyakarta.bps.go.id.

Safitri, B.R.A., Syafrudie, H.A., & Sutrisno. (2012). Relevansi program studikeahlianteknikbangunandenganpekerjaanlulusan [The relevance of the building engineering expertise study program with graduate work]. *Teknologidan Kejuruan, 35*(1), 29–36.

Wagiran. (2008). The importance of developing soft skills in preparing vocational high school graduates. *SEAVERN Journal, 1*(1), 1–9.

Widoyoko, E.P. (2014). *Evaluasi program pembelajaran* [Evaluation of learning programs]. Yogyakarta: PustakaPelajar.

Wiyanti, Sri. (2014). Dari 6,9 juta tenaga konstruksi, baru 159 bisa bersaing di ASEAN. Retrieved September 2018 on merdeka.com/uang/dari-69-juta-tenaga-konstruksi-baru-159-bisa-bersaing-di-asean.html.

Yuliana, C. (2009). Studi pemahaman dan penerapan standar kompetensi keterampilan kerja tenagakerja pada pelaksanaan proyek konstruksi [Study of understanding and applying work skills competency standards on the implementation of construction projects]. *Jurnal Teknologi Berkelanjutan, 10*(1), 83–91.

Yusof, N.A., Fauzi, S.N.F.M., Abidin, N.Z, & Awan, H. (2013). Improving graduates' employability skills through industrial training: Suggestions from employers. *Journal of Education and Practice, 4*(4), 23–29.

Zhao, Z., & Rauner, F. (2014). Competence research. In Z. Zhao & F. Rauner (Eds.), *Areas of vocational education research* (pp. 167–188). Berlin: Springer-Verlag Berlin and Heiderbelrg GmbH & Co. K.

Innovative Teaching and Learning Methods in Educational Systems – Retnowati et al. (Eds)
© 2020 Taylor & Francis Group, London, ISBN 978-1-03-224183-8

Elementary students' performance in mathematical reasoning

N. Andrijati, D. Mardapi & H. Retnawati
Universitas Negeri Yogyakarta, Yogyakarta, Indonesia

ABSTRACT: There are four competences to be mastered by learners in the 21st century: critical thinking and problem-solving, creativity, communication skills, and the ability to work collaboratively. Mathematical reasoning competence plays a crucial role in the development of critical thinking and problem-solving skills. However, current studies have not been able to define the actual mathematical reasoning competence of students. The paper and pencil test assessment method still dominates assessment of mathematical reasoning in elementary schools and the questions asked by teachers during assessment tend to be objective. The objective test used encouraged teachers to give similar training to students, as they have to answer questions in this format. To tackle limitations in the current assessment methods and provide teachers with thorough student profiles, performance assessments were conducted. The research subjects were fourth grader students from elementary schools in Tegal, Indonesia. Data was gathered through performance assignments using interpretation and integration methods. Data analysis was performed by a descriptive quantitative method. Research findings showed that students' mathematical reasoning performance was categorized as good but they had particular weaknesses in making and investigating assumptions, and developing arguments.

1 INTRODUCTION

The 21st-century students' learning framework (Mishra & Kereluik, 2011) and the needs of mathematics mean that the school mathematics curriculum emphasizes higher order thinking skills (HOTS). HOTS can be defined as a cognitive process involving analysis, synthesis, and evaluation (Moore & Stanley, 2010). In mathematics, HOTS can be defined as the ability to perform mathematical processes, complex assignments, and mathematical problems involving connection, problem-solving, and mathematical reasoning.

Mathematical reasoning involves collecting proofs, building assumptions, making generalizations, making arguments, and drawing logical conclusions. In other words, reasoning is a thinking process performed by connecting facts or evidence, leading to a conclusion or generalization. Generalization making in reasoning enables problem-solving, as generalization supports learners to analyze structures underlying problems or existing ideas.

Reasoning is an important aspect of mathematics and one of the aims of studying mathematics. According to The National Education Standards (RoI, 2013), the scopes of mathematical study materials in calculating, measuring, and algebra are intended to develop learners' thinking logic and competence. The National Council of Teachers of Mathematics (NCTM, 2000), refer to the general objectives of mathematical learning as competence in problem-solving, reasoning, communicating, building connections, and representation.

An international study on the ability of mathematical reasoning was conducted by the Trends in International Math and Science (TIMSS) study institution to monitor mathematics and science achievements by fourth grade and eighth grade students. In TIMSS 2015, the mathematics competence of fourth grade students was categorized as low. According to the cognitive domain, Indonesian elementary students have mathematical competences in the knowing, applying, and reasoning aspects at 32%, 24%, and 20%

respectively. The reasoning aspect at 20% is the lowest competence achieved by Indonesian students, far from the international average of 47% (Mullis, Martin, Foy, & Hooper, 2006). This clarifies that elementary students in Indonesia have low reasoning competence and are still unable to find non-routine solutions in an unfamiliar situation with complex contexts and multi-step procedures.

Reasoning competence can be improved in assessment activities. In reasoning assessments, teachers should consider the use of various assessment techniques that can include all fields or dimensions. The mathematics learning assessment is known as the term alternative or authentic assessment. Authentic assessment asks learners to use the same competence or combination of knowledge, skills, and attitudes required in a real-life situation.

At the elementary level, Indonesia has not developed an assessment that enables teachers to comprehensively observe mathematical reasoning competences. Mathematical reasoning competence assessments conducted so far tend to use written tests in the form of multiple choice or essays; but assessments using performance tasks have not been implemented. Therefore, this research aims to improve students' reasoning competence observed from assessments on performance tasks. It enables students to show what they have learned and how to solve a complex problem, either individually or in groups. Furthermore, this research aims to describe students' reasoning performance in elementary mathematical learning.

2 LITERATURE REVIEW

2.1 *Mathematical reasoning*

Reasoning is a basic skill of mathematics required in order to understand certain concepts, use mathematical ideas and flexible procedures, and construct mathematical knowledge. Reasoning is also defined as a thinking process, linking facts or evidence and leading to a conclusion (Keraf, 1982), or competences used to analyze connections and relationships among mathematical ideas then implement this understanding for solving new problems (English, 2004). Reasoning can be stated as an activity or thinking process used to draw a conclusion based on the relationship of available facts or evidence. The results of reasoning processes can be implemented to solve problems in new situations.

Standards of reasoning and NTCM verification (2000) state that mathematical reasoning occurs when learners 1) observe patterns and regularities, 2) formulate generalizations and assumptions related to the observed regularities, 3) test the assumptions, 4) build and assess mathematical arguments, and 5) explain (validate) logical conclusions about several ideas and relationships. Referring to the NCTM standards and customizing the reasoning definitions with elementary students' characteristics, the indicators used here are competencies in 1) analyzing problems, 2) making and investigating assumptions, and 3) providing evidence and arguments (Goos, Stillman, & Vale, 2007; Russel, 1999).

2.2 *Authentic learning and assessment*

Callison and Lamb (2004) emphasize that meaningful learning should involve students and transfer learning to a new situation. Focusing on the direct relationship between authentic learning and students' achievement, their research concludes that students in high-quality classes and authentic assessments gain higher scores in the standardized achievement test. Students' authentic learning assessments are measured using an authentic assessment, which focuses on measuring competencies in learning; while authentic learning focuses on involving students' activeness during the learning processes (Eddy & Lawrence, 2013).

Authentic assessment asks students to use knowledge competencies, skills, attitudes, and a combination of these (Lund, 1997). Students must use their knowledge, skills, and attitudes in the same way as professionals would in real-life situations (Gulikers, Bastiaens, & Kirschner, 2004).

2.3 *Mathematical reasoning performance assessment*

Attempts to improve students' mathematical reasoning require teachers to use various assessment techniques (such as methods, tasks, assignments, strategies, activities) in class, as teachers' assessment practices involve considering all methods to consider what students know and can do (Wilson, 1993; Adams, 1998). Class assessment is a process of information collection and is used by teachers to give scores for students' learning outcomes. In a curriculum, a students' learning outcome describes their process, activities, or performances. Performance is what is achieved by an individual or system (Williams, 2001) or performance is viewed as a process of something that is conducted. Therefore, performance measurement is observed from whether or not an activity will achieve the expected outcome.

Performance assessment development involves activities such as standard determination, authentic task determination, criteria making, and standard rubric making (Mueller, 2013). "Standard" constitutes statements on what should be known and done by learners, a competence that contains objectives achieved by learning processes. Selected authentic tasks must be in accordance with measured competencies and real-life situations, requiring problems or scenarios, authentic positioning and setting, and learning environment design. Criteria are required in authentic assessments to evaluate students' task qualities. Students' competences in a certain task are determined by matching their performance with a set of criteria to investigate to what extent they fulfill the task criteria. The level of learners' achievement or competence is determined using rubrics that usually contain important criteria and the level of criteria achievement used to measure learners' performances. The level of performance achievement is usually stated in numbers.

3 METHOD

This research applied a descriptive quantitative approach. The subjects were 27 fourth grade students at an elementary school in Tegal, Indonesia. The research instruments used for data collection were tasks and performance assessment rubrics on basic competences of estimating and rounding, referring to indicators of mathematical reasoning competence assessment, i.e. analyzing information, making and investigating assumptions, compiling evidence, and developing arguments. The instruments given to students had been validated by an experienced measurement expert, a mathematics lecturer, and a mathematics elementary teacher in Tegal.

Students' reasoning performances were analyzed using assessment data in the form of quantitative data categorized in accordance with qualitative scores based on the Standard Referred Evaluation as shown in Table 1.

Further analysis was conducted to gather reasoning performance information for each indicator.

4 RESULTS

The data on mathematical reasoning performance was gathered from 27 students' responses on performance tasks given to the domain of 'number content' with basic competences of

Table 1. Mathematical reasoning performance.

Score	Category
$6.6 < X \leq 8.0$	Excellent (E)
$5.2 < X \leq 6.6$	Good (G)
$3.8 < X \leq 5.2$	Rather Good (RG)
$2.4 < X \leq 3.8$	Bad (B)
$1.0 \leq X \leq 2.4$	Very Bad (VB)

interpretation and integration on whole numbers using the determined rubrics. Furthermore, the data were analyzed to establish the profiles of mathematical reasoning performance based on Table 1. The results are presented in Table 2.

For data/information analysis, performance data were used to reveal students' competences in connecting available information with mathematical ideas/objects. Two tasks were used to analyze the information: students were asked to 1) conduct distance integration by ordering distances from closest to farthest, and 2) determine the distance between two big cities. Table 3 summarizes the results.

For the assumption making investigation, performance data were used to reveal students' competences in making assumptions based on their learning experiences and investigating assumptions to obtain correct answers. In this task, students were asked to make an assumption of the distance between two cities that was a multiple of the distance of two other cities. The results are presented in Table 4.

In evidence and argument arrangement, performance data were used to show students' competencies in providing evidence by doing a calculation and drawing conclusions based on the calculation results. In this task, students were asked to prove the truth of a statement that after traveling a certain distance it was time to take a rest. The results are summarized in Table 5.

Table 2. Students' mathematical reasoning performance.

Score	Category	Number of students	
$6.6 < X \leq 8.0$	Excellent (E)	5	18.52%
$5.2 < X \leq 6.6$	Good (G)	4	14.81%
$3.8 < X \leq 5.2$	Rather Good (RG)	10	37.04%
$2.4 < X \leq 3.8$	Bad (B)	3	11.11%
$1.0 \leq X \leq 2.4$	Very Bad (VB)	5	18.52%

Table 3. Students' performance in information analysis.

Indicator1-1

Score	Number of students	Percentage
0	1	3.7
1	16	59.3
2	10	37.0

Indicator1-2

Score	Number of students	Percentage
0	8	29.6
1	1	3.7
2	18	66.7

Table 4. Students' performance in assumption making and investigation.

Score	Number of Students	Percentage
0	14	51.9
1	2	7.4
2	11	40.7

Table 5. Students' performance in evidence and argument arrangement.

Score	Number of students	Percentage
0	4	14.8
1	19	70.4
2	4	14.8

5 DISCUSSION

5.1 *Students' mathematical reasoning performance profile*

Based on the analysis of reasoning performance tasks in interpretation and integration materials given to fourth grade students, the information was obtained on students' reasoning performance. The percentages in each category indicated the same or relatively similar results of excellent and very bad performances and good and rather good performances. Meanwhile, most students managed mathematical reasoning performances within the good category (G). This confirmed that data of reasoning competence performances followed the normal curve distribution and that many students had mathematical reasoning performances within the good category.

5.2 *Students' performance in information analysis*

In analyzing the information in tasks 1 and 2, the score differences for students' performances were caused by the level of difficulty of the problems. Analyzing information on task 1 was relatively easier than that on task 2. This is evidenced by the score 0 on task 2 more than task 1. Scores 0, 1, or 2 were earned, respectively, if students did not answer the proposed question, gave information to show they understood but did an operation incorrectly, and gave information to show they understood and did the calculation correctly. Therefore, it could be stated that the performance analyzed information/data based on the level of performance task difficulties.

5.3 *Students' performance in assumption making and investigation*

In making and investigating assumptions, performance scores were described as follows: score 0, 1, and 2 were given if students did not write down their assumptions or make a calculation to investigate the assumptions, wrote down their assumptions and did the calculation incorrectly, and wrote down their assumptions and did the calculation correctly. Students' performance data revealed that 40% of students were able to write and investigate assumptions; while more than 50% of students were unable to write assumptions and do calculation correctly.

Writing or proposing assumptions requires the students to know mathematical concepts that underlie these assumptions. Students need the ability to determine the mathematical concepts to use in completing performance tasks. This is in accordance with the statement by Zulkardi (2003) that studying mathematics emphasizes the understanding of concepts, meaning that students must understand the concept first in order to be able to solve the problem/task and apply it in mathematics and the real-world learning. Students with moderate and high abilities are able to investigate assumptions by performing calculations involving the properties of operations in numbers. Around 50% of subjects have moderate and high mathematical abilities, so they are able to do calculations and find patterns to make generalizations or conclusions using inductive thinking. This is in accordance with the statement by Rochmad (2010) that inductive thinking is a better approach for students at elementary and junior high school levels.

5.4 Students' performances in evidence and arguments

Based on the results presented in Table 5, a minority of students were able to perform calculations and draw conclusions correctly, and most students were able to make the calculation but were unable to draw conclusions or gave the wrong conclusions. A few students were unable to either conduct calculations or draw conclusions. Most students were still unable to develop arguments; in other words, students' weaknesses in providing arguments or reasons are caused by a lack of ability to make generalizations or conclusions. This situation is in line with a statement by Aisyah (2016) that generalizations can be used to solve problems to prove the truth of a statement.

6 CONCLUSION

Based on the research findings and discussion, the general conclusion made was that students' mathematical reasoning performances on interpretation and integration materials could be categorized as good. In information/data analysis, students' performances were affected by the level of difficulty in the tasks. Most students had no ability to perform well in making assumptions and investigating. Students had no ability to write assumptions based on their learning experiences and perform calculation to investigate the assumption. In performances of evidence and argument arrangement, most students had no ability to write or draw conclusions to answer the truth of a statement based on a performed calculation.

REFERENCES

Adams, T.L. (1998). Alternative assessment in elementary school mathematics. *Childhood Education: ProQuest Professional Education*, 74(4), 220–224.

Aisyah, A. (2016). Studi Literatur: Pendekatan Induktif untuk Meningkatkan Kemampuan Generalisasi dan *Self Confident* Siswa SMK. *Jurnal Penelitian Pendidikan danPengajaran Matematika* 1(2), 83–94.

Brodie, K. (2010). *Teaching mathematical reasoning in secondary school classrooms*. New York, NY: Springer.

Callison, D. & Lamb, A. (2004). Key words in instruction: Authentic learning. *School Library Media Activities Monthly, Proquest Professional Education*, 21(4), 34–39.

Eddy, P., & Lawrence, A. (2013). Wikis as platforms for authentic assessment. *Innovative Higher Education*, 38(4), 253–265. doi: 10.1007/s10755-012-9239-7

English. (2004). Mathematical and analogical reasoning of young learners. London: Lawrence Erlbaum Associates Publisher.

Goos, M., Stillman G., & Vale, C. (2007). *Teaching secondary school mathematics: Research and practice for the 21st century* (1st edn.). Crows Nest NSW: CMO Image Printing.

Gulikers, J.T.M., Bastiaens, T.J., & Kirschner, P.A. (2004). A five-dimensional framework for authentic assessment. *The Journal of Educational Technology*, 52(3), 67–86.

Keraf, G. (1982). *Argumendannarasi: Komposisilanjutan III* [Arguments and narratives]. Advanced composition III]. Jakarta, Indonesia: Gramedia.

Lund, J. (1997). Authentic assessment: Its development & application. *Journal of Physical Education, Recreation & Dance*, 68(7), 25–28.

Mishra, P., & Kereluik, K. (2011). What 21st century learning? A review and a synthesis. *Proceedings of Society for Information Technology & Teacher Education International Conference 2011* (pp. 127–129). Chesapeake,VA.

Moore, B., & Stanley, T. (2010). *Critical thinking and formative assessments: Increasing the rigor in your classroom*. Larchmont, NY: Eye on Education.

Mueller, J. (2013). Authentic assessment. North Central College. Retrieved from http://jfmueller.faculty.noctrl.edu/toolbox/whatisit.htm

Mullis, I.V.S., Martin, M.O., Foy, P., & Hooper, M. (2016). *TIMSS 2015Internastional Result in Mathematics*. Chestnut Hill, MA: TIMSS & PIRLS International Study Center, Boston College.

NCTM. (2000). *Principle a standard for school mathematics*. Reston, VA: NCTM.

Rochmad. (2010). Proses berpikir induktif dan deduktif dalam mempelajari matematika. *Jurnal Kreano* 1(2), 107–117.

RoI. (2013). Peraturan Pemerintah Nomor 32 Tahun 2013 tentang Standar Nasional Pendidikan [Government Regulation No. 32 Year 2013 on National Standard of Education].

Russel, S.J. (1999). Mathematical reasoning in elementary grades. In L.V. Stiff & F.R. Curcio (Eds.), *Developing mathematical reasoning in grades K-12* (pp. 1–12). Virginia, USA: NCTM.

Williams, R.R. (2002). *Managinge mployee performance: Design and implementation in organizations.* London: Thomson Learning.

Wilson, P.S. (Ed.). (1993). *Research for the classroom: High school mathematics.* New York, NY: Macmillan Publishing Company.

Zulkardi. (2003). Pendidikan Matematika di Indonesia: Beberapa permasalahan dan upaya penyelesaiannya. Palembang: Universitas Sriwijaya.

Innovative Teaching and Learning Methods in Educational Systems – Retnowati et al. (Eds)
© 2020 Taylor & Francis Group, London, ISBN 978-1-03-224183-8

Evaluation model of the implementation of a quality management system of electrical skills in vocational high school

I.G.B. Mahendra
Graduate School Yogyakarta State University Yogyakarta, Indonesia

G. Wiyono
Electrical Engineering Education, Enginnering Faculty, Yogyakarta, Indonesia

ABSTRACT: This study aims to determine the following: (1) a proper evaluation model to evaluate the implementation of a quality management system for electricity expertise in vocational schools; (2) components of the evaluation model; and (3) the criteria for effectiveness of the evaluation program. The method used is a literature study with research data sources in the form of textbooks, journals, scientific articles, magazines, and newspapers. The results obtained are as follows. (1) A CIPP evaluation model is the most suitable model used in evaluating the implementation of a quality management of electricity expertise in vocational schools. (2) A CIPP evaluation model consists of four components: context, input, process, and product. (3) The criteria for the effectiveness of the evaluation program, on the four components of the CIPP evaluation model, were context effectiveness, input effectiveness, process effectiveness, and product effectiveness.

1 INTRODUCTION

Vocational high school is one of the levels of formal secondary education in Indonesia. Wulandari and Surjono (2013) stated that vocational high school is an advanced secondary education with the main objective of preparing a skilled, professional and highly disciplined workforce in accordance with the demands of the industry. The aim is to prepare graduates to work in certain fields that are in line with market needs, and in this case the business world and the world of work.

The implementation of vocational secondary schools is one of the policies by the Indonesian government to improve existing human resources. The needs of the world of work for quality human resources and skills influencing an educational institution is not unusual; vocational high schools should conduct quality assurance of the educational process to achieve positive values and customer satisfaction.

A good quality school will produce good quality graduates. With the demands of globalization, schools that have quality assurance will continue to improve their existing education services. The aspect that supports the improvement of the school education service quality needs to be improved. Implementing an ISO quality management system is one of the ways to carry out quality assurance of the educational process in an educational institution. Doherty (2003) stated that the implementation of ISO quality management system as an international quality standard will consistently improve efficiency in managing school resources and school quality.

In addition, it is hoped that there will be a continuous improvement process for school performance so that the quality and output of schools as an educational institution improve over time. ISO-based quality management systems aim to improve the competitiveness of an institution, and in this case an educational institution in order to improve the process of providing education to ensure customer satisfaction. One focus of the ISO principle is the customer focus or focus on the customer. Sallis (2014) stated that internal customers of educational institutions are teachers and staff in schools and external customers are parents of students and the community.

Lack of consistency in the performance of internal vocational high schools is characterized by the ups and downs of the performance of educators and education personnel. In addition, vocational secondary schools find it difficult to focus their work in accordance with ISO demands because there are many jobs in each vocational high school work unit. The most important obstacle in the application of ISO quality management systems in vocational high schools is the difficulty of teachers or employees adapting to the development of information and technology. Candra (2012) stated that common constraints experienced by vocational high schools are the problem of a school's culture.

The implementation of the ISO quality management system will be more successful if it is accompanied by a school vision that is the overall responsibility of the school community and not only the responsibility of individuals, especially leaders. On the other hand, the success of vocational high schools in obtaining ISO certificates requires high-quality culture in the management of leading organizations. The ISO certificate must be maintained, with the acculturation of the quality integrated into all policy actors in the organization within the school itself. Considering this, the commitment of vocational secondary schools in implementing the ISO quality management system in ensuring customer satisfaction requires evaluation.

This evaluation of the implementation of the ISO quality management system aims to provide benefits for the application of a quality culture in vocational secondary schools, so as to produce quality learning to prepare vocational high school graduates to work in certain fields in line with the needs of the workforce. Felestin and Triyono (2015) stated that one factor in the success of vocational education is an absorption of a graduate to the world of industries. It is important to study the world development quality in the vocational high school as a means of improving the partnership that produces vocational relevance of education to the labor market.

Aized (2010) stated that the evaluation process requires a suitable method to evaluate a program to produce a maximum evaluation. Michalska-Ćwiek (2009) stated that project quality management is the process that ensures the project meets the requirements and expectations of the beneficiaries of the project. Thus, this study aims to (1) ensure the right evaluation model is used to evaluate the implementation of the power management quality management system in vocational high schools, (2) determine components of the evaluation model and, (3) establish the criteria for effectiveness of the evaluation program on the implementation of the quality management system for electricity expertise in vocational secondary schools.

2 RESEARCH METHOD

This research is a– literature study that provides a logical description in accordance with the objectives of the study. This is done by giving meaning to the data that has been collected, which is carried out from beginning to end. This analysis and interpretation are done by referring to the theory or the results of the research in accordance with the problems to be solved.

To make it easier to see the results of the summary, a matrix was created. In the pattern of the matrix form, it is possible to see a complete picture of certain parts of the research results. On the basis of the pattern shown in the data presentation, it can be concluded that the data collected has meaning. As explained earlier in this study, the analysis process was carried out since the initial data was collected so that it was analyzed into a conclusion.

3 RESULTS AND DISCUSSION

Lazibat, Sutic, and Jurcevic (2009) stated that the ISO quality management system is recognized as an international standard on best practices in internal quality management. Bernik, Sondari, and Indika (2017) said that the quality management system is a formal system that documents processes, procedures, and responsibilities for achieving quality policies and objectives. ISO 9001: 2008 gives a series of general requirements that can be applied irrespective of an organization's activity, size or ownership.

The direct benefit that can be realized from the implementation of ISO 9001: 2008 is the combined alignment of the activities of internal processes that are focused toward the improvement of customer satisfaction, which will result in many other reimbursements, whether internal or external. Wiyono (2013) stated that eight principles of quality management are methods of how to lead, regulate, and control a school organization in accordance with the ISO 9001: 2008 quality management system.

These eight quality management principles (Point Development International, 2008) are as follows.

1 *Customer focus:* Organizations depend on their customers and therefore should understand current and future customer needs, meet customer requirements and strive to exceed customer expectations. Hoyle (2009) Successful vocational high school institutions are reliant upon their customers said El-Morsy et al (2014).

2 *Leadership:* Leaders of vocational high school institutions establish a unity of purpose and direction. They should create and maintain the internal environment in which people can become fully involved in achieving the institution's objectives. This goes beyond merely 'doing the work' to how people think and talk and behave in a 'quality culture'.

3 *Involvement of people:* El-Morsy et al (2014) state that vocational high school institutions should take full advantage of the staff's knowledge and experience; clarify their job and responsibility requirements, and teach them that reaching the institution's objectives is their objectives.

4 *Process approach:* A desired result is achieved more efficiently when the activities and the related resources are managed as a process, rather than as individual tasks. Managing these activities provides greater efficiencies through a clear view of what is happening.

5 *System approach to management:* The system approach to management includes identifying, understanding and managing, and interrelated processes as a system, and will contribute to the effectiveness and efficiency of the organization in achieving its objectives Gaspersz (2003). Management should view all activities and processes as parts of an integrated system. This will then contribute to the institution's effectiveness and efficiency in achieving its objectives.

6 *Continual improvement;* Continual improvement of the vocational high school institution's overall performance should be a permanent feature of the sector that really wishes to excel within the labor marketplace. The education institution can improve the quality system by managing the auditing periodically and continuously for daily tasks.

7 *Factual approach to decision-making:* In vocational high school institutions, effective decisions should be based on analysis of data and information that has been gathered via predetermined measures (Gaspersz, 2003).

8 *Mutually beneficial supplier relationships:* an organization and its suppliers are interdependent, and a mutually beneficial relationship will increase the joint ability to create added value. El-Morsy et al (2014) stated that vocational high school institutions and their customers are interdependent partnerships and a mutually beneficial relationship enhances the ability of both to create value.

Figure 1 covers all the requirements of this international sta ndard but does not show processes at a detailed level. In addition, the methodology known as 'Plan-Do-Check-Act' (PDCA) can be applied to all processes.

ISO 9001 (2008) PDCA can be briefly described as follows. Plan: establish the objectives and processes necessary to deliver results in accordance with customer requirements and the organization's policies. Do: implement the processes. Check: monitor and measure processes and product against policies, objectives, and requirements for the product and report the results. Action: take actions to continually improve process performance.

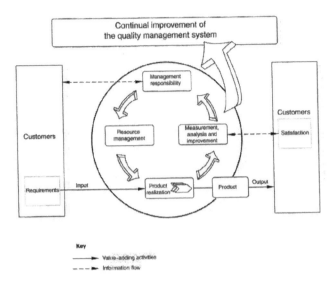

Figure 1.   Model of a process-based quality management system.

3.1   *A model to evaluate the implementation of the electricity management quality management system in vocational high schools*

Mahmudi (2011) stated that an appropriate evaluation is needed for the implementation of the quality management system. In this case, the electrical engineering expertise in the vocational middle class is the CIPP evaluation model.

CIPP is an evaluation model that uses a management-oriented approach referred to as a form of program management evaluation. Patton (1997) stated that the CIPP model rests on the view that the most important goal of program evaluation is not to prove but to improve. In this decision-oriented approach, program evaluation is defined as the "systematic collection of information about the activities, characteristics, and outcomes of programs to make judgments about the program, improve program effectiveness, and/or inform decisions about future programming." Stufflebeam (2003) stated that the CIPP evaluation model (see Figure 2) is a framework

Figure 2.   Components of Stufflebeam's CIPP Model.

for guiding evaluations of programs, projects, personnel, products, institutions, and evaluation systems.

Designed to assist administrators in making informed decisions, CIPP is a popular evaluation approach in educational settings (Fitzpatrick, Sanders, and Worthen, 2011; Zhang, et al., 2011) This approach, developed in the late 1960s, seeks to improve and achieve accountability in educational programming through a "learning-by-doing" approach. Its core concepts are context, input, process, and product evaluation, with the intention of not to *prove*, but rather *improve*, the program itself.

Stufflebeam (2003) state that an evaluation following the CIPP model may include a context, input, process, or product evaluation, or a combination of these elements. Mertens and Wilson (2018) state that the *context evaluation* stage of the CIPP model creates the big picture of both the program and evaluation. This stage assists in decision-making related to planning and enables the evaluator to identify the needs, assets, and resources of a community in order to provide programming that will be beneficial. Mertens and Wilson (2018) Context evaluation also identify the political climate that could influence the success of the program. To achieve this, the evaluator compiles and assesses background information, and interviews program leaders and stakeholders.

Mertens and Wilson (2018) state that key stakeholders in the evaluation are identified. In addition, program goals are assessed, and data reporting on the program environment is collected. Data collection can use multiple formats, including both formative and summative measures, such as environmental analysis of existing documents, program profiling, case study interviews, and stakeholder interviews. Throughout this process, continual dialogue with the client to provide updates is integral. To complement context evaluation, *input evaluation* can be completed. Mertens and Wilson (2018) In this stage, information is collected regarding the mission, goals, and plan of the program. Its purpose is to assess the program's strategy, merit and work plan against research, the responsiveness of the program to client needs, and alternative strategies offered in similar programs.

Fitzpatrick, et al. (2011) state that the intent of this stage is to choose an appropriate strategy to resolve the program problem. In addition to context evaluation and input evaluation, reviewing program quality is a key element to CIPP. *Process evaluation* investigates the quality of the program's implementation. In this stage, program activities are monitored, documented and assessed by the evaluator. Stufflebeam (2003) state that the primary objectives of this stage are to provide feedback regarding the extent to which planned activities are carried out, guide staff on how to modify and improve the program plan, and assess the degree to which participants can carry out their roles. Mertens and Wilson (2018) state that the final component to CIPP, *product evaluation*, assesses the positive and negative effects the program had on its target audience, assessing both the intended and unintended outcomes. Stufflebeam (2003) explain that both short-term and long-term outcomes are judged. During this stage, judgments of stakeholders and relevant experts are analyzed, viewing outcomes that impact the group, subgroups, and individuals. Applying a combination of methodological techniques assure all outcomes are noted and assist in verifying evaluation findings.

3.2 *Components of an evaluation model*

Mahmudi (2011) The CIPP model has four continuous elements.

1 Context evaluation primarily leads to identifying strengths and weaknesses of the organization and providing input to improve the organization. The main objective of context evaluation is assessing all organizational conditions, identifying its weakness, taking an inventory of the strengths that can be utilized to cover up its weaknesses, diagnosing problems faced by the organization, and finding solutions.
2 Input evaluation is specifically intended to determine the program to make changes required. Evaluate inputs look for obstacles and potentially available resources. The main goal is to help clients examine alternatives relating to organizational needs and goals.

3 Process evaluation basically checks the predetermined implementation plan. The goal is to give input for the manager and his staff regarding suitability between the implementation of the plan and the schedule that has been made previously and the efficient use of available resources. If the plan needs to be modified or developed, process evaluation provides instructions.

4 Product evaluation aims to measure, interpret, and assess program achievements. For further details, product evaluation aims to assess the success of the program in meeting program target needs. Assessments about the success of this program or organization are collected from people involved individually or collectively and then analyzed. That is, success or program failures are analyzed from various points of view.

3.3 *Criteria for the effectiveness of evaluation*

Evaluation activities are always related to predetermined criteria that are the objectives of the program. The basis of consideration makes it easy for evaluators to consider the value or price of the program components that they judge, whether they have been successful in accordance with the previous provisions or not. In this evaluation, the approach that will be used is a fidelity approach, where the criteria have been set before. In this approach the criteria approach refers to the success criteria for the implementation of the quality management program of electrical engineering expertise in vocational secondary schools, referring to the principles of the implementation of quality management systems.

Setiyana (2016) states that determining the success of a program requires an assessment criterion. Assessment criteria include the following:

3.3.1 *Context criteria*
Context criteria in the application of electrical engineering expertise quality management course at vocational high schools it is said to be effective if (1) the school has planned community needs and expectations through the field of activities, vision, mission and school objectives or school profile; and (2) the head of the school has determined the quality policy and school quality objectives.

3.3.2 *Input criteria*
Evaluation of inputs is intended for school readiness (principals, teachers, students, and facilities and infrastructure). The input criteria for the success of the implementation of quality management of electrical engineering expertise at vocational high schools are said to be effective if (1) teachers are involved in work mechanisms, quality plans, and training; and (2) there is control of documents and recordings as evidence of school activities.

3.3.3 *Process criteria*
Process criteria cover all activities of the process of applying the quality management of electrical engineering expertise in vocational high schools. Process components are said to be effective if (1) there are standard operating procedures and work procedures for learning procedures to control the learning process; (b) there is accurate data and information in making decisions such as survey data on the level of satisfaction of the working world of graduate users.

3.3.4 *Product criteria*
The product criteria cover all activities of the process of applying the quality management of electrical engineering expertise in vocational high schools, including the teaching and learning process. The product component is said to be effective if (1) the relationship between teachers and students, the relationship between the world of work and the school and students are well established and mutually beneficial; and (2) there is supervision and evaluation of quality policies and quality objectives as a preventive/corrective and corrective action.

4 CONCLUSION

The CIPP evaluation model is the most suitable model used in evaluating the application of quality management of electricity expertise to vocational high schools because it has a more holistic, detailed and broad approach of the object being evaluated. CIPP evaluation model consists of four components: context, input, process, product.

Criteria for the effectiveness of the evaluation program, based on the four components of the CIPP evaluation model are context effectiveness, input effectiveness, process effectiveness, and product effectiveness.

1 Context criteria: (1) The school has planned the needs and expectations of the community through the field of activities, vision, mission, and goals of the school. (2) The principal sets the quality policy and school quality objectives.
2 Input criteria: (1) The teacher is involved in the working mechanism, quality plan, and training; (2) there is control of documents and records as evidence of school activities.
3 Process criteria: (1) The existence of standard operating procedures and learning work instructions; and (2) the existence of accurate data and information in making decisions.
4 Product criteria: (1) The relationship between the teacher and the students, the relationship between the world of work and the school and students, are well established and mutually beneficial; (2) There are supervision and evaluation of quality policies and quality objectives as a preventive and improvement measure.

REFERENCES

Aized, T. (2010). *Total Quality Management and Six Sigma*. InTech. ISBN 978-953-51-0688-3, 306pp.
Bernik, M., Sondari, M.C., & Indika, D.R. (2017). Model of Quality Management System to Maintain Quality Consistency in Higher Education. *Review of Integrative Business and Economics Research*, 6(4), 235.
Candra, I.W. (2012). Studi Evaluatif Implementasi Penerapan Manajemen Berbasis Sekolah di SMK Negeri 3 Singaraja, Kabupaten Buleleng, Propinsi Bali [Evaluative study of school-based management implementation at Singaraja State Vocational School 3, Buleleng Regency, Bali Province]. *Thesis, Administrasi Pendidikan, Program Pascasarjana, Universitas Pendidikan Ganesha.*
Doherty, G. (Ed.). (2003). *Developing quality systems in education*. London, UK: Routledge.
El-Morsy, A., Shafeek, H., Alshehri, A., & Gutub, S.A. (2014). Implementation of Quality Management System by Utilizing ISO 9001: 2008 Model in the Emerging Faculties. *Life Science Journal*, 11(8), 119–125.
Felestin, F., & Triyono, M.B. (2015). The implementation of total quality management at vocational high schools in Indonesia. *REiD (Research and Evaluation in Education)*, 1(1), 13–24.
Fitzpatrick, J.L., Sanders, J.R., & Worthen, B.R. (2011). *Program evaluation: Alternative approaches and practical guidelines* (4 ed.). New York: Allyn & Bacon. Canadian Publisher: Pearson. ISBN: 978-0-205-57935-8.
Gaspersz, V. (2003). *ISO 9001: 2000 and continual quality improvement*. Jakarta: FT Gramedia Pustaka Utama.
Hoyle, D. (2009). *ISO 9000 quality systems handbook – using the standards as a framework for business improvement*. Butterworth-Heinemann: Oxford, UK.
ISO 9001. (2008). *Quality management systems-Requirements*. Switzerland.
Lazibat, T., Sutic, I., & Jurcevic, M. (2009). Quality management system at the faculty of economics and business. In: *31st Annual EAIR Forum in Vilnius, Lithuania.*
Mahmudi, I. (2011). CIPP: Suatu Model Evaluasi Program Pendidikan. *At-Ta'dib*, 6(1).
Mertens, D.M., & Wilson, A.T. (2018). *Program evaluation theory and practice: A comprehensive guide*. Guilford Publications.
Michalska-Ćwiek (2009). The quality management system in education – implementation and certification. *Journal of Achievements in Material and Manufacturing Engineering*. 37 (2).
Patton, M.Q. (1997). Utilisation-focused evaluation: the new century text. *Utilization Focused Evaluation: The News Century Text.*
Sallis, E. (2014). Total quality management in education. Routledge.
Setiyana (2016). Evaluasi Implementasi Sistem Manajemen Mutu ISO 9001:2008 di SMA Negeri 2 Yogyakarta dan SMA BOPKRI 1 Yogyakarta [Evaluation of the Implementation of the ISO 9001: 2008

Quality Management System at Yogyakarta 2 Senior High School and BOPKRI Yogyakarta 1 Senior High School]. *Thesis*, Universitas Negeri Yogyakarta.

Stufflebeam, D.L. (2003). The CIPP model for evaluation. In: *International handbook of educational evaluation* (pp. 31–62). Springer, Dordrecht.

Wiyono, G.v (2013). Pengembangan Model Sekolah Bermutu Total (Total Quality School) di SMK untuk Peningkatan Mutu Pendidikan Kejuruan [Development of a Total Quality School Model in Vocational Schools for Improving the Quality of Vocational Education]. *Seminar Nasional Pendidikan Teknik Elektro 2013. Universitas Negeri Yogyakarta.*

Wulandari, B., & Surjono, H.D. (2013). Pengaruh problem-based learning terhadap hasil belajar ditinjau dari motivasi belajar PLC di SMK. Yogyakarta [The effect of problem-based learning on learning outcomes viewed from the motivation of learning PLC in vocational schools]. *Jurnal Pendidikan Vokasi [Journal of Vocational Education]*, 3(2).

Zhang, G., Zeller, N., Griffith, R., Metcalf, D., Williams, J., Shea, C. & Misulis, K. (2011). Using the context, input, process, and product evaluation model (CIPP) as a comprehensive framework to guide the planning, implementation, and assessment of service-learning programs. *Journal of Higher Education Outreach and Engagement 15*(4), 57–84.

Formal & Informal Education

Innovative Teaching and Learning Methods in Educational Systems – Retnowati et al. (Eds)
© 2020 Taylor & Francis Group, London, ISBN 978-1-03-224183-8

Strengthening character education through the local wisdom: Indonesian folklore

T.A. Rini & P. Mahanani
State University of Malang, Malang, Jawa Timur

ABSTRACT: In the 21st century, strengthening character education becomes the center of attention for the betterment of a nation's character. One of the principles of strengthening character education is through the local wisdom. Local wisdom contains positive character values reflected in Indonesian folklore. This research is descriptive qualitative research using content analysis to describe and interpret the values of the local wisdom as the media of strengthening character education. The data of this research is obtained from Indonesian folklore texts, which are analyzed by using the socio-psychological approach. The result of the analysis shows that (1) character, personality, and events in the story represent the local wisdom of each region, (2) there are always social life values in each story, (3) each local wisdom has its own uniqueness in telling the story, and (4) the local wisdom values have five main characteristics as the media of strengthening character education.

1 INTRODUCTION

The degradation of character values in the 21st century has become a topic of much discussion during this decade. This degradation is supported by facts that show the results of the comparison of characters in Indonesia. Therefore, the revitalization of life values has been promoted through character education. It becomes one of the effective ways to improve the quality of national character, especially the learning process in schools (Sudrajat, 2011). Indonesian Government has promoted Penguatan Pendidikan Karakter (PPK) in 2016 as a foundation and a spirit of education. The sources of character education are continuously explored by adoption, adaptation, combination, and life-based creation. According to Walker, Roberts, and Kristnjanson (2013), character education means reviewing information in a more effective way to 'legitimate' a change through education training and policy.

One of the nine principles in strengthening life-based character education is learning through local wisdom. The local wisdom is defined as a form of public policy in using their mind to act and behave towards objects or events in a particular community (Mukhtar, et al, 2016:6). As a policy, local wisdom has been widely studied as a medium for character building that originates in regional culture. This kind of culture, according to Sauqi (2011:122) is believed as a universal value for humans at various levels adopted by all members of society. Learning through local wisdom is also supported by inheritance from generation to generation. One of the examples is form observation result, especially in Java, that the community highly respects the age difference in language, which is often not currently applied by the younger generation. Now, young people are more comfortable and used to talking informally to older people.

Based on the fact, culture as a source of local wisdom can be explored especially through Indonesian folklore. The local wisdom in Indonesian folklore has character uniquenesses or peculiarities, which can represent a nation's identity. Indonesian folklore is a form of literature built on space dialectics, beauty, and the value of ancestral cultural heritage for generations (Kurniawan, 2014: 4). Folklore can build children's character through messages implied or written in the story (Ardhyantama, 2017). In addition, according to Sarumpaet (2010: 22), folklore is also one of the most preferred kinds of stories by the community due to its noble values.

The local wisdom in Indonesian folklore has character peculiarities that can represent a nation's identity. These characters are portrayed through characters, personalities, and events in the story. The local wisdom as a medium for character building needs to be studied more deeply from various perspectives. The previous research was done by Sanjaya and Divayana (2015), which discussed character building through civic education proved that local culture can strengthen students' character. Sadjim, Muhadjir, and Sudarsono (2016) also conducted research, which showed that the revitalization of local wisdom values can be grown through community education to prevent social conflict. Furthermore, Riley and Prinston (2010) in their research also showed that folklore could have a positive influence on the community in having interaction with each other and the surroundings through myths and events, which have been passed down from generation to generation.

Based on the reasons mentioned in the previous paragraphs, the researcher formulates the purposes of this research to describe the values of the local wisdom in Indonesian folklore and its potential as a medium to strengthen character education. This research complements and updates the previous research by exploring more about the use of folklore in character strengthening. This research has an originality that different from other researches since in this research the values of local wisdom will be assessed in accordance with the five main values of the character as a medium for strengthening character education. Those values are stated in the guideline for strengthening character education (Kemendikbud, 2016: 8-10), which are religious (n1), nationalist (n2), independent (n3), mutual cooperation (n4), and integrity (n5). Moreover, the local wisdom in folklore will be examined for its peculiar or unique characteristics as the values, which represent Indonesian identity. Based on this, the results of the study are expected to contribute as a media for character strengthening especially in the education process by utilizing Indonesian folklore that is rich in local wisdom value.

2 RESEARCH METHOD

This research is defined as descriptive qualitative research using content analysis to examine the values of local wisdom in Indonesian folklore by utilizing a socio-psychological approach. This approach is used because the value of local wisdom is closely related to the values that are built from social interaction. The use of this approach also refers to the study results about folklore in Sarumpaet (2010: 46) as a classic work showing the symbolic representation of child development built in the story, the story in the story refers to the repeated patterns in human life and experience. The data in this study is descriptive data consisting of the results of examining Indonesian folklore and the local wisdom values.

The data source of this study is limited to only five Indonesian folklore taking from a book of Indonesian folklore entitled *Walansendow* and *La Laki yang Bijaksana* written by Y.B Suparlan (2015). This source was chosen because it has several stories from some different Indonesian regions so that it can represent national identity according to the values in the PPK program, as shown in Table 1.

Table 1. The research materials: Indonesian folklore.

No	Title	Province	Code
1	*Putri Nglirip*	East Java	*Pn*
2	*Ncuhi Mawo*	NTB	*Nm*
3	*Sutan Tanamoi*	Jambi	*St*
4	*Wahyu K. Mataram*	Yogyakarta	*Wk*
5	*Patih Senggilur*	South Sumatra	*Ps*

The instrument of the data collection in this study is a human instrument, which is the researcher herself who is equipped with supporting instruments in the form of data collection instruments of local wisdom values in the texts of Indonesian folklore. The value of local wisdom in Indonesian folklore is assessed through a socio-psychological approach by the following procedures (1) identifying events, characters and personalities that represent the local wisdom values, (2) identifying the local wisdom values outside related literary texts, (3) analyzing characteristics of Indonesian local wisdom, and (4) interpreting the local wisdom values as the media for strengthening character education.

3 RESULTS

The results of this study indicate that Indonesian folklore is full of local wisdom values, which can be used as the media to strengthen character education. The results of the study of local wisdom values in Indonesian folklore can be observed in Table 2.

The first folklore from East Java entitled Putri Nglirip tells about a woman named Dyah Kusumadewi, the youngest daughter of Adipati Tuban, who was very beautiful. Kusumadewi was always accompanied by his friendly and loyal maid *(Dayang)*. This maid served Kusumadewi to redeem the sins of her parents who hurt many people's hearts. Kusumadewi's sisters hated Kusumadewi's beauty so they always tried to harm her. Kusumadewi decided to leave the palace because she wanted to live peacefully. AdipatiTuban was furious and ordered his daughters to look for Dyah Kusumadewi wherever she was. They were not allowed to come back to the palace until they found Kusumadewi.

The second folklore was from West Nusa Tenggara, entitled Ncuhi Mawo (nw), which tells us about a regional leader named Ncuhi Mawo who was vicious because of his supernatural powers. As a result, Ncuhi Mawo got punishment from Sangaji Bojo. The

Table 2.   Characters, personalities, dan events.

Characters and Personalities	Events
Code pn • Kusumadewi– kind-hearted • Dayang– friendly • Adipati– wise • The daughters– jealous and envious	Adipati Tuban punished his two daughters who always hurt their sister, Dyah Kusumadewi, who was very beautiful until she decided to leave the palace.
Code nw • Ncuhi Mawo– arrogant • Ncuhi Jia– faithful • Baginda Sangaji Mbojo– wise	Ncuhi Mawogot punishment from Sangaji Bojo due to his vicious attitude. The punishment was carried out by his own brother, Ncuhi Jia who obeyed the king's order.
Code st • Sutan Tanamoi and Sanguiling – charitable and kind-hearted	SutanTanamoi and Sanguiling who were separated while wandering to find their sister could get together again.
Code wk • Ki A. Giring– careless • Ki Buluaji– spiteful • Ki A. Pemanahan– tricky	Ki Ageng Giring who broke his promise with Ki Buluaji failed to get revelation so that the revelation was taken by Ki Ageng Pemanahan
Code ps • PatihSangulir– wise	Patih Sangulir is a powerful community leader who is very concerned about the condition of his village

punishment was determined through a meeting attended by all Ncuhi (a designation for the regional leaders). Ncuhi Jia, Nuchi Mawo's brother, received an order to punish Ncuhi Mawo. Ncuhi Jia reluctantly carried out the orders of Sangaji Bojo as a form of loyalty and justice for his kingdom.

The third folklore was from Jambi entitled Sutan Tanamoi and Sutan Sanguiling, which tells us about two brothers who were sent by their parents to wander to find their separated brothers. Their father gave them incense charcoal that could be used to show them the direction where they should go. Following the smoke of the incense charcoal, they were separated and could not find each other. A few years later, Sutan Sanguling who had lived a prosperous life wanted to find his brother SutanTanamoi. After wandering, Sutan Sanguling and Tanamoi successfully found each other and lived together happily ever after.

The fourth folklore came from Yogyakarta entitled *Wahyu Kerajaan Mataram* (The Revelation of Mataram Kingdom), which tells about Ki Ageng Giring who received a revelation in the form of *Gagak Emprit* coconut. Ki Ageng Giring entrusted the coconut to Ki Buluaji. One day, Ki Ageng Giring took *Gagak Emprit* coconut without asking permission from Ki Buluaji. Ki Buluaji felt offended by Ki Ageng Giring's attitude because he felt that he had kept Ki Agen Giring's mandate well. At that time, Ki Ageng Pemanahan received a revelation to drink *Gagak Emprit* coconut milk due to the oversight of Ki Ageng Giring who made an agreement with Ki Buluaji.

The fifth folklore from South Sumatra, *Patih Senggulir,* tells about a community leader named Patih Sangulir who was powerful and wise. He was considered an influential figure in his village. When there were problems or activities carried out by the community, they would ask for his opinion and help. He was very concerned about the condition of his village and could manage his time very well.

Based on the description of descriptive data above, we can find the values of local wisdom in Indonesian folklore, which can be classified according to the five main personalities of the main characters included in the program of strengthening character education. Those values classification can be seen in Table 3.

Table 3 .   The Results of the Local Wisdom Values.

Code	Local Wisdom Values	Value
pn1	• Friendly and respectful	n5
pn2	• Good manners	n5
pn3	• Punishment for bad deeds	n1
pn4	• Compliance with parents	n5
nw1	• Compliance with laws and leaders	n2
nw2	• Deliberation in making decisions	n4
nw3	• Penalties for violations of customs/rules	n1
st1	• Depends on the natural surroundings	n1
st2	• Thankful and simple life	n1
st3	• Belief in mysticism	n1
st4	• Compliance with parents	n4
st5	• Hard work for prosperity	n3
st6	• Belief in parents' advice	n1
st7	• Living together as a happiness	n4
wk1	• Promises must be kept	n5
wk2	• Every act gets a reply	n1
ps1	• The obligation in maintaining order	n2
ps2	• Kindness and caring among others	n4
ps3	• Humble and caring	n5
Ps4	• Respectful to the wise figures	n2

4 DISCUSSION

4.1 *Characters, personalities, and event*

The results of the analysis show that folklore is full of local wisdom values, which are depicted in characters, personalities, and events in the story. Based on the results of the analysis it is known that Indonesian folklore always uses characterizations that differ in age or position. This characterization refers to exemplary and experience as a learning process. There is an inheritance of values from the previous generation for the next generation. It shows that the active role of the surrounding environment is needed in character education. Freeks (2015) concludes that people around students have an important role in character development and formation. This relationship forms mutual appreciation and respect in life.

In the characterizations, we can find that Indonesian folklore always uses the protagonist (Diah Kusumadewi, Sangatih, etc.) and antagonists (Ncuhi Mawo, Ki Bulujai, etc.). The interaction between the two figures shows a cause and effect that can be used as a role model in character education. The causal relationship shows social contact that involves cognitive, affective, and conative (Selasih and Sudarsana, 2018). The society believes that every good deed will get good and vice versa.

The events are always depicted with a progressive plot that begins with the introduction of characters to the development in the story. The events in the story tend to be described close to nature and the surrounding environment. This shows that folklore provides a story to respect and protect nature, both presented literally or symbolically (Sukmawan and Setyowati, 2017). Forests, community environments, private residences, and kingdoms are used to describe the life order of the past. People really appreciate and protect the natural surroundings.

4.2 *The local wisdom*

Analysis of meaning is done with a socio-psychological approach. The use of this approach refers to the results of Sarumpaet's (2010: 46) research on folklore as a classic work that represents the symbol of life development in the story built that refers to the repeating patterns and human experience in life. The first story (*pn*) represents a child's respect and responsibility for his parents. Parents become a figure who must be obeyed and respected. Moreover, a child must be willing to sacrifice for the sake of his parents. Children should maintain their attitudes and words towards their parents since the community believes that manners are closely related to position and age.

The second story (*nw*) represents the value of loyalty, justice, and high nationalism. All the orders of the leader and customary rules are considered as a form of wisdom and goodness, which should be highly respected. People really appreciate togetherness in life. If there is a problem, deliberation is held to make a decision on the problem. The third story (*st*) represents the indigenous people is a community that prioritizes togetherness and unity that is built on the basis of kinship. The happiness of life is defined as a blessing that can be enjoyed together especially in a family relationship.

The fourth story (*wk*) represents the sacredness of an agreement or promises so that those who make it must keep it and carry out carefully. In addition, the society believes that a revelation is a gift from God to certain people. A person's attitude and behavior will cause certain consequences in his life, which are related to the revelations received by those people and the society around them. Furthermore, the fifth story (*ps*) represents a view of magic and science as a noble power if they are used in a good way. The community leaders are respected based on their kindness and concern, which are then used as the role models.

4.3 *The peculiarities of the local values*

Based on the results of the analysis, the value of local wisdom in Indonesian folklore represents the Eastern culture. The local wisdom value as a part of the Eastern culture represents an agrarian value and culture that prioritizes life wisdom that comes from conscience and intuition by glorifying human intelligence (Soelaiman, 1990: 41). The results also show that deliberation, courtesy towards parents, as well as belief in good and bad, are found in the five stories. In addition, the depiction of life that is close to nature, the words, the attitudes, and the ways of depicting characters show the peculiarities of the East.

The events, characters, personalities, and the nature of society are much contemplated, abstracted and permitted and then meditated as a form of the local cultural wisdom of the Eastern community (Saryono, 2008: 32). Local wisdom in folklore has a specific meaning of obedience, togetherness, goodness in life and its impact. The harmony between people and their environment is very important to create happiness and prosperity. The results research conducted by Fatimah, Sulistyo, and Saddhono (2017) also found that the manifestation of character education can be seen through the values of local wisdom in the story of Sayu Wiwit, for instance, nationalism. The behavior and attitudes of the main characters reflect adherence and concern in the environment.

Another research by Sarman (2016) shows that folklore entitled Pinang Gading from Belitung represents honesty, responsibility, and deliberation in the community. It is followed by research conducted by Kanzunnudin (2017), which shows that folklore from Kudus represents the values of attitude, exemplary, heroism, and religiousness. From the values found and the results of previous studies, it can be concluded that Indonesian folklore has a peculiar or unique character that prioritizes social values in life.

4.4 *Media to strengthen character education*

Based on the results of the analysis, the local wisdom in Indonesian folklore can be classified as the five main characteristics of strengthening character education. From the five folklore, the spiritual value mostly came from other values. The belief in obedience and good behavior can be found in each story. Faith is the main focus in character education because we can clearly understand the concept of good and bad through the faith. This is in accordance with the opinion of Mailybaeva, Utegulov, Tazhinova, Assylova, and Shatyrbayeva (2015) that folklore can be a valuable tool for educating young people, unifying spiritual diversity, and moral education.

The society believes in the cause and effect of every behavior towards its surroundings, so keeping harmony around society is very important. It is different from living in the 21st century where people tend to be individualistic and apathetic. Many people are now more concerned with personal needs. In Indonesian folklore, it was found that people always prioritized togetherness, which was included in the value of mutual cooperation. Discussion among society members becomes the main activity in social communication as a symbol of deliberation and representation injustice.

Indonesian folklore also represents honesty, integrity, and tenacity in living life. The community is closely related to a philosophy that believes in what you sow you will reap (Mukhtar, et al, 2016: 17). These values can be used as a contextual learning process that many figures who currently have high integrity also succeed in achieving success and goodness in their lives. Comparison, examples, and internalization of meaning can be used as a way of using folklore as a medium to strengthen character education.

The five values of the main characters appear repeatedly in the local wisdom of Indonesian folklore. This shows that Indonesian folklore can be used as a medium to strengthen education through characters, personalities, events, and messages in the story. The Indonesian folklore can be used in formal classes at schools and home activities by involving teachers, parents, and the surrounding community to convey the diversity of the meaning of local wisdom through modeling and imitating the stories. This diversity, according to Anshori (2017), will develop and strengthen local wisdom that can provide an identity of the young generation as an Indonesian.

Folklore can be learned at school through language or literary appreciation activity. It can also be introduced as attractive reading material for young learners at home or in society. Related to the results of the previous discussion, the results of the comparison of this study with the previous one, which shows other folktales also contain local values of good faith for character strengthening. Teachers, parents, and society become the active 'storytellers' who tell the younger generation about the folklore. It also can be integrated with the learning process by selecting the appropriate material.

5 CONCLUSION

Indonesian folklore is full of local wisdom that can be used as a medium to strengthen character education. The results of the analysis of the five folklore show that characters, characterization, and events in the story describe life as a learning process that has passed from generation to generation. The local wisdom in Indonesian folklore that has been analyzed represents various values of social life in achieving harmony of life. Every local wisdom in folklore represents cultural distinctiveness as an Eastern nation. The local wisdom shows a polite and noble character of the East that prioritizes social values to realize a harmonious life. The values of local wisdom can be used as a medium to strengthen character education that integrates five main characters. The local wisdom in Indonesian folklore can be used in the learning process in schools and educational reading materials. Parents, teachers, and the surrounding community act as reviewers and the 'storytellers' in order to make sure that the values from the story are accepted and internalized in the character of the young generation.

REFERENCES

Anshori, I. (2017). Penguatan pendidikan karakterdi madrasah [Strengthening character education in Madrasas]. *Halaqa: Islamic Education Journal*, 1(2),63-74. doi: 10.21070/halaqa.v1i2.1243

Ardhyantama, V. (2017). Pendidikan karakter melalui cerita rakyatpada siswasekolahdasar [Character education through folklore in elementary school students]. *IJPE: Indonesian Journal Of Primary Education*, 1(2). doi: 10.17509/ijpe.v1i2.10819

Fatimah, F.M., Sulistyo, E.T., and Saddhono, K. (2017). Local wisdom values in sayuwiwit folklore as the revitalization of behavioral education. *KARSA: Journal of Social and Islamic Culture*, 25(1),1. doi: 10.19105/karsa.v25i1.1266

Freeks, F.E. (2015). The influence of role-players on the character development and character-building Of South African College Students. *South African Journal of Education*, 35(3). doi:10.15700/saje.v35n3a1086

Kanzunnudin, M. (2017). Menggali nilaidan fungsi cerita rakyatsultan hadirindan masjid wali at-taqwaloramkulon kudus [Exploring the value and function of the folklore of Sultan Hadirin and Wali At-TaqwaLoramKulon Kudus Mosque]. *KREDO:JurnalIlmuBahasadanSastra*, 1(1). doi:10.24176/kredo.v1i1.1748

Kemendikbud. (2016). *Konsepdan pedoman penguatan pendidikan karakter*[Concepts and guidelines for strengthening character education]. Jakarta: Kemendikbud.

Kurniawan, H. (2014). *Sastra anak dalam kajian strukturalisme, sosiologi, semioka, hinggapenulisankreatif* [Children's literature in the study of structuralism, sociology, semioka, to creative writing]. Yogyakarta: GrahaIlmu.

Mailybaeva, G., Utegulov, D., Tazhinova, G., Assylova, R., and Shatyrbayeva, G. (2015). Formation of moral values of school children by means of Kazakh Folklore. *Medwell Journal*, 10(6), 1076–1079. doi: 10.3923/sscience.2015.1076.1079

Mukhtar., Kahirun., Wianti, N.I., Umran. L.O., Karmila, W.O.N., Zalmina, S., Kete. S.C.R., and Jabuddin, L.O. (2016). *"Mecula&HaroaAnula"SuatutinjauankearifanlokalmasyarakatButon Utara* ["Mecula&HaroaAnula" A review of the local wisdom of the people of North Buton]. Yogyakarta: BudiUtama.

Riley, E.P., and Prinston, N.E. (2010). Macaques in farms and folklore: exploring the human-nonhuman primate interface in Sulawesi, Indonesia. *American Journal of Primatology*, 72(10), 848–854. doi: 10.1002/ajp.20798

Sadjim, U.M., Muhadjir, M., and Sudarsono, F.X. (2016). Revitalisasinilai-nilaiBhinneka TunggalIkadankearifanlokalberbasislearning society pascakonfliksosial di Ternate [Revitalizing the values of Bhinneka Tunggal Ika and local wisdom based on learning society after the social conflict in Ternate]. *Jurnal Pembangunan Pendidikan: Fondasidan Aplikasi*, 4(1),79–91. doi: 10.21831/jppfa.v4i1.7227

Sanjaya, D.B., and Divayana, G.H.D. (2015). An expert system-based evaluation of civics education as a means of character education based on local culture in the Universities in Buleleng. *IJARAI: International Journal of Advanced Research in Artificial Intelligence*, 4(12). doi: 10.14569/IJARAI.2015.041203

Sarman, S. (2016). Representasikearifanlokalmasyarakat Belitung dalamcerita Keramat Pinang Gading [Representation of the local wisdom of the Belitung people in the story of Keramat Pinang Gading]. *SirokBastra Journal*, 4(2). doi: 10.26499/sb.v4i2.85

Sarumpaet, R.H. (2010). *Pedomanpenelitiansastraanak*[Guidelines for children's literature research]. Jakarta: YayasanPustakaObor Indonesia.

Saryono, D. (2008). *Parasnilaibudaya* [Paras of cultural values]. Malang: Surya Pena Gemilang.

Sauqi, A., and Naim, N. (2008). *Pendidikanmulticultural* [Multicultural Education]. Yogyakarta: ArRuzz Media.

Selasih, N.N, and Sudarsana, I.K. (2018). Education based on ethnopedagogyin maintaining and conserving the local wisdom: A literature study. *Jurnal Ilmiah Peuradeun: The International Journal Of Social Sciences*, 6(2), 293–306. doi: 10.26811/Peuradeun.V6i2.219

Soelaiman, M.H. (1990). *Ilmu Budaya Dasar*[Basic Culture]. Bandung: PT Rosda Offset.

Sudrajat, A. (2011). Mengapapendidikankarakter [Why character education]. *JurnalPendidikanKarakter*, 1(1). doi: 10.21831/jpk.v1i1.1316

Sukmawan, S., & Setyowati, L. (2017). Environmental messages as found in Indonesian folklore and its relation to foreign language classroom. *Arab World English Journal*, 8(1), 298–308. doi: 10.24093/awej/vol8no1.21

Suparlan, Y.B. (2015). *Kumpulan ceritarakyatIndonesia: Walansendow*[Collection of Indonesian folklore: Walansendow]. Yogyakarta: PT KANISUS (Anggota IKAPI).

Suparlan, Y.B. (2015). *Kumpulan ceritarakyatIndonesia: La laki yang bijaksana*[Collection of Indonesian folklore: Wise man]. Yogyakarta: PT KANISUS (Anggota IKAPI).

Walker, D.I., Roberts, M.P., & Kristnjanson, K. (2013). Towards a new era of character education in theory and in practice. *Journal Educational Review*, 67(1), 79-96. doi: 10.1080/00131911.2013.827631

Innovative Teaching and Learning Methods in Educational Systems – Retnowati et al. (Eds)
© 2020 Taylor & Francis Group, London, ISBN 978-1-03-224183-8

Forming young citizen characters through youth organizations in Indonesia

Wellyana & Marzuki
Yogyakarta State University

ABSTRACT: Character as an outward characteristic possessed by each individual can be formed through formal or non-formal education. In non-formal education, character can be formed through participation in non-governmental organizations such as youth organizations. The purpose of writing this paper is to explain the role of youth organizations in building the characters of young citizens. The method used is a literature study on youth organizations, the concept of character education itself, and young citizens. The results of the literature review show that the activities of youth organizations indirectly support character building in young citizens even though there are still some obstacles.

1 INTRODUCTION

Every sovereign country has a major element in its state management: citizens. These countries have an obligation to educate and equip each of its citizens with a good education in order to form *good citizens* who understand their rights and obligations. However, the state cannot only rely on political policies for this but must involve other related parties. To create good citizens or good citizenship requires a strong education on 'character', so the issue of character education is much discussed and pursued by national governments.

In Indonesia, to support achieving optimal character education, the government makes related policies or regulations. The recently issued Presidential Regulation No. 87 of 2017 concerning character education (RoI, 2017) states that character education is not only obtained in schools or formal education institutions but involves various parties, including families (especially parents), the community and schools. According to Lickona (1991), the concept of character education for a noble character (good character) includes knowing about kindness, being committed to kindness and finally performing kindness. Thus, it can be said that character, includes several stages, starting from someone's knowledge (cognitive), attitudes, motivations, behaviors, and skills. Character education is not only teaching knowledge on character theoretically but is about how giving a real example to children about character as well so that the children can apply it in their daily lives.

Character education is not only a matter of formal education that children or students receive in school but also the non-formal education that children get in their families and the surrounding environment. However, there is certainly a view that only families and schools are required to provide character education. What is often forgotten is the community environment, yet this also has a significant contribution to the character education itself, and mainly character education for youths. In the community, there are many youth organizations that unwittingly become the place or tool in shaping youths into good citizens of young countries and understanding their rights and obligations. In Indonesia, examples of these include youth organizations (*Karang Taruna*), youth organizations in religious fields such as Anshor and IMM (Muhammadiyah Student Association), and other organizations. As stated by Widiatmaka, Pramusinto, and Kodiran (2016), the Anshor organization (a religious organization under Nahdatul Ulama (NU)) has a role in character education, especially in forming

77

religious character, responsibility, independence, cooperation, tolerance, and nationalism, obtained during activities basic education and training or namely *diklatsar*.

In relation to this, this article was written to determine through literature studies the role of youth organizations in forming young citizens' characters, especially the role of existing activities. The method used is a literature study conducted by analyzing descriptive qualitative data collected from books, journals, and previous studies on character education and youth organizations.

2   RESEARCH METHOD

The method used here is a literature study conducted by analyzing descriptive qualitative data from books, journals, and previous studies on character education and youth organizations.

3   RESULTS AND DISCUSSION

A school supervisor has an important role in solving problems in education quality because supervising is like a bridge between teachers and education quality. Teachers need help and solutions from a school supervisor to solve problems in the teaching-learning process. A clear learning process has become a factor for students to receive effective learning. Students who are given clear explanations in the learning process are more likely to receive high scores in examinations. Thus, the teacher is the key to successful learning for the students.

3.1   *Youth organization*

The youth organization is a forum that serves as a place for creativity and attitude and mental training. Wahjosumidjo (1992) stated that such organizations have a strong influence on social records. This strong influence is not only affected the community, but also its human resources as members of the organization that motivate active members in social change. In addition, Sudariya (2010) stated that youths have advantages related to leadership in an organization in the form of self-realization through work programs. Therefore, such organizations and youth are strongly interrelated, especially in developing and mentally training young people and maintaining social order.

In Indonesia, youth organizations exist in various forms: some formal and some informal. Formal organizations are usually certain agencies such as Scouts if at school, and a youth organization such as *Karang Taruna* in the village setting. Suharta (2009) described *Karang Taruna* as a social organization for developing young people in social welfare based on a sense of awareness and responsibility of young people in the developed village areas. Informal organizations are usually outside this, such as motorcycle clubs and soccer fans associations.

Zeldin, Gauley, Krauss, Kornbluh, and Collura (2015) stated that youth or community organizations encourage young people to explore or develop relationships between other youths or adults as well as the development of citizenship related to youth empowerment and relations with other organizations or communities. Therefore, it can be understood that youth organizations are a very important place in establishing character, because in the existing activity programs young people are forged both mentally and conceptually, can practice leadership skills and can exercise self-responsibility.

3.2   *Young citizens*

Karim (1992) explained that young people as a group are difficult to approach by political parties or other candidates for election, because of inadequate knowledge. In addition, de Vries and Wolbink (2018) started that more attention must be paid to this group. A collaboration of several parties is needed and with different innovations.

Law No. 10 of 2008 in Chapter II Article 19 paragraphs 1 and 2 and article 28 states that Indonesian citizens age 17 years or older on election day or those who are married have the right to vote (RoI (2008)). Hurlock (2011) states that at the stage of adolescence youths experience a significant change, both emotionally, physically, and through behavior patterns, and may have experienced many life problems.

Meanwhile, Cholisin (2004) defined 'citizen' as someone with the membership of political institutions, namely the state, and citizens as subjects as well as objects in the life of their country. Moreover, taking responsibility for the sustainability of a country automatically makes a person possess rights and obligations. In line with this, Syaukani, Gaffar, and Rasyid (2003) define citizens as a group of people who are part of the population that is an element in the forming of a nation.

Thus, it can be understood that young citizens are individuals who have not completely mature views and mindsets, but usually, in terms of politics and state, they have adequate knowledge so are not easy to provoke or mobilize. They have begun to understand the rights and obligations of a citizen.

3.3 Contribution of youth organization to the character formation of young citizens

The concept of youth organizations in forming the character of young citizens is currently an interesting topic with lots of research. Character education not only relies on family or parents and formal institutions such as schools, but non-formal institutions in society also have an influence in shaping the character of children, especially youths.

Felice and Solheim (2011) stated that youth organization is considered a meaningful learning program for young people, especially in learning skills and attitudes for citizenship. Gay, Hjorth, Penderson, and Roelsgaard (2018) said that youth organizations are a place to hone or train skills especially related to one's character. Thus, each youth organization has its own role in forming the characters of young people, and this can be seen as non-formal education, which in practice is not only related to theories but direct practice in daily life.

Moore (2015) stated that organizations play an important role in the vision of peace and forming one's ideal character through each of its process and activities. Otero (2016) stated that youth organizations are able to improve youth's skills because they are considered as non-formal education. Kolano and Davila (2019) stated that young people need special non-formal institutional spaces (youth organizations) that can maintain identities, experiences, and social awareness, which will encourage leadership attitudes and good character.

From several theoretical studies of youth organizations in forming the characters of young citizens, it can be stated that youth organizations are not only beneficial to social life, but they also have a large share in forming a person's character. Lickona (1991) stated that noble character (good character) encompasses knowledge of kindness, creates a commitment to kindness, and finally performs kindness. So it can be said that character includes several stages, starting from someone's knowledge (cognitive), attitudes, motivations, behavior, and skills. Therefore, character education is not only that which provides theoretical knowledge about the character, but is also based on how the character education agent is able to educate, by giving children a direct understanding of character so they can apply it in their daily lives. From some of these concepts, it can be concluded that character education is not only related to formal education in schools but also non-formal education in families and community environments, such as youth organizations.

Youth organizations are one of the agents of character education outside formal education that favored by young people as places for forming skills, knowledge that will shape good behavior, and characters that will create good citizen attitudes and behaviors. The existing studies show that youth organizations have agendas or activities that unwittingly train and sharpen youth skills, for example, training or workshops and social service activities. This is in accordance with RoI (2017), who explains that character education is not only the obligation of one party, but various parties or agents of character education, both formal and non-formal education, starting from parents or families, schools, and the community environment.

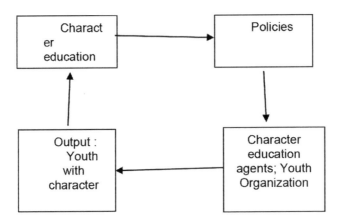

Figure 1.   The character education process through youth organizations.

Thus, we should provide opportunities for young people to develop their interests and talents by participating in youth organizations, even though they must remain under the supervision of other relevant parties. Figure 1 depicts the explanation above.

4   CONCLUSION

Character education is not only formed in the family and school environment as formal education but can be formed in the community, for example in youth organizations. Young people should be given the opportunity to develop their interests and talents by participating in youth organizations that aim to shape character. Families and schools continue to play a role in providing supervision to children or young people in each of their organizational activities.

REFERENCES

Cholisin. (2004). *Pendidikan kewarganegaraan* [Civic Education]. Diktat. Yogyakarta, Indonesia: Fakultas Ilmu Sosial dan Ekonomi UNY.

de Vries, M. & Wolbink, R. (2018). Transition and transformation in youth care in the Netherlands: Emergent challenges for leadership and management in the youth sector. *International Journal of Public Leadership*, *14*(2), 96–108. doi: 10.1108/IJPL-07-2017-0028

Felice, C.D. & Solheim, L. (2011). Youth organisations: Exploring special contributions and challenges. *Development in Practice*, *21*(8), 1094–1108. doi: 10.1080/09614524.2011.606892

Hurlock, E.B. (2011). *Psikologi perkembangan: Suatu pendekatan sepanjang rentang kehidupan* [Developmental psychology: An approach throughout the range of life]. Jakarta, Indonesia: Erlangga.

Karim, M.R. (1992). *Pemilu demokratis kompetitif* [Competitive democratic election]. Yogyakarta, Indonesia: PT. Tiara Wacana Yogya.

Kolano, L.Q. & Davila, L.T. (2019). Transformative learning of refugee girls within a community youth organization serving Southeast Asians in North Carolina. *Journal of Research in Childhood Education*, *33*(1), 119–133. doi: 10.1080/02568543.2018.1531447

Moore, M. E. (2015). Building a nonviolent organization: Religious leadership in a violent world. *The Religious Education Association*, *110*(40), 435–450. doi: 10.1080/00344087.2015.1063966

Otero, M.S. (2016). Young people's views of the outcomes of non-formal education in youth organisations: Its effects on human, social and psychological capital, employability and employment. *Journal of Youth Studies*, *19*(7), 938–956. doi: 10.1080/13676261.2015.1123234

Gay, P.U.H., Hjorth, T.L., Penderson, K.Z., & Roelsgaard, A.O. (2018). Character and organization. *Journal of Cultural Economy*, *12*(1), 36-53. doi: 10.1080/17530350.2018.1481879

RoI. (2017). *Peraturan Presiden (Perpres) Nomor 87 Tahun 2017 tentang Penguatan Pendidikan Karakter* [Presidential Regulation Number 87 of 2017 concerning Strengthening Character Education]. Jakarta, Indonesia: Republic of Indonesia.

Sudariya, N. (2010). *Gaya kerja kepemimpinan situasional* [Situational leadership work style]. Bali, Indonesia: UNDIKSHA.

Suharta. (2009). *Pengembangan organisasi kepemudaan* [Youth organization development]. Yogyakarta, Indonesia: Universitas Negeri Yogyakarta.

Syaukani, H. R., Gaffar, A., & Rasyid, R. (2003). *Otonomi daerah dalam negara kesatuan* [Regional autonomy in a unitary state]. Yogyakarta, Indonesia: Pustaka Pelajar.

Lickona, T. (1991). *Educating for character: How our school can teach respect and responsibility.* New York, NY: Bantam Books.

RoI. (2008). Undang-Undang Nomor 10 Tahun 2008 tentang Pemilihan Umum [Law Number 10 of 2008 concerning General Elections]. Jakarta, Indonesia: Republic of Indonesia.

Wahjosumidjo. (1992). *Kepemimpinan dan motivasi* [Leadership and motivation]. Surabaya, Indoensia: Ghalia Indonesia.

Widiatmaka, P. Pramusinto, A., & Kodiran. (2016). *Peran organisasi kepemudaan dalam membangun karakter pemuda dan implikasinya terhadap ketahanan pemuda (Studi pada pimpinan cabang gerakan pemuda Ansor di Kabupaten Sukoharjo Provinsi Jawa Tengah)* [The role of youth organizations in building youth character and its implications for youth resilience (Study on branch leaders of the Ansor youth movement in Sukoharjo Regency, Central Java)]. *Jurnal Ketahanan Nasional, 22*(2), 180-189. doi: 10.22146/jkn.12002

Zeldin, S., Gauley, J., Krauss, S. E., Kronbluh, M. & Collura, J. (2015). Youth-adult patnership and youth civic development cross-national analyses for scholars and field profesionals. *Youth and Society,* 49(7),851-878. doi: 10.1177/0044118X15595153

Innovative Teaching and Learning Methods in Educational Systems – Retnowati et al. (Eds)
© 2020 Taylor & Francis Group, London, ISBN 978-1-03-224183-8

Revitalization in vocational training centers for improving the quality of human resources

I.A. Manalu & R. Asnawi
Universitas Negeri Yogyakarta

ABSTRACT: This research is related to revitalization at a Vocational Training Center (VTC). VTCs conducts training programs based on guidance from the Ministry of Labor. The VTC requires revitalized to maintain the quality of training. Revitalization was carried out in relation to training components, namely training facilities and infrastructure and training instructors. The components were rearranged according to standards and needs to improve the quality of training. Thus, the method of this research was a literature study based on books, journals, articles, and government regulations, to find out information about training programs. This study discussed training facilities and infrastructure and training instructors according to standards and in accordance with the needs of training programs. This component greatly affects the success of training. VTC revitalization is required in order to meet the needs of the training program so that its objectives can be achieved.

1 INTRODUCTION

The unemployment rate in Indonesia tends to be high. Based on data from the Central Statistics Agency (BPS, 2018) as of February 2018, the total number of unemployed people was more than 6.8 million. This is the overall unemployment in Indonesia at different levels of education: never going to school, graduates of elementary, middle school, high school, vocational school, diploma level, or bachelor level education. High school graduates contribute the most in unemployment level. Vocational school and Junior high school graduates are in the second and the third place respectively. This data shows the level of readiness of educational institutions in preparing students. Educational institutions have not been able to optimally prepare students to be competitive in the world of work.

Based on RoI (2003), one of the goals of the vocational school is to prepare students to be productive people, able to work independently and fill job vacancies in accordance with the competencies in their expertise programs. Vocational education institutions are designed to prepare graduates who are ready to work and compete in the world of work. At present, the number of unemployed people with vocational education is still relatively high. This means that vocational schools have not achieved the specific goals set.

Based on previous data, it can be gauged that the high number of unemployed persons is not a weakness of only higher educational institutions, but all levels of education, including those who never went to school. This education basically does not prepare the students for employment. Graduates tend to be prepared to continue their education. Graduates of this level of education do not have special skills that have prepared them for this level of education. The open unemployee who do not have the skills need to be given special education or training. This step is to prepare them to be able to work according to their interests and have competitiveness.

Vocational Training Centers (VTC) carry out special education and training programs. VTC are expected to produce human resources that are ready to work and able to compete in the world of work. The Ministry of Manpower states that Indonesia aims to be the seventh largest economy in the world by 2030. To achieve this, Indonesia needs 113 million skilled workers yet in April 2017 the number of skilled workers in Indonesia was 57 million people. The Ministry of Manpower must meet the needs for skilled workers that are still lacking, with a target of 3.7 million people annually. To achieve this, the ministry implemented a training program that was implemented by the VTC. The main target of this training program is working-age people who do not have a job. With the target number and quality of training graduates, VTC need to be revitalized to help achieve its goals and maintain its function and purpose. Revitalization relates to facilities and infrastructure, instructors, and the process of implementing training.

Revitalization of facilities and infrastructure is done to ensure the quality of facilities and infrastructure used in the training process. Facilities and infrastructure, in general, are based on standards in the draft directorate of course and training coaching. Revitalization of the instructor is related to the qualifications and competence listed in Minister of Education and Culture regulations 90 of 2014 (RoI, 2014a). Instructors must have an undergraduate or diploma education qualification from an accredited university and have a certificate of expertise according to competence in the training program. Instructor competency standards are in the form of pedagogic, personal, social, and professional competencies. Revitalization of the training process is related to the training process, which begins with the selection of participants, and the process of implementing training activities for evaluation. Training programs are designed before training occurs. So, the expected training process is in accordance with the initial planning.

The purpose of this study was to determine the quality of facilities and infrastructure and a good training process. The research was conducted to examine the importance of revitalization in training institutions.

The research method used was a literature approach related to the revitalization of VTCs. This literature study was carried out by reviewing regulations, books, research journals on facilities and infrastructure, instructors, and the training process in the VTC.

2 DISCUSSION

The VTC is a venue for training programs. Training is a form of non-formal education. The implementation of training programs is focused on one particular area and implements practically the skills needed by the participants (Agrawal, 2013). A participant is given training on one type of special skill. Participants need skills that can be used in the workforce, so they can be responsible for their work.

Training at the VTC prioritizes the mastery of work skills that include knowledge, skills, and attitudes according to standards and needs in the workforce (Depnakertrans). The VTC creates a training plan that aims to create or improve the skills of participants and this plan will guide the entire implementation of the training toward the target and training objectives. The training includes all activities in giving, obtaining, improving, and developing competencies, productivity, discipline, attitudes and work ethics.

The quality of the training program must be a priority for the person in charge of the program and the VTC needs to carry out revitalization in order to maintain its quality. Revitalizing VTCs as non-formal education institutions also needs to be done for several reasons. (1) Preparing people to face regional and global competition. At present, the community is faced with wider competition; in addition to national competition, the community must also face competition from the ASEAN Economic Community (AEC). The training program must be updated according to the needs of the community and the level of competition that exists. RoI (2016) stated that the MEA agreement provides the possibility of open employment being as high as 14 million until 2025. (2) Defining steps to fulfill needs. Indonesia has the potential to become the seventh largest economy in the world, but to achieve this, 113 million competent workers are required. Revitalization of VTCs is expected to produce competent workers to fill the needs of the workforce. (3) Creating employment opportunities for people who do not have an education. The VTC offers a job training program in working-age communities with varying levels of education. People who only have primary school education or even did not go to school can attend the training and those who do not have special competencies have the opportunity to learn new skills and enter the workforce. Revitalization carried out at the VTC includes facilities and infrastructure as well as training instructors.

Revitalization of the VTC was officially carried out in February 2017 through the Decree of the Minister of Manpower No. 23 of 2017. This program is called the 3R (reorientation, revitalization, and rebranding) program, which covers all parts of the implementation of training programs. The VTCs implementing this program are in Bandung, Serang, Bekasi, Semarang, and Medan. Revitalization is deemed necessary because the VTCs needs a lot of improvements, especially with regard to the facilities and infrastructure that support training programs and the needs of instructors.

Facilities are the equipment needed directly in the process of conducting training. These facilities referred to under the tender regulations include machinery/equipment, hand tools, health, and safety equipment, and supporting equipment. Infrastructure is equipment that is indirectly needed in the implementation of training. Infrastructure standards at the VTC, include the availability of office buildings, theory rooms, practice rooms, and supporting infrastructure. Supporting infrastructures are toilets, worship rooms, parking lots, libraries, training material warehouses, dining rooms/canteens, security posts, archive rooms, service rooms, dormitory buildings, etc.

The next standards related to facilities and infrastructure are discussed by Soendjojo et al. (2017). The facility standards include furniture and supporting facilities. The furniture in question is a strong, stable and easy to move desk and instructor chair with a standard size to be able to sit comfortably. Supporting facilities include archives and documents etc., a clock, a trash can, a first aid box, internet connection, fire extinguishers, water and a fan or air conditioner. Other facilities needed are related to learning media, namely blackboards, LCD projectors, computers/laptops, and posters according to the subject material. The place for the implementation of the training program also requires supporting facilities, including a table and chair according to the room, filing cabinets, photos of the leaders of the state, facilities for learning spaces (means of learning theory, media of learning theory, teaching materials, means of practical learning), and facilities for supporting spaces (such as a leadership room, instructor room, reading room, administration, guest/public rooms, and storage of tools and goods).

The standard of infrastructure relates to land, buildings, learning spaces, supporting rooms (as listed above). The building land used must give property rights or a minimum of 3 years lease, avoid potential hazards that threaten health, and avoid river pollution and noise. Buildings used in training must be at least 100 m^2, have easy and fast evacuation access, and have air ventilation. The learning room requires adequate facilities for practical learning, a minimum learning area of 54 m^2 with a minimum width of 6 m, capacity for 25 people, and lighting that suits the needs.

Instructors in training institutions can be called trainers. RoI (2017) states that instructors are someone who has technical and methodological competence. This competence is proven by a competency certificate.

RoI (2014a) describes the qualifications of instructors. Instructors in training must have a minimum education qualification of undergraduate (S1) and diplomae-4 (D4) from accredited universities. In addition, the instructor must have a vocational competence certificate, as in the Minister of Manpower Ministry's Regulation of 2017. The Minister of National Education Regulation No. 41 of 2009 also mentions a third qualification, namely that the instructor has experience in the relevant field.

Competency standards used for instructor competency assessment in carrying out training, are standards of pedagogic competence, personality, social, and professional competence. Pedagogic competency standards include eight things: (1) understanding the characteristics of students; (2) mastering learning theory and the principles of learning in training; (3) mastering the concepts, principles, and procedures needed in curriculum development or expertise programs; (4) mastering the theories, principles, and strategies of learning in training; (5) being able to create active, interactive, communicative, effective and fun learning conditions and provide effective guidance for students; (6) mastering the use of media, technology, information and reflective actions in improving the quality of learning; (7) mastering the principles, concepts and assessment strategies; (8) understanding the process and results of training for students.

The next competency is personality competence, which is having good character and being a role model for students and the general public. The instructor acts in accordance with the norms, religion, law, social, and culture of the Indonesian people. Instructors should be honest, friendly, humane, tolerant, stable, wise and authoritative, and have good manners. A good instructor is the one who have enthusiasm, attitude and democratic behavior.

Training instructors should have social competence and be able to display an open, familiar, empathy and sympathy attitude to trainees and the community. The instructor is responsible for his role as a trainer, with responsibility, a good work ethic, and strong confidence. An instructor must be committed and understand the instructor's code of ethics, have an open attitude, be objective and not discriminatory. The instructor also has the responsibility to build effective, sympathetic, empathic and polite communication toward students, other instructors, and the community. Instructors must be able to tolerate and respect the culture of the surrounding community.

Finally, the instructor must have professional competence covering several things: (1) mastering the concept and scientific mindset in accordance with the vocational school; (2) mastering basic competencies in the field of expertise in accordance with the vocation; (3) being able to develop training materials; (4) developing professionalism in a sustainable manner; (5) and utilizing information and communication technology in developing professional capabilities.

Every vocational training program must have at least two instructors. Instructors have an important role to play in implementing the training program and are given the responsibility and authority to carry out training activities. The learning process is fully the responsibility of the instructor. Hidayat (2016) stated that instructors had a large influence on the quality of graduates, at around 58.6%.

Dawaous (2013) examined the influence of facilities and infrastructure in a training institution. Completeness of facilities and infrastructure is one of the principles of service quality in the education and training institutions and they help streamline the process of conducting training. In addition, complete facilities and infrastructure will optimize training so that trainees can achieve competency standards according to their vocational skills.

Hidayat (2016) conducted a study on the influence of infrastructure and instructors in VTCs. His research results show that high-quality instructor performance tends to produce high-quality graduates, and vice-versa. Hidayat stated that the completeness of facilities and infrastructure affected the enthusiasm of the trainees to learn. Furthermore, the quantity and quality of facilities and infrastructure affected the motivation, discipline and learning independence of trainees.

The training program conducted at the VTC must be competency-based. Competency-based training focuses on mastering work skills, including knowledge, skills, and attitudes of graduates in accordance with work needs. This means that training focuses on training participants in mastering one area of competence according to their needs at work.

The principles of the implementation of competency-based training according to the Minister of Manpower and Transmigration, are as follows. (1) Training is carried out based on the results of the identification of training needs and/or competency standards. This training program at the VTC is for people who are not working to equip them with certain competencies. (2) The recognition of the competencies that they possess. Prospective trainees are selected through tests in accordance with the chosen competencies. (3) Centering is on individual trainees. (4) Participants may start and end training programs at different times and levels according to their ability. (5) Participants are assessed in accordance with competency standards. (6) Competency-based training is carried out by registered or nationally accredited training institutions.

The training model applied is off-the-job training and on-the-job training and trainees are assessed in each of these methods. At the end of the training, participants receive training certificates. The training program is carried out for different time durations based on competence or vocational training. However, the duration of the on-the-job training is usually uniform for each vocation. Off-the-job training is carried out in a special place designed as an illustration of the workplace. This model is applied to train participants appropriately and they are equipped with the knowledge and work skills needed in the workplace. After the off-the-job training is complete, the participants continue to on-the-job training. This model aims to provide the skills needed when working. The implementation of the job training is industry/company/employment. In this stage, participants get special assistance from the company (Hamalik, 2007).

The training program was carried out with the steps defined by the CDC (2008): (1) analyze training needs; (2) design training programs; (3) develop training courses, curriculums, and materials; (4) implement training programs; (5) evaluate training results; and (6) improve training based on evaluation results.

3 CONCLUSION

Facilities and infrastructure are important components and are very influential in the implementation of a training program. These components will determine the level of success of the training program. One indicator of the success of a training program is its graduates. The training program is said that to be successful if there are significant changes to the competencies of the trainees. In this case, the instructor provides a competency test for trainees after they in a training program. The training program requires revitalization to improve and maintain its quality. Facilities and infrastructure can experience quality deterioration over time. This can be influenced by age and usage. Instructors must be sensitive to and updated on the competencies needed in training and still be able to use the learning tools properly.

REFERENCES

Agrawal, T. (2013). Vocational education and training programs (VET): An Asian perspective. *Asia-Pacific Journal of Cooperative Education, 14*, 15–26.

BPS. (2018, July 30). *Pengangguranterbukamenurutpendidikantertinggi yang ditamatkan 1986–2018* [Open unemployment according to higher education ended 1986–2018]. Retrieved from https://www.bps.go. id/statictable/2009/04/16/972/pengangguran-terbuka-menurut-pendidikan-tertinggi-yang-ditamatkan-1986—2018.html

CDC. (2008). *CDC unified process project office: Training planning.* Retrieved from https://www2.cdc. gov/cdcup/library/practices_guides/CDC_UP_Training_Planning_Practices_Guide.pdf

Dawaous, G.G. (2013). *Pengaruhmanajemensaranadanprasaranaterhadapmutulayanansaranadanprasaranadiklat di Pusatpendidikandanpelatihan (pusdiklat) geologi Bandung* [The influence of facilities and infrastructure management on the quality of service facilities and infrastructure training in the Bandung geological education and training center]. (Undergraduate's Thesis, UniversitasPendidikan Indonesia, Bandung, Indonesia). Retrieved from http://repository.upi.edu/2894/

Hamalik, O. (2007). *Manajemenpelatihanketenagakerjaanpendekatanterpadu* [Management of integrated employment training approaches]. Jakarta, Indonesia: Bumi Aksara.

Hidayat, R. (2016). *PengaruhkinerjainstrukturdansaranaprasaranaterhadapmutululusanbalailatihankerjadanpengembanganproduktivitasProvinsi Daerah Istimewa Yogyakarta* [The effect of instructor performance and infrastructure on graduate quality of vocation training and productivity development in Yogyakarta Special Territory] (Master's thesis, Universitas Negeri Yogyakarta, Yogyakarta, Indonesia). Retrieved from: https://eprints.uny.ac.id/30784/

Menteri Tenaga Kerjadan Transmigrasi [Minister of Manpower and Transmigration]. Pedoman Penyelenggaraan Sistem Pelatihan Kerja Nasional di Daerah [Guidelines for Implementing National Work Training Systems in the Regions].

RoI. (2003). *Undang-undangNomor 20 Tahun 2003, SistemPendidikanNasional* [Law Number 20 Year 2003, National Education System]. Jakarta, Indonesia: Ministry of Education, Republic of Indonesia.

RoI. (2014a). *Peraturan Menteri Pendidikandan Kebudayaan Nomor 90 Tahun 2014 tentang Standar Kualifikasidan Kompetensi Instrukturpada Kursusdan Pelatihan* [Regulation of the Minister of Education and Culture number 20 of 2016 about standard qualifications and instructor competencies in courses and training]. Jakarta, Indonesia: Ministry of Education and Culture, Republik of Indonesia.

RoI. (2014b). *Peraturan Menteri Tenaga Kerjadan Transmigrasi Republik Indonesia Nomor 8 Tahun 2014 tentang Pedoman Penyelenggaraan Pelatihan Berbasis Kompetensi* [Regulation of the Minister of Manpower and Transmigration of the Republic of Indonesia number 8 of 2014 about the guidelines for the implementation of competency-based training]. Jakarta, Indonesia: Ministry of Manpower and Transmigration, Republik of Indonesia.

RoI. (2016). *Revitalisasipendidikanvokasi* [Revitalization of vocational education]. Jakarta, Indonesia: Ministry of Education and Culture, Republik of Indonesia.

RoI. (2017). *Peraturan Menteri Ketenagakerjaan RI Nomor 8 Tahun 2017 tentang Standar Balai Latihan-Kerja* [Regulation of the Minister of Manpower and Transmigration number 81 of 2017 about the standard VTC]. Jakarta, Indonesia: Ministry of Manpower and Transmigration, Republik of Indonesia.

Soendjojo, R.P., Dhieni, N., & Gunarti, W. (2017). *Standarsaranadanprasaranalembagakursusdanpelatihan* [Standard of infrastructure for courses and training institution]. Jakarta, Indonesia: Ministry of Manpower and Transmigration, Republik of Indonesia.

Learning Models

Innovative Teaching and Learning Methods in Educational Systems – Retnowati et al. (Eds)
© 2020 Taylor & Francis Group, London, ISBN 978-1-03-224183-8

Theatrical stage of technology and humans in relation to education

T. Öztürk
Ankara University, Turkey

ABSTRACT: Since primitive times, humans have used technology in changing their lives for the better. Even small technological developments have led to new civilizations or other key historical moments and, in turn, humans have developed technology to serve a variety of purposes, mainly aimed at effective and productive results. It is important to understand the role of the 'design' element of the technology, which is fed by positive and social sciences. From the educational science perspective, this study will discuss current technological advancements and perspectives, drawing on historical experiences of humans in relation to technology and how to make sense of these technologies from the angle of economy-politics in all levels of education and teaching praxis. New technologies in education will be introduced, and some models and approaches will be discussed in practice, such as the Technological Pedagogical Content Knowledge model and computational thinking.

1 INTRODUCTION

In ancient times, acting as 'extensions of limbs and senses' (Emerson, 1904), technology helped humans cope with the problems they faced in nature and altered the format of reciprocal relationships between humans and nature. Even small technological inventions have changed civilizations, together with mass stimulators such as religion. In his book *Guns, Germs and Steel*, Diamond (1999) refers to the dire consequences of the invention of a transportation technology – the steam train – on changing social structures of a civilisation, and remarks that invention of the steam train is the lasting symbol of the triumph of European guns, germs and steel.

Technology acceptance had an impact on societies, leading to the industrial age, but there is also a similar but reverse effect on societies if technology is rejected. For instance, the Ottoman Empire initially rejected the arrival of the printing press. As a result, literacy was passed on by oral transmission from generation to generation and this caused falsification and fabrication of information. Hence, the level of education remained much lower than in the western societies that accepted printing press and learned from accessible and reproducible printed materials. It is believed that while science improved in Europe thanks to the printing press, which made information and techniques spread rapidly, the Ottoman Empire significantly fell behind the Europeans. There were reasonable excuses to reject this invention in the Empire: the printing press caused calligraphers to lose their jobs; there were religious concerns, such as one spelling mistake in the Quran could have brought about misinterpretations that would spread among societies, which eventually might cause social disorder. There is now a Turkish phrase that translates as "missing the train" and means signifying falling behind because of indifference to developments. From a technological point of view, "train" could represent a technology – like the steam train leading to the start of the industrial age – and "missing" it could represent the 'time' that cannot be captured again.

Coming back to the 21st century, technology is still associated with concepts of development, advancement and employment, and industry 4.0 is detailed in discourses of educational components. It is asserted that education should not fall behind industry, and industry 4.0 is now an issue for education 4.0. Today, automation and smart industrial technologies fed by artificial intelligence algorithms threaten the loss of thousands of jobs. According to Frey and Osborne (2017), "about 47 percent of total US employment is at risk" due to computerization. In line with Frey and Osborne's findings, Ra, Chin and Liu (2015:2) report on "imbalances between the supply of skills and the demand for skills in the world of work".

Drawing on a study by Dobbs, Madgavkar, Barton, Labaye, Manyika, Roxburgh, Lund, and Madhav (2012), Ra, Chin and Liu (2015) referred to the global view of skills mismatch and educational qualifications along with tendencies in levels of development:

i. Low-income countries tend to have employment concentrated in low-skill jobs overall, but face a surplus of workers with low-level skills alongside a shortage of workers to fill critical (if relatively limited) jobs requiring mid-level skills;
ii. Mid-income countries tend to have roughly balanced supply and demand for workers with low-level and mid-level skills but face significant shortages of highly skilled workers to fill emerging occupations requiring advanced skills; and
iii. Advanced countries tend to principally face skills mismatch characterized by having a surplus of workers with mid-level skills alongside a shortage of workers with advanced skills.

Overall, the study predicts that by 2020, there will be global shortages of high-skilled (advanced) and medium-skilled workers, and a surplus of low-skilled workers.

The quotation above demonstrates the tight relationship between industry 4.0 and education 4.0, whereas this match distracts our attention from the role of education in serving up the "ideal human".

The scope of the developments is not limited to the issues of employment and technology in relation to education. Now, information and communication technology (ICT) is embedded in society – so embedded that we could say 'surrounded' instead of embedded – in ATMs, at airports, in political propaganda through social media. So, one must possess the knowledge of them first and be 'literate' for democracies for better convenience.

We now mention "multiple literacies" to describe a minimum level to be active citizens. In order to have quality of life, literacy as only knowing how to read and write means nothing compared to multiple literacies, such as media literacy, technology literacy, financial literacy, and so on. As one of the aims of education is to make individuals that are ready to live in society compatible with existing environmental and social structures, it could be suggested that knowledge and skills of digital technologies should be taught at first at the schools.

In the light of the discussions presented, from the times dating back to the early civilizations technology and education have taken the stage together, by transforming each other, and sometimes by one dominating more than the other, and this is the basis if the argument of this paper. This relationship could be disclosed through an emphasis on the practices of educators related to "technology education" or "technology in education". The discussions revolving around the aforementioned relationship could help us uncover whether technology is a means rather than an end or an end rather than a means. In fact, these discussions are highly relevant to the critical educators' points on technology and education. Sometimes, technology becomes the main purpose of education (technology education) in a situation where technology is actually a means for effective education (technology in education). For instance, while technology could be used in STEM education to promote better understanding of the topics for students, STEM education is put aside in teachers' pedagogical practices and technology is taught. These kind of implications are harshly critcized today and over the years. Sometimes, coding in its 'mechanical terms' is taught even in early childhood education to help students live in technology embedded environments, whereas coding is the topic of computer programmers or engineers. Here, the curriculum could be problematized because of the state of thinking of a child with their adulthood stage based on future predictions and based on 'wishful' intentions of a curriculums' vision.

In order to make this more concrete, the remainder of this paper will refer to the practical implications on *technology in education* and *technology education*.

2 TECHNOLOGY IN EDUCATION: TEACHERS' PEDAGOGICAL APPROACHES TO TECHNOLOGY

Although teachers' experiences with technology vary significantly from the global perspective, it is apparent that the use of technology in education is not naturally occurring. In particular, for teachers with more than 10 years' exprience, hardware technologies are introduced to the 'classroom' by an outsider such as a technology firm, a policy maker, an administrator, etc. Considering that there is no one technological tool developed especially for education – for instance, the internet, computers, projectors were invented and/or first used for military purposes – it is challenging for these educators to adopt these tools to use in education. In these situations, teachers should not be forced to use technology and alienate their occupation.

Regarding meaningful use of technology in education, two frameworks put forward by Mishra and Koehler (2006) and ISTE, a leading community in the field, might help educators. Briefly, the Technological Pedagogical Content Knowledge (TPACK) model developed by Mishra and Koehler (2006) could inspire internalization of technology (Figure 1).

The TPACK framework focuses on the interplay of content, pedagogy and technology and reviews technology integration through the intersections of these dimensions as pedagogical content knowledge, technological content knowledge, technological pedagogical knowledge and in the middle of all these intersections *technological pedagogical content knowledge* (TPACK) (Koehler, 2012). The model could be utilized by teachers from any specialization in delivering a wide variety of content.

Regarding use of technology in teachers' practices, the International Society for Technology in Education (ISTE, 2017), a community of practitioners whose expertise or interests revolve around technology and education, sets elaborative standards for educators with the potential to guide the practices on technology in education. The ISTE standards for educators are presented in Table 1. The main contribution of these frameworks to the teachers' practices is that the frameworks do not force teachers to use technology just because they are the "necessities of the 21st century" but they help teachers to connect the essential educational components together and also have the inspiration to emerge pedagogical practices such as teachers as facilitators, analysts and collaborators.

3 TECHNOLOGY EDUCATION IN GENERAL AND THE CASE OF 'CODING'

Today's post-industrial societies heavily relying on technological investments expect to be impressed by the "mesmerizing" effect of technology coming true *in education* and *human thinking*.

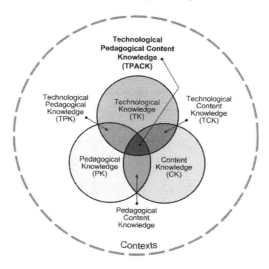

Figure 1.   The TPACK framework and its knowledge components (2012, tpack.org).

Table 1. ISTE standards for educators (ISTE, 2017).

Teacher as a...	Description
Learner	Educators continually improve their practice by learning from and with others and exploring proven and promising practices that leverage technology to improve student learning.
Leader	Educators seek out opportunities for leadership to support student empowerment and success and to improve teaching and learning.
Citizen	Educators inspire students to positively contribute to and responsibly participate in the digital world.
Collaborator	Educators dedicate time to collaborate with both colleagues and students to improve practice, discover and share resources and ideas, and solve problems.
Designer	Educators design authentic, learner-driven activities and environments that recognize and accommodate learner variability.
Facilitator	Educators facilitate learning with technology to support student achievement of the ISTE Standards for Students.
Analyst	Educators understand and use data to drive their instruction and support students in achieving their learning goals.

In terms of "education" aspect, the actual outcome often fails the "mesmerizing" anticipations. For example, an OECD (2015) report featuring data on 64 countries and economies, drawing on internationally comparative analysis of the digital skills of the students, highlights that

"The results also show no appreciable improvements in student achievement in reading, mathematics or science in the countries that had invested heavily in ICT for education. And perhaps the most disappointing finding of the report is that technology is of little help in bridging the skills divide between advantaged and disadvantaged students. Put simply, ensuring that every child attains a baseline level of proficiency in reading and mathematics seems to do more to create equal opportunities in a digital world than can be achieved by expanding or subsidizing access to high-tech devices and services."

In terms of the "human thinking" aspect, *computational thinking* is a further example of something enchanting, in educators' eyes, put forward in the changing curricula. Previously, students were taught that computer hardware resembles human organs, such as RAM working like the memory of a human, CPU as a brain, etc. A further example is that it thought a Turing machine could replicate "any symbolic algorithm executed by a human" (Stanford Encyclopedia of Philosophy, 2015). Now, we are working on how humans can think like a computer, which raises the idea of computational thinking, described as "the thought processes involved in formulating a problem and expressing its solution(s) in such a way that a computer—human or machine—can effectively carry out" (Wing, 2014). Özçinar (2018) summarizes the historical development of computational thinking in education agenda as follows:

"In the 1960s, when Alan Perlis suggested that computer programming should be taught to everyone, the main reason behind this was the idea that computer programming would enable the students to adapt to a new way of thinking, and this way would be beneficial in solving the problems faced in different disciplines or in daily life. Likewise, Papert (1980) and Wing (2006) had the same opinion about the main benefits of teaching programming at the level of primary school: solving the problems encountered in different areas with the mental tools provided by the computers."

One of the main problems with this development is that educators and related experts do not agree on what to teach, on which level of education, and for which outcomes. Below are some reflections on a possible solution of this complexity.

In early childhood education, coding should have minimal place in the classroom; and nature, real materials and other child development issues must be prioritized. Within this short time and in order to question the educational value of coding practices, the criteria should be whether the activities and equipments used to teach coding are developmentally appropriate (Öztürk & Calingasan, 2018). For instance, are the self-confidence and imagination of children supported? Is there aconnection with surroundings of the child? Is there a child-friendly design? Is there a meaningful interaction between the kit and child? One of the important questions is, does the kit support emotional stimulations?

Furthermore, computational thinking should be integrated in a way that facilitates teachers to promote the student's problem-based learning, inquiry-based learning and learning from trial and error (Öztürk & Calingasan, 2018). Activities and kits should support the Positive Technological Development (PTD) framework (Bers, 2010), emphasizing fostering the positive aspects of children's characters rather than "correcting their wrong behaviors". The model is summarized and visualized in Figure 2.

Given that computer tools are not developmentally appropriate for small children, using digital materials in abstract contents in acceptable frequencies and time spans is more appropriate than teaching children about the hardware.

In primary education, as in early childhood education, it is more important to focus on proper use of technology than teaching students about the hardware. At these ages, situations such as addiction and cyber loafing could be experienced, and therefore, social and psychological parameters in relation to the use of ICT are more important. At this point, the curriculum should also include parents, as the professional approach to children is crucial and overcoming these issues could be possible through collaboration with families. At this level of education, after school activities such as robotics could be popular.

In secondary education, ICT teaching can take place in the curriculum in order to advance the students from the level of "awareness and basic level of ICT knowledge and skills" to more "advanced level of ICT use". Again, after school clubs such as robotics, coding etc could be organized if there is students demand.

In high school education, undesired social and psychological situations could be more prevalent than at any other level of education. As the students at high schools are entering the "adulthood" stage, they must also be introduced to ICT as political tools, through topics such as ethics, surveillance, censorship, etc. Therefore, students must become more conscious about the likely damaging effects of technology use through the help of ICT teachers, subject teachers and psychological counselors at school. ICT could also be used for social responsibility projects. For instance, splint could be designed and produced through 3D printers for a street animal with a broken leg; a smart system could be designed from recycled materials; green energy solutions could be tested by Ardunio, and so on.

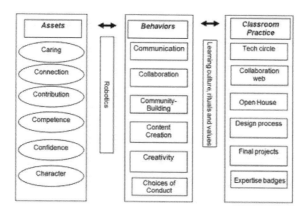

Figure 2.   Positive Technological Development (PTD) framework (Bers, 2010).

4 CONCLUSION

Revisiting *Guns, Germs, and Steel,* along with some variables such as biology and geography, technology is an important element of today's world: humans created the technology, technology changed the habitats and social structures of human life, and in return, humans changed the technology. This interplay continues even 'now', at a time when we discuss the dominance of one over the other. Although the developments in artificial intelligence and robotics seem to be in favor of "technique" over "humans", the most benefits could be gained in "ideal human" terms if we, educators, can clarify the vague borders between technology in education and technology education.

As a conclusion, rather than focusing on the mesmerizing effect of technologies and the knowledge and skills of techniques in relation to employment or to any kind of pragmatist and instrumentalist view, the questions of "what" "when" "why" and "how" to teach or use technology in education should guide the educators and policy makers.

REFERENCES

Bers, M.U. (2010). The tangible K Robotics Program: Applied computational thinking for young children. *Early Childhood Research & Practice, 12*(2).
Diamond, J.M. (1999). *Guns, germs, and steel: The fates of human societies.* New York, NY: W. W. Norton.
Emerson, R.W. (1904). *The complete works of Ralph Waldo Emerson.* Boston, MA: Houghton, Mifflin and Company.
Frey, C.B. & Osborne, M.A. (2017). The future of employment: How susceptible are jobs to computerisation? *Technological Forecasting and Social Change, 114*(C), 254–280. doi:10.1016/j.techfore.2016.08.019
ISTE. (2017). *ISTE standards for educators.* Retrieved from https://www.iste.org/standards/for-educators.
Koehler, M. (2012). *TPACK explained.* Retrieved from http://www.tpack.org/
Dobbs, R., Madgavkar, A., Barton, D., Labaye, E., Manyika, J., Roxburgh, C., Lund, S., & Madhav, S. (2012). *The world at work: Jobs, pay, and skills for 3.5 billion people.* Paris, France: McKinsey & Company.
Mishra, P. & Koehler, M.J. (2006). Technological pedagogical content knowledge: A framework for teacher knowledge. *Teachers College Record, 108*(6), 1017–1054. doi:10.1111/j.1467-9620.2006.00684.x
OECD. (2015). *Students, computers and learning: Making the connection.* PISA, OECD Publishing. doi:10.1787/9789264239555-en
Ozturk, H.T. & Calingasan, L. (2018). Robotics in early childhood education: A case study for the best practices. In H. Ozcinar, W. Gary & H.T. Ozturk (Eds.), *Teaching Computational Thinking in Primary Education.* Hersey, PA: IGI Global.
Özçınar, H. (2018). A brief discussion on incentives and barriers to computational thinking education. In H. Ozcinar, W. Gary & H.T. Ozturk (Eds.), *Teaching Computational Thinking in Primary Education.* Hersey, PA: IGI Global.
Papert, S. (1980). *Mindstorms: Children, computers, and powerful ideas.* New York, NY: Basic Books.
Ra, S., Chin, B., & Liu, A. (2015). *Challenges and opportunities for skills development in Asia changing supply, demand, and mismatches.* Mandaluyong City, Philippines: Asian Development Bank.
Stanford Encyclopedia of Philosophy. (2015). *The computational theory of mind.* Retrieved from https://plato.stanford.edu/entries/computational-mind/
Wing, J.M. (2014). *Computational thinking benefits society.* Retrieved from http://socialissues.cs.toronto.edu/2014/01/computational-thinking/on 17 September 2018.
Wing, J.M. (2006). Computational thinking. *ACM SIGCSE Bulletin, 39*(1), 195–196. doi:10.1145/1227310.1227378

Innovative Teaching and Learning Methods in Educational Systems – Retnowati et al. (Eds)
© *2020 Taylor & Francis Group, London, ISBN 978-1-03-224183-8*

Improving students' critical thinking abilities in probability problems through problem-based learning

Rauzah & Kusnandi
Universitas Pendidikan Indonesia, Bandung, Indonesia

ABSTRACT: This study aims to determine the improvement of students' critical thinking skills through the implementation of a problem-based learning (PBL) model for probability problems. This was a quasi-experimental research with a non-equivalent pre-test-post-test control group design, involving 28 students of grade X in one of the senior high schools in Banda Aceh, selected using a purposive sampling technique. The research instrument used in this research was critical thinking about probability test problems used to measure the students' critical thinking abilities. The results of the test were analyzed with quantitative methods using statistical analysis, paired-samples t-tests and using gain scores to determine the effectiveness. The results of this study indicate that the application of a PBL model to the probability problems had a positive impact on improving students' critical thinking skills. We conclude that the implementation of PBL can improve student's critical thinking abilities.

1   BACKGROUND OF THE STUDY

The concept of critical thinking has received a great deal of attention in recent years (Kuhn, 1993; Ennis, 1987). It is not only of interest to academic psychologists but has also been emphasized in the practical context of further and higher education in both the United States and the United Kingdom (Anderson, *et al.*, 2012). The term 'critical thinking' refers to the use of those cognitive skills or strategies that increase the probability of a desirable outcome In the long term, critical thinkers will have more desirable outcomes than 'noncritical' thinkers (Helpern, 1998). When people think critically, they are evaluating the outcomes of their thought processes, and how good a decision is or how well a problem is solved (Helpern, 1996, 1998). Indeed, the various skills that are collectively termed 'critical thinking' are regarded as an important component of the so-called 'transferable skills' accrued during higher education (Anderson, et al., 2012).

It is generally accepted that producing students who are capable of critical thinking and problem-solving is an important objective of the education system (Pikkert & Foster, 1996). In mathematics, mastering critical thinking is required for understanding concepts, analyzing problems and determining the exact solution for a problem. However, studies show that the ability of mathematical students in Indonesia to think critically is still low and unsatisfactory (Fristadi & Bharata, 2015; Suprapto, 2016; Stacey, 2011). The low performances in mathematics were revealed in Trends in Mathematics and International Science (TIMSS) 2011: Indonesian students were 38th out of 42 countries (Jupri, Drijvers & van den Heuvel-Panhuizen, 2014). In addition, the Program for International Student Assessment (PISA) 2012 placed Indonesian students in 64th out of 65 countries, lower than Vietnam, Thailand, and Malaysia in South-East Asia (OECD, 2012).

Previous research shows that many mathematics teachers in Indonesia still adhere to the old paradigm: the transfer of knowledge, teachers as information and students as recipients of information (Fristadi & Bharata, 2015; Hartati & Sholihin, 2015). Richard (2009) reported that teachers often ask students to retell, define, describe, decipher and register rather than analyzing, drawing conclusions, linking, synthesizing, criticizing, creating, evaluating, thinking and rethinking. As a result, many schools graduate students who think superficially, only swimming on the surface of the problem, rather than students who are able to think deeply.

To improve students' critical thinking skills, we provided learning experiences by designing a learning process called the problem-based learning (PBL) approach. Nurun and Suryanto (2000) defined PBL as a learning model that uses real problems as a context for learners to improve their critical thinking and problem-solving skills, and to acquire essential knowledge and concepts from the subject. It makes a positive contribution to critical thinking skills that are part of the cognitive and metacognitive sub-dimensions of self-regulation and resource management strategies (Hake, 1999). According to this theory, Richard (2009) stated that PBL activities are carried out as the core activities of learning, which consists of five learning stages: 1) giving orientation on the problem; 2) organizing students to learn; 3) guiding the students' inquiry independently as well as in groups; 4) developing and presenting the work; and 5) analyzing and evaluating the problem-solving process.

2   RESEARCH METHODS

To determine the improvement of students' critical thinking skills through the implementation of the PBL model for probability problems, a pre-test post-test control group experiment was set up. The study involved 28 students of grade X (16–17 years old) in one of the senior high school in Banda Aceh. Students were selected using a purposive sampling technique: two classes were chosen out of eight, which the mathematics teachers felt had equivalent characteristics and academic ability. This was part of a pilot study to design learning materials for improving the critical thinking of senior high school students. The documents for the purpose of the analysis included all written solutions by the students for solving two pretest probability problems, as shown in Figure 1.

The post-test document for determining any improvement in the students' critical thinking skills is shown in Figure 2.

---

1. In a sports equipment store, there are 20 red strings, 15 green strings, and 25 blue strings for the same price. If the blue strings have been sold, determine the probability of selling blue strings for the second time!
2. A lottery is held in a supermarket. Buyers who spend above Rp. 50,000.00 get 2 coupons. Fifty people spend above Rp. 50,000.00, including Rudi who gets 2 coupons. Supermarkets provide 3 prizes for the lucky winners. If Rudi has been won the first prize, what is the probability he wins the second prize?

Figure 1.   Two pre-test probability problems used in the study.

---

1. In a coop there are 40 chickens. There are 21 roosters and 15 among them are black-feathered. Andi wants to catch a hen, so determine the probability of a white-feathered hen being caught if there are 19 black-feathered chickens.
2. Budi has 3 sets of school uniforms: gray, white and scout uniform; 3 pairs of shoes: vantopel shoes, scout shoes, and sports shoes; and 2 hats: animal hat and scout hat, Budi always wears uniform and a hat to school, and wears a scout hat on Saturdays. Every week the uniform schedule is Monday = white, Tuesday and Wednesday = gray, Thursday and Friday = free choice, and Saturday = scout uniform. Determine the probability of Budi wearing a gray uniform with an animal cap and vantopel shoes.

Figure 2.   Two post-test probability problems used in the study.

There were three steps for the research.1) Preparation – in the preparation stage, the syllabus, lesson plans, instruments, which had already been validated, were prepared, and the instruments were tested. 2) Implementation – a pre-test was conducted to determine the students' initial critical thinking skills, and then the normality test was carried out to determine the normal distribution of the data; a post-test was conducted after the treatment had finished, and the normality test was then performed on the data of the post-test. 3) The data were analyzed using a paired t-test, and a normalized gains score test, to answer the research problems as described in the background to this research. The instrument of critical thinking used an essay test and a rubric of critical thinking with the scale 0–5. Increases in students' critical thinking skills were calculated using a normalized average score (N-gain) data with the following formula (Hake, 1999):

$$<g> = \frac{post - test\ score - pre - test\ score}{maximum\ score - pre - test\ score}$$

Research activities had been conducted for five meetings. At the first meeting, all students were given the pre-test to determine the students' early critical thinking skills. In the second, third and fourth sessions, the students in the experimental class received problem-based learning (PBL) and students in the control class received conventional learning activities. The lesson lasted for 90 minutes and consisted of three respective parts in the experimental class. First, a worksheet activity was completed and included posing problems and a whole-class discussion. In this activity, the teacher introduced the concept of probability through posing problems and guided class discussion, while the students paid attention. Then, during the group work, students were asked to solve a series of tasks on a worksheet activity under the teacher's guidance. Finally, the students and the teacher reflected the activities by making conclusions from the results of the discussions. All students in the experimental class and control class were given a final test (post-test) to determine students' final critical thinking abilities.

3   RESEARCH RESULTS

A summary of written student work for pretest and post-test is shown in Table 1. The table presents the mean of pre-test, post-test, and N-gain values of critical thinking for both the experimental class and control class.

After giving PBL to the experimental group and conventional learning to the control group, there was an increase in critical thinking skills in both groups. As shown in Table 1, the experimental group's pre-test mean of 45.71 increased to 61.27 in the post-test, whereas the control group's pre-test mean of 37.96 increased to 49.62. To see the magnitude of the improvement from the pre-test to post-test, the normalized gain score was calculated. Based on Table 1, it shows that the gain score of the experimental group was higher than the gains score in the control group. Within these classes, the students in the experimental group improved more than the students in the control group. The overall result shows that the implementation of the PBL model effectively improved critical thinking skills for about half of the participants.

Table 1.   Pre-test, post-test, and N-gain mean of critical thinking skills.

Score	Class	
	Experimental	Control
Pre-test	45.71	37.96
Post-test	61.27	49.62
%N-Gain	44	27
Category	Medium	Medium

The hypothesis testing used a paired t-test with the criteria of hypothesis testing. H_0 was rejected if $t_{count} > t_{table}$ or significance level sig < 0.05, meaning that H_1 was accepted. Thus, the implementation of the PBL model could improve critical thinking, and H_0 was accepted if $t_{count} < t_{table}$ or significance level sig > 0.05 meaning that H_1 was rejected, and thus the implementation of the PBL model could not improve critical thinking. The test of the difference in the mean value of the improvement of student's critical thinking ability (N-gain), with significant level $\alpha = 0.05$ and df = 54, from t distribution table, obtained $t_{(0.95)(54)} = 1.67$. Based on the calculation results $t_{count} = 3.0$ which means H_0 rejected and H_1 accepted. This means that the implementation of the PBL model can improve critical thinking skills.

Critical thinking improvement can also be shown as representative of comparison between written student work (S1) pre-test and post-test in Figures 3 and 4. Figure 3 shows a representative example of written student work in the pre-test. In this case, the results of the pre-test show that S1 made a mistake in determining the number of events to get a second prize if he had won the first prize, so reached an incorrect solution.

Post-test results in Figure 4 show that, after receiving PBL, S1 was able to identify the problem easily and completed the problem systematically based on critical thinking indicators, so that the correct solution was obtained.

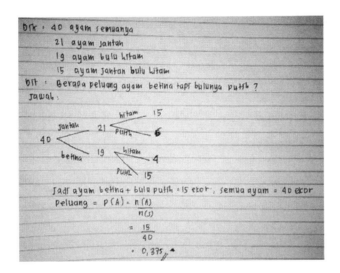

Figure 3.   A representative example of written student work on the pre-test.

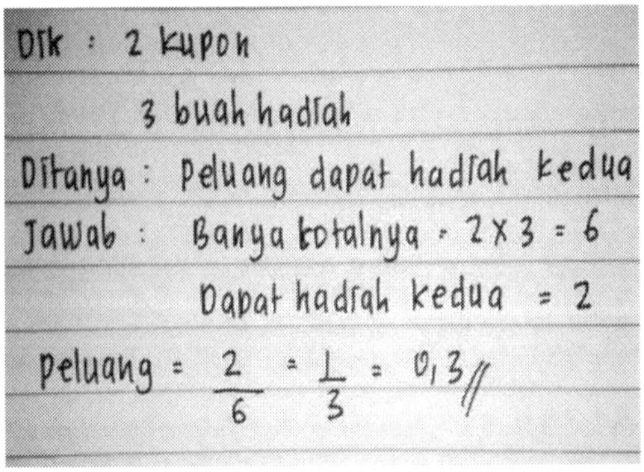

Figure 4.   A representative example of written student work on post-test.

The critical thinking ability measured in this study consists of four indicators adopted from Fisher (2008): identifying problems, gathering relevant information, developing a number of problem-solving alternatives, and interpreting solutions. Each item is adapted to an indicator of problems that refers to the mastery of critical thinking skills and is a separate part of one problem so it is not a prerequisite for the solution of the next question.

A = identifying the problem; B = gathering relevant information; C = arranging alternative problem-solving; D = interpreting the solution.

The comparison of improvements in each indicator of critical thinking ability between the experimental and control classes is presented in Table 2. Based on Table 2, it appears that the experimental class predominates for the overall improvement of the mastery of measured critical thinking indicators. This can be seen from the percentage distribution of students' critical thinking ability, at around 22.56% – 57.71%. The dominance of the increase in the mastery of critical thinking is clearly seen in the indicators 'arranging alternative problem-solving', with a percentage increase to 57.71%.

Increasing critical thinking ability in the indicator 'arranging alternative problem solving' in the experiment class is high enough to indicate that the students' ability in the experimental class to understand the problems presented, determine what solution that will be used as an alternative, and choose the criteria to be used in the evaluation of alternative solutions to problem-solving, have improved after applying the PBL model to the probability problems. In addition, mastery of critical thinking skills in other indicators also shows that the PBL model contributes positively to the mastery of students' critical thinking skills. It shows that the learning activity has been in accordance with one of the goals of PBL, which is to develop students' thinking skills to a high level, or the ability to think critically and scientifically (Richard, 2009).

This indicates that PBL models give students more opportunities to optimize their critical thinking skills than conventional learning models. In PBL models, students are required to explore, inquire, discover and solve problems, while in the direct learning model students only accept problems presented by the teacher. This is consistent with research from Newel & Simon (1972) and Hmelo & Silver (2004), which shows that PBL trains students' ability to analyze, and think critically and metacognitively. It is also supported by the research from Shepherd, (1998), Gallagher, (1997) and Hmelo, (1998), that PBL significantly increases critical thinking ability.

An essential principle that focuses on building students' understanding of mathematics is "make the subject problematic." Learning should give students the opportunity to think "why," do inquiry, find solutions, and verify solutions (Ericson, 1999). This means learning should start with problems, dilemmas, or questions for students. On that basis, mathematical learning must be packed into a process of constructing rather than receiving knowledge. In the process of learning, mathematics students are expected to build their own knowledge through active involvement in teaching and learning (Confrey, 1995).

Table 2. Improving students' critical thinking skills on each indicator.

Indicator	% N-gain	
	Experimental	Control
A	33.25	20.47
B	22.56	15.06
C	57.71	46.65
D	42.90	37.43
Sum	156.42	119.61
Maximum score	57.71	46.65
Minimum score	22.56	15.06

As mentioned earlier, there is an increase in critical thinking skills through applying a PBL model, meaning that students are encouraged to analyze a problem and consider alternative analysis because the starting point of PBL is a problem. Based on the philosophical foundation of constructivism adopted by the PBL model, student-centered learning activities are placing students as the main actors in learning and thinking skills.

Students are trained to think independently and develop self-confidence and appreciate the activities that are happening. In addition, students are also required to work in groups to achieve results together. Starting from the definition of the problem, the students conduct a discussion to equate perceptions about the problem and set goals and targets to be achieved. Next, the students look for materials from sources in the library, the internet, and through personal knowledge or observation. The teacher acts as a facilitator who guides the students in group activities to identify problems, to make hypothesize, and to search for data, conduct investigations, formulate solutions and determine the best solution to the condition of the problems presented in the student worksheet at each lesson. Finally, this research provides evidence that the implementation of the PBL model can create an atmosphere that supports students in improving their critical thinking skills.

4 CONCLUSION

From results described in the previous section, it can be concluded that the application of a PBL model to probability problems had a positive impact on the critical thinking abilities of students in grade X in one of the senior high school in Banda Aceh. Potential for further research includes whether the implementation of PBL could improve students' hard skills.

REFERENCES

Anderson, T., Howe, C., Soden, R., Halliday, J & Low, J. (2001). *Peer interaction and the learning of critical thinking skills in further education students Instructional Science*, 29(1), 1–32.

Confrey, J. (1995). *A theory of intellectual development journal for the learning of mathematics*, 15(1), 38–47.

Ennis, R.H. (1987). A taxonomy of critical thinking dispositions and abilities. In J.B. Baron & R.J. Sternberg (Eds.), *Teaching Thinking Skills Theory and Practice*. New York: W.H. Freeman and Co Google Scholar.

Erickson, D.K. (1999). *A problem-based approach to mathematics instruction in the mathematics teacher*. Reston: VA NCTM.

Fisher, A. (2008) *BerpikirKritis* [Critical thinking]. Jakarta: Erlangga.

Fristadi, R. & Bharata, H. (2015). Meningkatkankemampuanberpikirkritissiswadengan problem-based learning [Improve students' critical thinking skills with problem-based learning]. *In Prosiding Seminar Nasional Matematika Dan Pendidikan Matematika UNY*, 597–602.

Gallagher, S.A. (1997). Problem-based learning: Where did it come from, what does it do, and where is it going? *Journal for the Education of the Gifted*, 20(4), 332–362.

Halpern, D.E. (1996). *Thought and knowledge an introduction to critical thinking 3rd ed Mahwah*. NJ: Erlbaum.

Halpern, D.F. (1998). Teaching critical thinking for transfer across domains Disposition skills structure training and metacognitive monitoring. *American Psychologist*, 53(4), 449.

Hartati, R. & Sholihin, H. (2015). Meningkatkankemampuanberpikirkritissiswamelaluiimplementasi model problem based learning (PBL) padapembelajaran IPA terpadusiswaSMP. *Prosiding Simposium Nasional Inovasidan Pembelajaran Sains* 2015, 505–508.

Hmelo, C.E. (1998). Problem-based learning: Effects on the early acquisition of cognitive skill in medicine. *The Journal of the Learning Sciences*, 7(2), 173–208.

Hmelo, E. & Silver. (2004). Problem-based learning: What and how do students learn? *Educational Psychology Review*, 16, 235–265.

Jupri, A., Drijvers, P. & van den Heuvel-Panhuizen, M. (2014). Student difficulties in solving equations from an operational and a structural perspective. *Mathematics Education*, 9(1), 39–55.

Kuhn, D. (1993). Connecting scientific and informal reasoning. *Merrill-Palmer Quarterly*, 39, 74–103.

Newell A. & Simon H. (1972) *Human problem-solving*. Englewood Cliffs, NJ: Prentice Hall.

Nurun, Y. & Suyanto, W. (2000). Penerapanmodel problem-based learning untukmeningkatkanketeram-pilanberpikirkritisdanhasilbelajarsiswa [The application of problem-based learning model to improve critical thinking skills and student learning outcomes]. *JurnalPendidikanVokasi*, *4*(1), 125–142.

OECD (2014). *PISA 2012 result what students know and can do-student performance in mathematics, reading and science* (volume 1, revised Edition, February 2014). Paris: OECD Publishing.

Pikkert, J.J. & Foster, L. (1996). Critical thinking skills among third year Indonesian English students. *RELC Journal*, *27*(2), 56–64.

Hake (1999). *Analyzing change/gain score Indiana*. Indiana: University.

Richard I.A. (2009). *Learning to teach ninth edition*. New York, NY: McGraw Hill.

Shepherd, H.G. (1998). The probe method: A problem-based learning model's effect on critical thinking skills of fourth-and fifth-grade social studies students. *Dissertation Abstracts International Section A Humanities and Social Sciences*. September 1988, *59*(3–A).

Stacey, K.(2011). The PISA view of mathematical literacy in Indonesia. *Journal on Mathematics Education*, *2*(2), 95–126.

Suprapto, N. (2016). What should educational reform in Indonesia look like? Learning from the PISA science scores of East-Asian countries and Singapore in Asia-Pacific. *Forum on Science Learning & Teaching*, *17*(2).

Cultural map media as an innovation to overcome cognitive learning difficulties in Social studies at elementary school

L. Fatmawati, V.Y. Erviana, D. Hermawati, I. Maryani, M.N. Wangid & A. Mustadi
Ahmad Dahlan University, Yogyakarta, Indonesia

ABSTRACT: This research aims to (1) develop a cultural map media, (2) discover the feasibility of the product, and (3) determine the effectiveness of the product in overcoming cognitive learning difficulties in Social studies for fourth grade elementary school students. The type of research was RnD, adopting the ADDIE model. The result of the research is the development of cultural map media to be used in the learning process in an elementary school. The experts validation tests showed that the mean score was very feasible (score: 80.89); the teachers' response stated that the assessment was very good (score: 82.67); and the students' response stated that the assessment was very good (score: 90.39). The results of the operational field tests research showed that based on the pre-test and post-test scores

1 INTRODUCTION

The curriculum is a very important tool in the achievement of educational success. Without an appropriate and proper curriculum, educational programs that have been planned and implemented to achieve educational goals and objectives will likely fail. Education is seen as a benchmark in educating a nation and in shaping and improving the quality and competitiveness of human resources to achieve national development goals and educate a nation.

Social sciences are an embodiment of social values developed through study (Bauto, 2016). The scope of the social sciences is included in Social studies education. Social studies education is one part of an education program and takes the form of subject matter.

Research suggests that there are several problems in the process of Social studies learning in the classroom. 1) Many students think that learning is boring (Wibawa, 2017). 2) There is a lack of optimal learning media for the learning process in the classroom (Faruk, 2014). 3) Social studies subjects tend toward accumulating various facts and material that is more cognitive (Agung, 2011). Social studies is considered a difficult and tedious subject because the scope of the material being studied is very broad, and the application of learning methods do not stimulate students (Effendi, Soetjipto, & Widiati, 2016).

The elementary school students are at the concrete operational stage. Piaget explained that the concrete operational stage is between 7 and 12 years old, when children's understanding and thinking are developed, and logical structures are created (Šramová, 2014). The learning process in the concrete operational stage is concrete or real learning, and also involves students directly. Learning success can be determined by two main components: learning method and learning media (Ramdhani & Muhammadiyah, 2015). For this reason, there is a need for media to support students' learning processes, and interactive multimedia is a solution that can be used by both teachers and students in the learning process. Interactive multimedia based on Lectora Inspire is effective software used to develop learning media.

Observations and interviews with students and teachers at Muhammadiyah Kleco II Elementary School, Muhammadiyah Kleco III Elementary School, and Muhammadiyah Purwodiningratan 2 Elementary School showed that there were several problems in learning. The use of learning media was perceived as still not optimal and concrete in improving the quality of learning. Learning materials contained in learning resources (textbooks) are limited and minimal. One of students' difficulties is related to concept comprehension, one of which is Indonesian cultural map. Other problems include teachers' difficulties in developing learning resources, their lack of ability and creativity in developing learning resources, and students' enthusiasm for operating computer technology.

This data showed that media is needed to support student learning. One learning media that can be used by both teachers and students in the learning process is interactive multimedia. Lectora Inspire is an effective software used to develop learning media. The aim of this research was to develop learning media using Lectora Inspire for cultural map material, discover the usefulness of the product, and determine the effectiveness of the Lectora Inspire based learning media on a cultural map to overcome learning difficulties in Social studies.

2 LITERATURE REVIEW

2.1 Social studies education

Social studies education is one part of an education programs in the form of subject matter. In general, social studies is described as a science whose components come from social sciences, humanities, and natural sciences that are oriented as educational goals (Samawi, 2017). Social sciences are an embodiment of social values developed through study (Bauto, 2016). The scope of the social sciences is included in Social studies education. Social studies is an educational program where the material covered is from various social and humanities disciplines, including sociology, history, geography, economics, politics, law and culture. In other words, Social studies is a combination of various social and humanities studies to produce individuals who can participate in various social problems (Agung, 2011).

Students are expected to be able to explore information, skills, attitudes, and sensitivities when dealing with various issues and social and environmental challenges. This is in line with the purpose of Social studies education that is to explore values, develop the ability to solve social problems, moral education, independence, responsibility, and discipline.

2.2 Learning difficulties

Learning difficulties are often referred to as learning disorders and are closely related to the achievement of learning outcomes and daily activities. Learning difficulties refers to obstacles in learning, in which students cannot learn well. There are several factors that influence learning difficulties (learning disabilities): emotional (internal factors); environmental factors (i.e. family and society) (Morrison & Cosden, 1997); development and behavioral problems (Shaw-Smith et al., 2004); and a lack of visual representation strategies in solving problems (Garderen, Scheuermann, & Poch, 2014). Learning difficulties will affect children's cognitive abilities.

2.3 Learning media

The word "media" comes from the Latin language, which is *medium*, meaning "middle", "intermediary" or "agent"(Anwariningsih & Ernawati, 2013). Media is one of the determinants of students' learning success (Buchori & Setyawati, 2015) and is considered to be an important component in supporting the learning process.

Media is important in the learning process because it has roles: 1) A means to raise desires and interests, increase motivation, and also stimulate learning activities (Buchori & Setyawati, 2015). 2) Learning will be more interesting and learning materials will be easily understood. It can control and achieve learning objectives, and learning methods will be more varied (Anwariningsih & Ernawati, 2013); 3) It creates two-way interaction (Faruk, 2014). 4) It improves students' learning and supports interaction activities in the learning process (Wibawa, 2017); 5) It can improve clarity in the material delivery and provide feedback on learning outcomes (Ramdhani & Muhammadiyah, 2015). 6) It can be a means of overcoming common obstacles such as boredom, limited learning hours, etc. (Herdini et al., 2018).

The selection criteria for media used in the learning process are: 1) learning objectives, 2) learning methods, 3) students' condition, 4) efficiency, and 5) availability (Ramdhani & Muhammadiyah, 2015). Well-planned media will affect multimedia success.

2.4 *Lectora Inspire*

The term multimedia means "means of communication through several media" (Krstev & Trtovac, 2014). One interactive multimedia based learning media is the Lectora Inspire software, an effective program to create and develop learning media (Faruk, 2014). Lectora Inspire is e-learning software that is relatively easy to use and implements because the available programming languages are easy to understand and are sophisticated (Wibawa, 2017).

3 METHOD

The method used in this study was RnD, referring to the ADDIE model. The phases of product development were as follows:

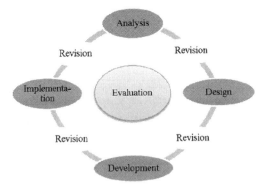

Figure 1.   ADDIE model (Aldoobie, 2015).

The data consisted of complementary qualitative and quantitative data. Data collection techniques used were questionnaires, multiple choice tests and short answers, and interviews. The subjects used were divided into two: the product validation tests and the user trials. Validation tests were conducted to determine the feasibility of the product through the validation of Social studies by a materials expert, an IT-based learning media expert, and an elementary school learning expert. User trials involved fourth grade elementary school teachers and students in three elementary schools to find out the users' response to the media. The trial was limited to ten students at Muhammadiyah Purwodiningratan II Elementary School and 30 students at Muhammadiyah Kleco II Elementary School. After determining the response of students and teachers, an effectiveness test was carried out on 30 students at Muhammadiyah Kleco III Elementary School to find out the effectiveness of cultural map media in overcoming cognitive learning difficulties in Social studies classes for fourth grade elementary school students.

Hypothesis testing on the results of One Group of Pre-test-Post-test Design data was calculated using a paired sample t-test. The proposed hypothesis refers to Riadi (2016) as follows.

H_0: $\mu_1 = \mu_2$ (There is no significant difference between pre-test and post-test)

H_1: $\mu_1 \neq \mu_2$ (There is a significant difference between pre-test and post-test)

4 RESULTS

The result of this research is an interactive learning media product based on Lectora Inspire, which contains Indonesian cultural map material for fourth grade elementary school students. The product development process went through several phases and referred to the ADDIE development model. The first phase was needs analysis and three things were analyzed: the curriculum, the students' needs, and the materials. The results of the curriculum analysis revealed that the three schools used the 2013 curriculum with an integrative thematic approach.

The results of the analysis of students needs are that the learning process had not yet optimized the use of IT-based media. This is especially true for the Indonesian cultural map material, which is seen as difficult to understand because of the great diversity of Indonesian cultures, and it is still abstract in the students' minds. Thus, learning resources are needed that make it easier for students to understand Indonesian cultural map material.

The results of the material analysis are that students also found it difficult to understand Indonesian cultural map material, especially on the indicators of identifying Indonesia regional cultural diversity. This is because the material contained in learning resources (textbooks and students worksheets) is still limited.

The second phase was design, which was compiling the design concept of cultural map media. This involved creating a flowchart and storyboard to facilitate the determination of the components and systematics of the media.

The third phase was the development of an interactive cultural map product with the Lectora Inspire software. The media components were media user instructions, core competencies, basic competencies, teaching materials, evaluations, bibliography, and biographies. The scope of the teaching materials is the Cultural Map of West Sumatra, Yogyakarta, Bali, West Kalimantan, South Sulawesi, Maluku, and Papua Provinces. It discusses traditional musical instruments, traditional clothing, traditional weapons, local languages, folk songs, and traditional dances from each province.

The product was then validated by a Social studies material expert, an IT-based learning media expert, and an elementary school learning expert. The mean results from the experts' validation indicate that the product developed is in the 'very good category' and can be used in elementary school learning. The results of the assessment in quantitative terms can be seen in Table 1:

Qualitative advice was also provided by experts. There are several parts of the media product to be revised in accordance with the comments and feedback given. These are described below.

Feedback from IT media expert: (a) In the main menu, each group of buttons and menu images needs to be given a box-shaped layer. (b) There is a few explanation about regional cultural diversity in the media comparing to what really happened. (c) Adding user instruction for operating signs before and after the page of Indonesia's map to make it easier to see regional cultural diversity in each province. (d) Increasing the quality of the provincial map image to be more interesting and detailed. Examples of added menu items can be seen in Figure 2.

Table 1. Results of assessment of product feasibility by the experts.

Assessment	Score	Category
Materials expert	79	Good
Media expert	78.33	Good
Learning expert	85.33	Very good
Total	242.33	
Mean	80.89	Very good

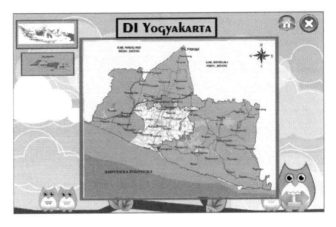

Figure 2 The sample of adding back and next menus and improving map quality.

Feedback from elementary learning expert: (a) Improvement of lesson plan components, i.e. learning indicators adjusted to the basic competencies, compatibility of indicators with learning objectives in closing activities that need to be added with the learning evaluation to measure students' understanding of the material that has been delivered. (b) Adding introductory pages in the form of conversation or questions and answers, by involving several figures including a teacher whose purpose is to explore students' initial understanding (reinforcement). An example of the improvements made on the advice of the elementary learning materials expert can be seen in Figure 3.

Feedback from a material expert: (a) Adding video material on regional cultural diversity, i.e. traditional dances, traditional musical instruments, and local songs: a video of a *kipas pakarena* dance from South Sulawesi, musical instrument *talempong*, drum *tabuik*, and *bansai* from Minangkabau, West Sumatra, and the local song *Sue ora Jamu* from Yogyakarta. (b) Checking the spelling and paying more attention to upper and lower case in the material and questions. (c) Adding a bibliography, pictures, animations, sounds, videos, songs, and material references. An example of the improvements made on the advice of the materials expert can be seen in Figure 4.

After the product was revised according to the experts' inputs, then it entered the fourth phase of implementation. This trial included two stages: limited group testing was conducted to grade students at Muhammadiyah Purwodiningratan II Elementary School and asked for

Figure 3. The reinforcement addition sample.

Figure 4.   The addition of traditional musical instruments material.

the response of one classroom teacher. The next stage was a large group trial for 30 fourth grade students of Muhammadiyah Kleco II Elementary School and asked for a response from the classroom teacher. In the trials, the teachers and students were given a response question-naire to measure the users' assessment of product quality. The analysis of students and teachers assessment questionnaires is summarized in Table 4.

After conducted a large group trial, an effectiveness test was then conducted on 30 fourth grade students at Muhammadiyah Kleco III Elementary School to measure the effectiveness of the product by applying the design model One Group Pre-test-Post-test Design. The results of the descriptive analysis of pre-test and post-test are presented in Table 5.

Correlation analysis was conducted to determine the analysis of the correlation before and after using the cultural map media. The results of the correlation analysis with paired t-tests are shown in Table 6.

The test results above show that the *p-value* < 0,05 caused H0 to be rejected which implies that there is a significant difference between the results before (pre-test) and after (post-test) the learning. Therefore, the developed product is effective at overcoming cognitive learning

Table 4.   Data of the results of trials on students and the teacher responses.

No.	Assessment	Score	Category
1.	Limited group test	90.55	Very good
2.	Large group test	89.97	Very good
	Mean of the trial on students	90.26	Very good
3.	The limited group of teacher responses	84	Very good
4.	The large group of teacher responses	81.33	Very good
	Mean of teacher responses	82.67	Very good

Table 5.   The analysis results of the description of students' scores before and after using the cultural map media.

	Paired sample correlations				
	N	Min score	Max score	Mean	Std.Deviation
Pre-test	30	40	60	56	11.11789
Post-test	30	70	90	77.78	15.13015
Valid N	30				

109

Table 6. Analysis of correlation before and after using cultural map media.

		Paired differences				
		Mean	Std.Dev	Std.Error Mean	Lower	Upper
Pair 1	Pre-test and post-test data	14.96233	16.75917	3.05979	8.70436	21.22031
		T		Df	Sig. (2 tailed)	
Pair 1	Pre-test and post-test data	4.890		29	.000	

difficulties in the Indonesian cultural map material for fourth grade elementary school students.

Finally, an evaluation phase was carried out in every phase of development, starting from the analysis, design, development, and implementation phases. In this evaluation phase, researchers also used summative evaluation to determine the feasibility of the product in the learning process.

5 DISCUSSION

The 2013 curriculum is the current one in Indonesia. At the elementary school level, the main characteristic of the 2013 curriculum is a thematic approach, which integrates various learning content to create a more active, interesting and meaningful learning atmosphere (Chon, Mat, Mohd, & Nazri, 2012). To realize this goal, a teacher must have professional competence and a good personality. Teachers must be organized and be able to manage and interact with students(Moreno Rubio, 2009).

In the implementation of thematic learning there are several problems: the teacher's ability in using the learning method is still minimal, and teacher-centered learning results in students being passive (Agung, 2011), which causes many students to experience cognitive learning difficulties.

Based on the results of the needs analysis, cognitive learning difficulties were discovered in fourth grade students regarding the material of Indonesian cultural maps. One effective way to overcome this problem is through multimedia applications. There is still a lack of interactive learning media for Indonesian cultural map material, so it was necessary to develop a media map based on Lectora Inspire.

The development of this Lectora Inspire based cultural map media was in line with the ADDIE model, with five phases: analysis, design, development, implementation,and evaluation (Aldoobie, 2015). To discover the usability of the cultural map media developed, a prior feasibility assessment was carried out by an IT-based media expert, a Social studies material expert, and an elementary school learning expert. The mean of the results of the assessment by experts wass 80.89, which is in the "very good" category. The experts' feasibility assessments were both quantitative and qualitative assessments, giving feedback for improvement. Feedback from the experts was used for revision of the media product, so that the media could improve the quality of the learning process and help students absorb material more deeply and completely.

The next phase was a usage test. In this trial phase, students and teachers were given a questionnaire to assess the response to the cultural map media: the students' assessment received a mean score of 90.26 in the "very good category" and the teachers' assessment received a mean score of 82.67 in the "very good category". Feedback from the experts showed that a strong point of the developed cultural map media is the interactive quality, which is in accordance with the requirements of IT development (Troseth, Russo, & Strouse, 2016).

After conducting extensive group trials, the effectiveness test was then conducted to measure the effectiveness level of the product. A comparison of the pre-test and post-test scores showed a significant increase of 21.78. This shows that the cultural map media helps students understand the material so that the learning difficulties with Social studies content can be minimized. The result of the statistical test showed that the product can be effectively used to overcome cognitive learning difficulties in the Indonesian cultural map material for fourth grade elementary school students.

The application of Indonesian cultural map media is effective to overcome learning difficulties which are influenced by the variation of the active learning method i.e. example non example. The Cooperative Learning Model type Examples Non Examples is a learning model that uses examples and not examples. Examples can be obtained from cases/pictures that are relevant to basic competencies. The advantages of this learning model are (1) training students to think critically in analyzing pictures or teaching aids, (2) students know the application of the material, (3) the learning process is more interesting and interactive, and (4) efficiency in time and energy, which also can improve the quality of students' learning outcomes.

6 CONCLUSION

A cultural map media product has been successfully developed according to ADDIE procedures. Based on assessments by experts, it can be concluded that the product based on Lectora Inspire can be used for thematic learning, especially Social studies content for fourth grade elementary school. The teachers and students in the trial group also gave a "very good" assessment. From the results of the effectiveness test, it can be seen that the cultural map media is effective at overcoming cognitive learning difficulties when using Social studies content. This is evident from the increase in students' scores before and after using the media.

REFERENCES

Agung, L. (2011). Character Education Integration In Social Studies Learning. *International Journal of History Education, XII*(2), 392–403. Retrieved from http://akhmadsudrajat.wordpress.

Aldoobie, N. (2015). ADDIE model. *American International Journal of Contemporary Research, 6*(2015), 68–72.

Anwariningsih, S.H., & Ernawati, S. (2013). Development of Interactive Media for ICT Learning at Elementary School Based on Student Self Learning. *Journal of Education and Learning (EduLearn), 7* (2), 121–128. Retrieved from http://journal.uad.ac.id/index.php/EduLearn/article/view/226/pdf_8

Bauto, L.M. (2016). Socio-cultural values as community local wisdom katoba muna in the development of learning materials social studies and history. *Jurnal Pendidik Dan Peneliti Sejarah, 14*(2), 195–218.

Buchori, A., & Setyawati, R.D. (2015). The development model of charactereducation through e-comic in elementary school. International Journal of Education and Research. *International Journal of Education and Research, 3* (9). Retrieved from www.ijern.com

Chon, K., Mat, M.A., Mohd, R., & Nazri, I. (2012). Teachers' Understanding and Practice towards Thematic Approach in Teaching Integrated Living Skills (ILS) in Malaysia. *International Journal of Humanities and Social Science, 2*(23). Retrieved from www.ijhssnet.com

Effendi, A., Soetjipto, B.E., & Widiati, U. (2016). The Implementation of Cooperative Learning Model Tsts and Carousel Feedback to Enhance Motivation and Learning Outcome for Social Studies. *Journal of Research & Method in Education (IOSR-JRME), 6*(3), 131–136. doi:10.9790/7388-060304131136

Faruk, A. (2014). Development of Interactive Learning Media Based Lectora Inspire in Discrete Method Course. In *International Conference on Research, Implementation and Education of Mathematics and Sciences* (pp. 18–20). Palembang: Universitas Sriwijaya. Retrieved from http://eprints.unsri.ac.id/4803/

Garderen, D., Scheuermann, A., & Poch, A. (2014). Challenges students identified with a learning disability and as high-achieving experience when using diagrams as a visualization tool to solve mathematics word problems. *ZDM Mathematics Education, 46*(1), 135–149. doi:10.1007/s11858-013-0519-1

Herdini, H., Linda, R., Abdullah, A., Shafiani, N., Darmizah, F. 'Alaina, & Dishadewi, P. (2018). Development of interactive multimedia based on Lectora Inspire in chemistry subject in junior high school or madrasah tsanawiyah. *Journal of Educational Sciences, 2*(1), 46–55. Retrieved from https://ejournal.unri.ac.id/index.php/JES/article/view/4895

Krstev, C., & Trtovac, A. (2014). Teaching Multimedia Documents to LIS Students. *The Journal of Academic Librarianship, 40*(2), 152–162. https://doi.org/10.1016/J.ACALIB.2014.02.006

Moreno Rubio, C. (2009). Efective teachers-profesional and personal skills. *Ensayos: Revista de La Facultad de Educación de Albacete, ISSN 0214-4842, ISSN-e 2171-9098, N°. 24, 2009, Págs. 35-46*, (24), 35–46. Retrieved from https://dialnet.unirioja.es/servlet/articulo?codigo=3282843

Morrison, G.M., & Cosden, M.A. (1997). Risk, Resilience, and Adjustment of Individuals with Learning Disabilities. *Learning Disability Quarterly, 20*(1), 43–60. doi:10.2307/1511092

Ramdhani, M.A., & Muhammadiyah, H. (2015). The Criteria of Learning Media Selection for Character Education in Higher Education. In *Proceeding International Conference of Islamic Education: Reforms, Prospects and Challenges* (pp. 174–182). Malang: UIN Sunan Gunung Jati. Retrieved from http://digi lib.uinsgd.ac.id/5118/1/Ramdhani%26Muhammadiyah. The Criteria of Learning Media Selection for Character Education in Higher Education.pdf.

Riadi, E. (2016). *Statistika penelitian (analisis manual dan IBM SPSS)*. Yogyakarta: Penerbit Andi.

Samawi, A. (2017). Inclusive Education Management in Social Studies Course of Children with Special Needs. *Journal of ICSAR, 1*(2), 155–158.

Shaw-Smith, C., Redon, R., Rickman, L., Rio, M., Willatt, L., Fiegler, H., … Carter, N.P. (2004). Microarray based comparative genomic hybridisation (array-CGH) detects submicroscopic chromosomal deletions and duplications in patients with learning disability/mental retardation and dysmorphic features. *Journal of Medical Genetics, 41*(4), 241–248. doi:10.1136/JMG.2003.017731

Šramová, B. (2014). Media literacy and Marketing Consumerism Focused on Children. In *Procedia - Social and Behavioral Sciences* (Vol. 141, pp. 1025–1030). Bratislava, Slovak Republic: University in Bratislava. doi:10.1016/j.sbspro.2014.05.172

Troseth, G.L., Russo, C.E., & Strouse, G.A. (2016). What's next for research on young children's interactive media? *Journal of Children and Media, 10*(1), 54–62. doi:10.1080/17482798.2015.1123166

Wibawa, S.C. (2017). The Design And Implementation Of An Educational Multimedia Interactive Operation System Using Lectora Inspire. *Elinvo (Electronics, Informatics, and Vocational Education), 2*(1), 74. doi:10.21831/elinvo.v2i1.16633

Learning innovations in citizenship education for strengthening digital and ecological citizenship

K.E.R. Marsudi & S. Sunarso
Universitas Negeri Yogyakarta, Indonesia

ABSTRACT: There is worldwide concern over the issues of citizenship and the misuse of information technology such as cybercrime and also environmental damage due to global warming. The main factors that cause these problems are the characteristics of humans themselves, and the weakness of digital and ecological citizenship, which ultimately leads to ignorance of the negative effects of technology abuse and also excessive exploitation of nature and environmental destruction without being able to maintain balance. These problems have a very large impact on the survival of humankind. Therefore, it is imperative that appropriate solutions are found. This paper aims to find a solution that has a programmed and sustainable structure in the face of weak digital and ecological citizenship. A suggested solution is innovative learning in civic education (Civics) as a curricular program in schools to shape digital and ecological citizenship for young citizens. Digital citizenship is formed using a value-based learning model often known as the value clarification technique (VCT) by considering nine elements of digital citizenship. Ecological citizenship is strengthened via eco-literacy using a project-based ecological citizen learning model.

1 INTRODUCTION

Currently, two global issues are key topics of discussion in many countries: *digital citizenship* and *ecological citizenship*. These two issues urgently require attention from groups that have a close relationship with various lines of human life. Digital technology is developing rapidly in the 21st century and this time is often referred to as the digital era. Various findings have emerged about the ease of information technology. These conditions have various positive and negative impacts on human life. With rapid digital development, it is easier to access information from anywhere and communicate with anyone, it is easier to express opinions and criticisms of anything from government policy to social care, and there are many other benefits. But undoubtedly there are negative impacts that are difficult to avoid by digital citizens.

Some examples of massive negative impacts found in various fields of human life in the 21st century include rampant hoaxes with information on politics, economics, and even health. The various *cyber crimes* range from defamation, *bullying*, electronic data theft, fraud, buying and selling illegal goods such as drugs, human trafficking and many other detrimental misuses of information technology.

Another that has caught the world's attention in this global era is ecological or environmental issues. Humans live on this earth as one component of an ecosystem. Humans do not live alone and are unable to live alone but must coexist in interdependence with various other components in the ecosystem. Human life will not last long if we are only utilizing livelihood resources around us to meet and enrich our needs and desires without properly maintaining the balance of the other components.

To protect the survival of human life in the future, humans must understand how living resources on this earth can be maintained undamaged. Current reports indicate decreases in ecological sustainability due to irresponsible human activities: excessive use of resources, massive areas of forest clearing for economic purposes using destructive methods such as forest fires, large-scale plastic waste disposal, and water and air pollution. These and much other ill-treatment of nature have led to *global warming*, which is already at a very concerning level.

It is now necessary to find solutions to minimize the negative impacts of these two issues, both in digital development and ecological problems. The correct solution will be based on the root cause of the problem, which naturally arises because of human beings themselves, who must be educated to reduce actions that can cause negative effects, from the rapid development in the digital era and environmental sustainability.

If examined in more depth, we can see that some of the problems caused by rapid digital development and environmental damage in various parts of the world, including Indonesia, arises from low *digital character citizenship* and *ecological citizenship*.

There is a lack of awareness of the use of digital media in a moral and responsible manner no sense of responsibility from the community toward environmental sustainability, and indifference to the effects of environmental damage, especially for the long term.

A number of parties have made various efforts over the years to improve *digital citizenship* and *ecological citizenship* in the community, using various approaches to solve issues related to the problem of digital abuse and environmental destruction. There are parties who use a general approach to *digital citizenship* and *ecological citizenship* and others who use thematic approaches, for example, "anti-hoax" or "*internet security*" activities, or focus on certain communities such as "teenagers" and "children" or "students" and "housewives".

In the formation of *ecological citizenship,* there are countless communities that focus on the environment, the *go-green* movement, campaigns on the importance of protecting the environment, and many other activities aimed at the formation of digitally literate citizens and ecological literacy.

Increasing digital literacy and ecology communities is not an easy task that can be realized through one or two approaches given the rapidly growing digital problems and *global warming,* which are increasingly worrying. This must be carried out through various aspects and levels of human life achieve fast and precise goals.

For this reason, further movement is needed to find ways, models or approaches that target the situation today and for the long term. This requires a program that is truly structured, carried out through curricular programs included in civic education (Civics) in schools. This will effectively create Indonesian citizens who are digitally and ecologically literate, involved in the digital world and responsible for environmental sustainability, and as far as possible also morally and responsibly in accordance with national identity contained in the values of Pancasila.

Strengthening *digital citizenship* and *ecological citizenship* through curricular programs in schools will have long-term effects. This early education will be given to students through the Civics curriculum, which in Indonesia runs from basic primary education to universities. The questions posed in this study include how to effect learning innovations in civic education in schools and higher education in order to realize both the digital citizenship and ecological citizenship communities effectively and efficiently.

The study will provide recommendations for strengthening the *digital citizenship* and *ecological citizenship* movements in Indonesia to meet an unfulfilled agenda. The recommendations will also be useful for compiling a civic education module based on *digital citizenship* and *ecological citizenship* with an Indonesian context in accordance with the values of Pancasila.

2 LITERATURE REVIEW

2.1 *Digital citizenship concept within an Indonesian concept*

The concept of *digital citizenship* arises when very rapid developments occur in the field of technology and information, particularly regarding the internet, websites and social media such as Facebook, Twitter, Instagram, and YouTube, etc. These developments have provided certain benefits, including fast global communication not hindered by space and time, as we can easily interact with others through cyberspace. Digital actors join and form communities to share information and utilize various content through digital media, usually in the form of photos or images, videos, and documents, etc. The perpetrators or users of various digital technologies in accordance with the context of digital citizenship are called digital citizens.

Citizens of digital or *digital citizens* according to Norris and Bimber in Mossberger, Tolbert, and McNeal (2007) are those who use the internet regularly and effectively every day. Mossberger et al. (2007) define digital citizens as those who often use technology, those who use it in relation to politics to fulfill their civic duties, and those who use technology at work for economic gain. Digital citizens are current users of information technology.

Hintz, Dencik, and Wahl-Jorgensen (2017) defined digital citizenship as the application of self from the role of people in society through the use of digital technology. Ribble and Bailey (2007) described it as the norms and responsible behavior related to the use of technology. Digital citizenship is a concept related to the formation of healthy and safe relationships to interact in the internet environment by using conscious digital technology. In fact, digital citizenship is also a concept of what can or cannot be done using digital technology. As the 21st century is the digital era via technology and the internet, individuals should have certain techniques to avoid any negative impacts from technology (Kuzu, Odabasi, & Gunuc, 2013). It can be concluded that what is meant by digital citizenship is a concept or idea that can be used in educating digital citizens about ways that are moral, good, right and responsible for using information technology.

Digital citizenship is usually determined through actions taken by people, not their formal status and the rights and responsibilities that follow. This indicates that citizens create and carry out their role in the world community. Isin and Ruppert (2015, p. 43) stated that *"we are enacting ourselves in cyberspace"* and Isman & Gungoren (2014) discussed how in this technological era everything can quickly change and develop. Given these technological advancements, schools have a responsibility to shape and create good digital citizens who are literate in the development of information technology. Digital citizens must have knowledge, skills, and ethics in relation to internet use.

Digital citizens must be continually taught provisions to deal with this digital era. The younger generation will face new world challenges: daily interactions with individuals with diverse ethnicities, genders, languages, races and socioeconomic backgrounds from around the world. They may experience related problems such as health, injustice, environmental damage, population explosion, transnational migration, ethnic nationalism, and the decline of nation-states (Murdiono, 2014). Smart and strong digital citizenship behaviors are keys to ensuring that citizens can positively contribute to digital life. When digital citizens are not smart or good in their activities, it will have a negative impact on life in society both at the individual and group level.

Indonesian national education was designed not only to form intelligent and skilled students but also to create citizens who can uphold positive values in line with their identity, culture, and character that has been inherited by the nation. The main objective here is describing a preventive measure for all nations in the face of globalization.

The key objective is that the process of globalization does not cause citizens to be uprooted from their own nation's culture and character. In order to contribute positively in the midst of globalization, a citizen requires competence in the form of knowledge, skills, and attitudes that are integrated across various aspects in life, but at the same time maintains and deepens their sense of identity and not forget their integrity as part of a country's citizens. Digital citizenship that is in accordance with the Indonesian character adheres to the noble values of Pancasila as its basis. This is absolute, considering that Pancasila is a way of life for the nation and state, especially in the face of globalization and the digital era in the 21st century.

2.2 *Ecological citizenship concept within the Indonesian context*

The concept of *ecological citizenship* refers to environmental care by citizens of various interdependent components of the earth's ecosystems. Humans have a responsibility to protect the other components for the balance and survival of the ecosystem. *Ecological citizenship* is an idea that caught public attention and triggered the emergence of various environmental care movements both individually and publicly.

Lummis, Morris, Lock, and Odgaard (2017) discuss ecological citizenship in terms of global community citizenship, where citizens are bound by the responsibility to preserve our ecology for future generations. As ecological citizens, individuals have a responsibility to impose daily lifestyle behaviors without damaging the surrounding environment. Biagi and Ferro (2011) discuss ecological citizenship as a new concept emerging from the political ecology in response to the need for theoretical reformulation of democracy in the context of global environmental risk. Ecological citizenship does not look at place boundaries and emphasizes individuals' responsibilities to protect the environment. Curtin (2002) describes *ecological citizenship* as a new idea that contains the principles of establishing awareness to form citizens who have direct involvement in pursuing environmental conservation.

Direct involvement by citizens must also be supported by other elements, including government policies. This is reinforced by Barry (2006) who states that the policies made by the state in relation to the environment will be very effective in forming ecological citizenship in a sustainable manner without neglecting other dimensions of economic, political, and social life. However, on occasional national policies related to the environment are defeated by the economic interests of capitalism, as evidenced by illegal logging carried out by large economic corporations and the lack of waste management from the rapidly growing industries.

In addition to government policies, another important role that determines the formation of environmentally literate citizens is early education. In many countries, education is still focused only on children's academic development: children are formed to become good citizens but rarely as ecological citizens. Jordan, Singer, Vaughan, & Berkowitz (2008) state that ecological education must be a major part of the learning experience of citizens. This must be taught in the classroom environment and continue through learning experiences to adulthood, with the aim of improving ecological education for students of all ages and the general public.

The importance of learning ecological citizenship is also discussed by Kelly and Abel (2012) who argue that environmental education provides important insights into learning experiences, especially environmental service-learning, and encourages students to see they have civic responsibilities to play an active role in critical environmental problems. This is supported by Crane, Matten, and Moon (2008) who state that ecological citizenship comes from the desire to connect and relate emotionally and spiritually to the natural environment.

According to Porto (2018), global citizenship is based on normative values and a responsibility to solve world problems by using three related dimensions: moral, institutional, and political. Ecological citizenship blends with the traditional conception of citizenship in several ways. First, it is not only related to rights but also to obligations and responsibilities. Second, it is not only related to the public domain but also the private domain. Third, it is not limited to the nation but is more territorial. In Indonesia, *ecological citizenship* in the context of Indonesia means ecological literacy that is understood and applied in the lives of citizens where it cannot conflict with the basic laws and ideology of the Indonesian nation itself, Pancasila. The description of the points of Pancasila has its own relevance to the field of ecology.

Civics courses are compulsory various levels of education and have tremendous influence on the formation of ecologically literate citizens. In secondary education, the Civics curriculum includes learning points that will form citizens who love and care for the environment, because the strengthening of *ecological citizenship* must be done in an innovative, effective and efficient way using civic education.

3 RESEARCH METHODOLOGY

The method used by here was library research with an inductive analysis model. It is called library research because the data or materials needed to complete the research come from the library in the form of books, encyclopedias, dictionaries, journals, documents, magazines, etc. (Harahap, 2014). The inductive analysis will uncover a variety of themes, categories, and patterns of relationships between these categories. Library research uses library resources to obtain research data, meaning that activities are limited to library collection materials without the need for field research (Zed, 2004). A literature study is very important in conducting research because the research is not separated from the scientific literature. This research was conducted by examining various studies and research results related to *digital citizenship, ecological citizenship* and Civics learning in schools. The analysis shows that the learning innovations that can be used by citizenship education teachers to strengthen the character *digital citizenship* and *ecological citizenship* are in accordance with the noble values of Pancasila.

4 RESULTS AND DISCUSSION

4.1 *Strengthening digital citizenship through citizenship education*

Citizenship education is learning used to form good citizens. The learning target of the Civics curriculum is the integration of the three aspects of competence: understanding (cognitive), attitude (affective) and skills (psychomotor). It is expected that the learning material (values) is manifested in attitudes and behavior according to good citizens or good manners.

Facing the current digital developments, people still require to maintain their national identity and the values adopted by each country. In the context of Indonesia, the values and identity of the nation that must be maintained are Pancasila, which is a manifestation of the moral character that has lived in society since time immemorial. As explained earlier, strengthening human character that is moral and responsible in accordance with the values of Pancasila must be an educational priority. This is especially key in civic education given the development of increasingly globalized information technology in this digital era, which has a variety of negative impacts that potentially lead to the disintegration of the nation and raises various conflicts for Indonesian people.

According to researchers, the appropriate learning model to strengthen *digital citizenship* is the value clarification model, better known as the *values clarification technique* (VCT). VCT is a method of value clarification in which students do not memorize and are not "fed" with the values that others have chosen, but are helped to find, analyze, account for, develop, choose, take a stand, and practice their own values. Value clarification is one method that seeks to foster students' intellectual abilities to form moral decisions full of responsibility, especially in dealing with the digital era today. This method teaches students to overcome problems, even in dilemmas of high moral values.

Based on some of the effectiveness of learning through VCT delivered by Toyibin and Djahiri (1997), the relationship with learning *digital citizenship* will correlate with the research as follows. First the process of student learning activities that are *clarifying*, where students (through their various potentials) seek and study moral values in the digital era, so they will know what to do and what not to do. Second, the learning activities are *spiritualized* and are valued through the heart of *digital citizens.* Third, along with the process *valuing*, there is also a process of applying the knowledge and values of *digital citizenship* in real life.

The *values clarification technique* can train students to assess the values of life that exist in society in this digital era and establish morals and responsibilities in performing the role of a digital citizen. PKn and its development cannot be separated from the challenges of strengthening *digital citizenship*. In accordance with its objectives, Civics must create citizens who understand and behave according to norms, who are responsible in their use of technology, who respond to issues in the digital era, and who have strong digital native ethics.

According to the researcher, VCT can be implemented in Civics learning to strengthen characters *digital citizenship* as follows. Ribble and Bailey (2007) identified nine elements of digital citizenship: *digital access, digital commerce, digital communication, digital literacy, digital etiquette, digital law, digital rights and responsibilities, digital health and wellness,* and *digital security.*

For *digital access*, in Civics classes, students are invited to respect the right of every individual to have access to technology. But they must also understand that every digital citizen should be aware of the constraints to technological achievements, such as infrastructure factors, and their respective customs and cultural factors. *Digital commerce* discusses economic or commercial issues through the digital world, not only through using technology but also in terms of risk such as fraud. Students in Civics classes must know how to be good online buyers and sellers and have good relationships with others. Digital communication has many benefits in various fields of life. Civics learning must direct students on how to use digital communication morally, with adjustments to values and norms in accordance with national culture. The fourth element is *digital literacy*, and it is expected that through Civics learning each individual will have the ability to use digital technology and know when and how to use it appropriately. Misusing digital media should be avoided for the sake of national integration. *Digital etiquette* shapes citizens' behavior in accordance with the standards and values of people's lives. Digital ethics are used to maintain feelings, remain in harmony with other users, and grow a community as responsible digital citizens. It is very important to teach these characteristics through early Civics lessons in school. Educational psychology states that it is easier to form someone's character early in life. *Digital law* provides guidance to digital citizens on how to use digital media legally, because if they make a mistake they can be subjected to moral and legal sanctions as stipulated in *undang-undangnomor 11 tahun 2008* concerning Information and Electronic Transactions. This section teaches students how they can adhere to the law because of awareness as digital citizens. The seventh element is rights and responsibilities in the digital world. Digital citizens or individuals have rights that must be respected by others and must in turn respect the rights of others. Before demanding their rights, an individual must fulfill their obligations first. This teaches tolerance and tolerance. Regarding *digital health and fitness*, behind the various benefits of digital technology, there are several mental and physical health problems caused by improper or excessive use.

For this reason, in Civics classes students are always taught to use technology appropriately and not excessively. The ninth element is digital safety (*digital security*). Among the many digital citizens in the world, it is not uncommon to experience cases of digital abuse and cheating, such as damaging the work of others and plagiarism, etc. We should not be mistaken for other digital users, but every digital citizen must remain vigilant to maintain their safety. PKn forms students into someone who is critical and moral according to the Pancasila corridor so that they do not commit or become victims of misuse of digital technology. The development of the nine elements of digital citizenship through Civics learning can be adapted to the students' developmental level and the situations and conditions in the learning environment, such as facilities and infrastructure in schools.

4.2 *Strengthening ecological citizenship through citizenship education learning*

Innovations in civic education that includes ecological education to create ecologically literate citizens can be done by applying *ecoliteracy* through project-based ecological citizenship learning models. *Ecoliteracy* is an abbreviation of the term *ecological literacy*, which can be interpreted as environmental literacy or ecological literacy. Ecology itself can be interpreted as the science of how to care for and maintain the universe where living creatures live. Keraf (2014) states that ecological literacy or *eco-literacy* describes humans who have a high level of awareness about the importance of the environment.

Ecology literacy (*ecoliteracy*) can be defined as the ability to use an understanding of ecology, thinking and habits of mind to learn to care for the environment (Berkowitz, Ford, & Brewer, 2005). *Ecoliteracy* is fast becoming a major priority in various parts of the world, given the increasingly alarming natural conditions caused by human activities. At present, governments, the private sector, and the world community are working on ways to minimize environmental damage. One effective and long-term effect is *eco-literacy*, which can be practiced in the field of formal education as *ecological citizenship* taught through civic education.

Various components and material studies in the field of citizenship education are appropriate for the development of learning about *ecoliteracy*. This is key to reconstructing the understanding and behavior of citizens so they realize the importance of maintaining the global ecology in order to create a balance between the needs of humans and the ability of the earth and nature to support their needs because all elements need each other to survive.

Education is a significant way to change one's mindset and behavior. *Ecoliteracy* knowledge can be improved through education on maintaining the environment, through various studies on ecology, and through the direct practice of how to behave. Sustainable and continuous formal *ecoliteracy* education across the generations will reap long-term results. This is consistent with Turner and Donnelly (2013), who argues that the classroom is a good place to change a person's behavior to positive. Education has the potential to transform beliefs and understanding of the world.

There are three pathways that can be used to implement the idea of *eco-literacy*, as conveyed by Prasetiyo and Budimansyah (2016). The first is education through the provision of learning materials that address ecological or environmental issues. The second is using the surrounding environment with real-life examples as a learning resource. The third is ecological or environmental education that does not only develop knowledge and skills but also the values and ethics of ecological citizenship. The goal is not only to know but truly to have the character of the ecology of citizenship.

In line with the three paths of implementing *eco-literacy* detailed above, citizenship education itself contains three citizenship competencies. The first *civic knowledge* or knowledge of citizenship means anything that should be known by citizens. The second is *civic skills* or citizenship skills, which mean the ability of citizens to participate directly in the life of a country. The third is *civic disposition* or citizenship character, in which a citizen really has the character, values, and ethics of good citizenship.

The thinking above illustrates a strategic correlation and the importance of *ecoliteracy* in reconstructing ecological citizenship, where civic education carries a value and attitude mission in three domains, cognitive, affective, and psychomotor, to achieve instructional objectives and companion goals (Winataputra, 2012). The idea of *eco-literacy* is appropriate for civic education by describing basic competencies that have already contained various subjects related to ecology, the living creatures are elaborated more detail in order to create ecologically literate citizens. Goleman (2009) suggests that there are five points to develop the character of *ecoliteracy*. First, *develop empathy for all forms of life*, and learning must focus on awareness and attitude (empathy) for the environment. Basically, every child has empathy for his environment and this attitude can be seen when students feel sorry for living things when they are hurt. This attitude of empathy must be developed by the teacher in the classroom, so that the students' sense of empathy is stronger, not only to fellow beings but also to nature. Through this sustainable practice, children can assess and reflect on how what they is good or bad for the environment. Second, students should *embrace sustainability as a community practice* in group learning, and ask questions with their friends or other people related to strengthening the ecology of citizenship. Students will understand how ecological or environmental balance is the responsibility of each individual, including themselves. Third, it is necessary for students to *make the invisible visible*, and experience real-life learning about environmental preservation. Students will be closer to and animate each learning process if it is truly practiced directly. They will carefully follow the steps and procedures in ecological activities. So, they will experience the goals of learning *ecological citizenship*. This will make learning more meaningful. Students can experience directly how learning takes care of the environment. Fourth, *anticipating unintended consequences* will teach students to take full responsibility for their role as humans in protecting the environment. Fifth, *understanding how nature sustains life* will bring students directly into the evaluation stage. Students will realize what happens when the environment is not maintained properly and understand that life is the responsibility of the people who manage it. Good management will have positive effects on the environment and vice versa. This will provide a separate experience for students and will further strengthen the character of *ecological citizenship* in students.

One model of learning suited to learning *ecoliteracy* in civic education is a project-based learning model, which will take students directly into the field to carry out real activities related to ecological citizenship, for example, tree planting projects, environmental extension projects and many other activities that are concrete manifestations of *ecological citizenship*. According to Trianto (2012), using a project-based learning model has considerable potential to make learning more interesting and challenging and has a real positive impact both for the students and for the environment. Project-based learning models find problems or issues around students related to the environment, provide learning experiences that are more interesting and meaningful to students, produce tangible products, and have an evaluation process carried out continuously in the real world rather than only theory in class.

Some of the advantages of using project-based learning in ecological citizenship learning in Civics subjects are as follows. First, it increases students' motivation to face challenges and encourages real work. Second, it improves students' ability to solve the ecological problems that surround them. Third, it improves students' ability to work together and their communicating skills, so that ecological issues can be easily resolved between many parties. Fourth, it provides experience to students in organizing projects, allocating time, and managing resources such as equipment and materials to complete tasks and display real ecological citizenship in real life.

5 CONCLUSION

Based on the results of the analysis of various literature, it can be concluded that there is much scope for innovation in Civics learning to strengthen the digital citizenship and ecological citizenship characters considered crucial in the 21st century. First, in strengthening digital citizenship, citizenship education uses the learning values clarification values (VCT) model by considering nine digital citizenship elements: *digital access, digital commerce, digital communication, digital literacy, digital etiquette, digital law, digital rights and responsibility, digital health and wellness, and digital security.* That way, it is hoped that strengthening digital citizenship through the curricular program of Civics can be a structured, sustainable, effective and efficient program for the formation of digital citizens who are moral and responsible, in accordance with the noble values of Pancasila, in their role as users of digital technology in the 21st century.

Second, the ecological strengthening of citizenship in citizenship education is carried out through ecoliteracy, which uses a project-based model of ecological citizenship learning. That way, it will create young citizens who are ecologically literate and able to utilize and manage the environment in order to maintain the balance of ecosystems and human survival and various other components of the universe, given that global warming is a key issue. With formal learning through Civics, strengthening the digital character of citizenship and ecological citizenship can be well structured and have a sustainable long-term effect.

REFERENCES

Barry, J. (2006). Resistance is Fertile: From Environmental to Sustainability Citizenship. In A. Dobson, & D. Bell (Eds.), Environmental Citizenship (pp. 21–48). Massachusetts, US: MIT Press.

Berkowitz, A.R., Ford, M.A., & Brewer, C.A. (2005). A framework for integrating ecological literacy, civics literacy, and environmental citizenship in environmental education. In E.A. Johnson, & M.J. Mapping (Eds.), *Environmental education and advocacy: Changing perspectives of ecology and education* (pp. 227–266). Cambridge: Cambridge University Press.

Biagi, M., & Ferro, M. (2011). Ecological citizenship and socialrepresentation of water: Case study in two Argentine cities. *SAGEOpen, 2*(1), 1–8.

Crane, A., Matten, D., & Moon, J. (2008). Ecological citizenship and the corporation: Politicizing the new corporate environmentalism. *Organization & Environment, 21*(4), 371–389.

Curtin, D. (2002). Ecological citizenship. In E.F. Isin, & B.S. Turner (Eds.)., *Handbook of Citizenship Studies* (pp. 293–304). New Delhi: SAGE Publications.

Goleman, D. (2009). *Ecological intelligence: How knowing the hidden impactsof what we buy can change everything*. New York, US: Broadway Books.

Harahap, N. (2014). Penelitian kepustakaan. *Iqra': Jurnal Perpustakaan dan Informasi, 8*(1), 68–74.

Hintz, A., Dencik, L., & Wahl-Jorgensen, K. (2017). Digital citizenship and surveillance society. *International Journal of Communication, 11*(1), 731–739.

Isin, E.F, & Ruppert, E. (2015). *Becoming digital citizens*. London, UK: Rowman& Littlefield International.

Isman, A., & Gungoren, O.C. (2014). Digital citizenship. *Turkish Online Journal of Educational Technology - TOJET, 13*(1), 73–77.

Jordan, R., Singer, F., Vaughan, J., & Berkowitz, A. (2008). What should every citizen know about ecology? *Frontiers in Ecology and the Environment, 7*(2), 495–500.

Kelly, J.R., & Abel, T.D. (2012). Fostering ecological citizenship: the case of environmental service-learning in Costa Rica. *International Journal for The Scholarship of Teaching and Learning. 6*(2), 1–19.

Keraf, S. (2014). *Filsafat lingkungan hidup: Alam sebagai sebuah sistem kehidupan* [Philosophy of living environment: Nature as a system of life]. Yogyakarta, Indonesia: Kanisius.

Kuzu, A., Odabasi, H.F., & Gunuc, S. Evaluation of a social network activity within the scope of the digital citizenship. *World Journal on Educational Technology, 5*(2), 301–309.

Lummis, G.W., Morris, J.E., Lock, G., & Odgaard, J. (2017). The influence of ecological citizenship and political solidarity on Western Australian student teachers' perceptions of sustainability issues. *International Research in Geographical and Environmental Education, 26*(2), 135–149.

Mossberger, K., Tolbert, C.J., & McNeal. R.S. (2007). *Digital citizenship: The internet, society, and participation*. London, UK: MIT Press.

Murdiono, M. (2014). Pendidikan kewarganegaraan untuk membangun wawasan global warga negara muda [Civic education to build young citizens' global insight]. *Cakrawala Pendidikan, 33*(3), 349–357.

Porto, M. (2018). 'Yo antes no reciclaba y esto me cambiopor completo la consiencia': Intercultural citizenship education in the English classroom. *International Journal of Primary, Elementary and Early Years Educaion, 46*(3), 317–334.

Prasetiyo, W.H., & Budimansyah, D. (2016). Warga negara dan ekologi: Studi kasus pengembangan warga negara peduli lingkungan dalam komunitas Bandung berkebun [Citizen and ecologists: Acase study of developing citizens' environment care among the members of Bandung gardening community]. *Jurnal Pendidikan Humaniora, 4*(4), 177–186.

Ribble, M., & Bailey, G. (2007). *Digital citizenship in schools*. Portland, OR: International Society for Technology in Education (ISTE).

Toyibin, M.A., & Djahiri, A.K. (1997). *Pendidikan Pancasila* [Pancasila education]. Jakarta, Indonesia: Rineka Cipta.

Trianto. (2012). *Model pembelajaran terpadu* [Integrated learning model]. Jakarta, Indonesia: BumiAksara.

Turner, R., & Donnelly, R. (2013). Case studies in critical ecoliteracy: A curriculum for analyzing the social foundations of environmental problems. *Educational Studies, 49*(5), 387–408.

Winataputra, U.S. (2012). *Pendidikan kewarganegaraan dalam perspektif pendidikan untuk mencerdaskan kehidupan bangsa (gagasan, instrumentasi, danpraksisi* [Civic education in the perspective of improving nations' life (ideas, instrumentation, and practice)]. Bandung, Indonesia: WidyaAksara Press.

Zed, M. (2004). *Metode penelitian kepustakaan* [Library research method]. Jakarta, Indonesia: Raja Grafindo Persada.

Needs analysis for an electronic module (e-module) in vocational schools

S. Oksa & S. Soenarto
Universitas Negeri Yogyakarta, Yogyakarta, Indonesia

ABSTRACT: The research objective was to determine the need for electronic modules (e-modules) employing project-based learning for a Basic Graphic Design course in Vocational Schools. This research is descriptive research. Data collection techniques were interviews, questionnaires, and documentation. The data were analyzed descriptively. The results of this study indicate that: (1) the Basic Graphic Design course had not applied a massive project-based learning model; (2) 92.54% of students complained about the difficulties in practical sessions of learning Basic Graphic Design because of the unavailability of learning materials; (3) 80.60% of students needed learning materials that covered theory, working procedure, and video tutorials. Based on the study results, it was necessary to develop an electronic module (e-module) employing project-based learning that could be used as learning materials effectively and efficiently for vocational education students.

1 INTRODUCTION

The importance of 21st-century learning that integrates knowledge, skills, and attitudes, as well as mastery of ICT has been formulated into the revised Indonesian curriculum 2013 to 2017. Revised curriculum 2013 to 2017 is applied at all levels of education from primary and secondary education. Vocational secondary education, in the form of educational units of Vocational High Schools (SMK), prioritizes the development of professional skills and attitudes, as well as the preparation of students for certain types of employment (RoI, 1990). In addition, Vocational Schools require students to have high productivity to create graduates with competence and skills. The revised curriculum-2013 structure for 2017 in vocational secondary education consists of core competencies and basic competencies (KI-KD) of national content subjects (A), regional content (B), basic expertise (C1), basic expertise programs (C2), and skill competence (C3) (RoI, 2017a).

Vocational high schools (SMK) have several expertise programs, including a Computer and Information Engineering expertise program, which includes Software Engineering, Computer, and Network Engineering, Multimedia and Information Systems, Networks and Applications. The competence of multimedia skills requires students to be able to apply the principles of graphic arts and animation in presenting and producing multimedia products. This encompasses several subjects, and the expertise program (C2) taught in Vocational Schools is Basic Graphic Design.

Basic Graphic Design is a productive course that is classified as new in the revised 2017 curriculum and taken by students in the odd semester or even semester class X. Basic Graphic Design course aims to develop students' basic knowledge and skills in delivering visual communication based on art and technology. The learning process in Basic Graphic Design course is supported by the use of information and communication technology, thus requiring students to be able to operate graphic design learning software such as CorelDraw, Photoshop, Illustrator, etc. One of the basic competencies in Basic Graphic Design is vector image processing using CorelDraw learning software (RoI, 2017b). This requires a cognitive understanding of the use of interface functions, tools, and features in CorelDraw. But in reality, students' graphic design abilities in real-life situations are still low, because these subjects are taken by students in class X who have never received graphic design training before. In addition, there are no specific books, modules or worksheets (LKS) for the Basic Graphic Design course.

Basic Graphic Design in the revised curriculum-2013 to 2017 also demands that student-centered learning involves students actively and directs them to explore their potential. However, in the field students complained about difficulties completing practical learning activities using graphic design learning software because there were no books, modules or student worksheets (LKS) that directly referred to the steps of graphic design. This led to the implementation of practical learning activities in which students were only guided and instructions were given by the teacher.

The objectives of this study include the following: (1) analyzing the Basic Graphic Design learning that has been carried out; (2) analyzing students' needs for electronic modules (e-modules) based on project-based learning; and (3) analyzing learning difficulties experienced by students. The results of the analysis are expected to become the basis for researchers in developing electronic modules (e-modules) based on project-based learning to create effective and efficient teaching materials for vocational school students.

1.1 *Characteristics of vocational students*

The characteristics of vocational students are reviewed through their cognitive, psychomotor, and affective development. Cognitive development of vocational students is included in the formal operational development stage because vocational students are in the age range 16–18 years. Characteristics of students at this stage include the ability to think abstractly and logically. were, students also have been able to think scientifically in drawing conclusions, interpreting, and solving a problem.

Psychomotor development of vocational students occurs in several stages: the cognitive stage, where student movements are still stiff and slow in the learning process; the associative stage, where students need a short time in thinking about the movement; and the autonomy stage, where students do not need instructions in making movements. The affective development of vocational students in mastering learning materials can take the form of maintaining an ego, an increased sense of anxiety such as frustration and worry, courage in taking risks and having characteristics related to individual involvement with the feelings of others.

Based on the analysis of the characteristics of class X students with multimedia vocational skills competence, it can be concluded that the students' ability in using graphic design learning software was low because they had not learned any graphic design at the previous educational level. Students' claim to be more motivated to learn through productive learning or practice activities rather than theoretical learning activities. Students can only practice graphic design during the learning process of the Basic Graphic Design course, because of the limited learning resources that can be used outside of school hours. Learning systems independently and face-to-face are both in demand by students in the learning process.

1.2 *Basic Graphic Design course*

Basic Graphic Design includes productive subjects consisting of compulsory practice and theoretical learning activities that must be taken by students of the Computer and Informatics Engineering Skills Program in class X, namely in odd or even semesters (RoI, 2017a).

This subject aims to develop basic knowledge and skills of students in delivering visual communication based on art and technology. In addition, Basic Graphic Design also provides basic competencies of graphic design science, graphic design components, and graphic design principles. Basic Graphic Design course is supported by the use of information and communication technology, thus requiring students to be able to operate learning software. In accordance with the core and basic competencies that must be achieved by students, the Basic Graphic Design course can teach students the basics of graphic design.

One of the subjects taught is vector image processing; students are expected to know the functions of vector image processing applications, interfaces, and tools used in vector image processing applications; and how to operate vector image processing applications for graphic design. The vector image processing software used is the CorelDraw application.

1.3 *Electronic module (e-module)*

Learning through information technology for the presentation of teaching materials will add to simple teaching materials such as a whiteboard, chalk, drawings, or model tools (Sukiman, 2012). Thus the presentation of teaching materials is not only limited to print technology but makes use of audiovisual technology, computer-based technology and multimedia (Warsita, 2008). One presentation of teaching materials that utilize audiovisual technology, computer-based, and multimedia is an electronic or e-module, where the teaching materials using print modules can be transformed into electronic or digital forms.

Sugianto, Abdullah, Elvyanti, and Muladi (2013) suggested that electronic modules are a form of presentation of independent teaching materials designed to assist students in achieving certain learning goals that are systematically arranged into the smallest learning unit and presented in an electronic format. Electronic-based modules not only have text, but can be animations, audio, images, video, and interactive, which are processed into a learning material (Putra, Wirawan, & Pradnyana, 2017). Wijayanto and Zuhri (2014) described the electronic module or e-module as a display of information or messages in a book format, presented electronically using hard disk, diskette, CD or flash disk storage media that can be read using a computer or electronic book reader.

A digital-based module or electronic module is one of the independent learning media that utilizes computer technology in presenting learning materials so that students can use interesting, creative, and practical learning materials that can lead them to discover concepts independently (Bakri, Permana, & Siahaan, 2016). Serevina, Sunaryo, Astra, and Sari (2018) stated that an e-module is interactive learning material, where students not only read the text but also see the actual process so as to facilitate understanding. The above statement can be explained as using an electronic module can facilitate the student learning process, where students not only learn by reading text, but also can see animations or even simulations of processes that resemble actual events so as to improve students' understanding.

2 METHOD

This type of research is a qualitative descriptive study. It began on 1 and 2 March 2018 as a pre-survey and the study continued until 7 August 2018. The research sample was 67 students of class X Multimedia and two classes of Basic Graphic Design at SMK 2 Sewon.

Data collection techniques for collecting primary data and secondary data were interviews, observation, and documentation. Primary data was obtained through needs analysis interviews with Basic Graphic Design teachers related to the learning processes that occurred. Observations were made to collect data by distributing needs analysis questionnaires to students related to the learning activities in Basic Graphic Design course.

3 RESULT AND DISCUSSION

3.1 *Result*

The electronic module needs analysis (e-module) was divided into two: the needs analysis for subject teachers and for students.

3.1.1 *Analysis of subject teacher needs*

The need analysis for teachers in Basic Graphic Design course covered their perception of students' activities, interests, and motivations towards learning Basic Graphic Design, applied learning models, learning media used, material considered difficult to learn with proof of completeness of student learning outcomes, as well as identification of learning resources needed by students in Basic Graphic Design courses.

The results of the analysis of the needs of the Basic Graphic Design teachers are as follows. (1) Students were less active in finding other learning resources for Basic Graphic Design. (2) Students were more interested in practical learning than theory. (3) Student learning motivation was low and only a few students were eager to learn. (4) Teachers commonly used tutorials as learning models. (5) No media was used in learning. (6) Material that is difficult for students to learn, including other vector image processors and processing bitmap images using learning software; this was evidenced by the completeness of student learning outcomes in the material at below 50%. (7) Learning resources should be developed to facilitate students' understanding of multimedia in the form of electronic modules (e-modules), which include the theory and work steps and learning through video tutorials.

3.1.2 *Student needs analysis*

The analysis of the requirements for electronic modules based on project-based learning provided to students was undertaken by distributing questionnaires that contained activities, the interests and motivation of learning, the constraints faced, the availability of learning resources that can be used, the material that is considered difficult to learn, and the learning resources desired by students.

The learning process of Basic Graphic Design at SMK 2 Sewon is a productive learning activity or practical learning, which begins by manually designing something using paper, stationery, and coloring. After the manual design is complete the item is designed using graphic design learning software, CorelDraw.

The results of the analysis of student needs in the Basic Graphic Design course were as follows (see Table 1). (1) 83.58% of students were less interested in Basic Graphic Design learning materials because the learning process that took place only guided the delivery of material or instructions given by the teacher. (2) 92.54% of students complained about learning difficulties in practical learning activities using graphic design learning software because there were no books, modules or student worksheets (LKS) that directly referred to the steps of graphic design. (3) Only 23.88% of students had the ability to operate learning software because students had never received graphic design learning in previous education levels. (4) 14.93% of students wanted a guidebook, 4.48% of students wanted a printed module, and 80.60% of students wanted a module electronics (e-module) as learning resources that could contain work steps and learning tutorial videos.

Based on Table 1, 83.58% of students were less interested in Basic Graphic Design. 92.54% of the students stated there were no learning resources available, and 76.12% of students did not have the ability to operate graphic design learning software. So it can be concluded that students need teaching materials that can be used in learning Basic Graphic Design.

Table 1. Results of the analysis of e-module need based on project-based learning for students.

Number	Statement	Answer	
		Yes	No
1	Students' opinions that they are less interested in learning Basic Graphic Design	83.58%	16.42%
2	Students' opinion that there is no learning source used for Basic Graphic Design	92.54%	7.46%
3	Students' opinion that their ability to operate graphic design learning software is low	76.12%	23.88%

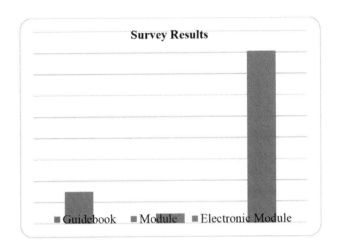

Figure 1.   Survey results of the learning resources needed.

Figure 1 shows that 14.93% of students needed a guidebook as a learning source, 4.48% of students needed a print module, and 80.60% of students needed an electronic module (e-module). So it can be concluded that students need learning materials in the form of electronic modules (e-modules) that can be used for Basic Graphic Design.

3.2   *Discussion*

Graphic design is a creative visual art discipline that covers many fields including art, typography, layout, information technology, and other creative fields (Ambrose & Harris, 2009). Basic Graphic Design includes productive topics consisting of compulsory practice and theoretical learning activities that must be taken by students of the Computer and Informatics Engineering Skills Program in class X, in odd or even semesters (RoI, 2017a).

This subject aims to develop basic knowledge and skills in delivering visual communication by integrating art and technology. In addition, Basic Graphic Design also provides basic competencies to achieved in graphic design science, graphic design components, and graphic design principles. Meanwhile, one of the basic competencies taught is vector image processing, which requires a cognitive understanding of the use of interface functions, tools, and features in CorelDraw learning software.

Productive learning activities or practices should be carried out by emphasizing the process of giving experience to students in integrating their initial knowledge with appropriate knowledge of the scientific concepts learned. The process of integrating this requires an intermediary or media (channel) so that messages or knowledge can enter into students' understanding. Intermediaries needed in 21st-century learning include learning media that are interesting, effective, and efficient by utilizing computer technology.

Learning media is everything that is used in channeling messages or information and can stimulate students' thoughts, attention, feelings, and willingness so that it can encourage the deliberate, purposeful, and controlled learning process (Miarso, 2009). The sophistication of computer technology at this time makes it possible to support the creation of multimedia-based learning media. Multimedia learning can be defined as a combination of various media that are packaged in an integrated and interactive manner of presenting certain learning messages (Warsita, 2008).

The use of multimedia as interactive teaching material in the learning process can present a learning material that contains multimedia components such as text, graphics, images, sound, video, and animation. However, the multimedia component must be processed and manipulated and integrated digitally using a computer or electronic device (Surjono, 2017).

Through multimedia learning, the learning process takes place in more interesting and interactive way, the quality of student learning can be improved, and the learning process can be carried out wherever and whenever, so that student learning attitudes can be improved (Daryanto, 2013). Students can also learn according to their abilities without being bound by time (Kustandi & Sutjipto, 2011).

Multimedia learning includes non-printed teaching materials that require computer assistance to display messages or information for learning (Prastowo, 2014). Digital-based modules or electronic modules (e-modules) are one of the multimedia learning tools that utilize computer technology in the presentation of learning materials in the form of text, graphics, images, sound, video, and animation. Thus, students have more interesting, creative and practical learning materials that can guide students in discovering the concept of learning independently (Bakri et al., 2016). The use of electronic modules (e-modules) can condition students to always be involved in meaningful learning experiences and always think about what they are doing so that active learning occurs. Active learning can be realized through project-based learning, which is integrated into the preparation of electronic modules (e-module). In addition, Maswan and Muslimin (2017) emphasize that the presentation of messages or knowledge in this information era will always use media, both electronic and non-electronic. Furthermore, Mukarom and Rusdiana (2017) stated that a module consisting of printed material, audio material, and audio-visual material is a system of delivering messages or information using various types of learning materials arranged to form a learning package.

Project-based learning is a teaching method that links technology with problems of daily life that are familiar to students, or through a school project (Warsono & Hariyanto, 2017). In the project-based learning process, students will design, solve problems, make decisions, and conduct their own investigation activities when completing a project. The end result of the project is in the form of artifacts, which include scientific work, a model, film, video, CD, DVD, and so on.

Basic Graphic Design course needs practical learning in understanding the concepts and principles, therefore the right multimedia learning source is in the form of project-based learning (e-module). Multimedia learning is needed in the Basic Graphic Design course because it can improve students' metacognitive understanding of the concepts and principles of graphic design through their own experience. Furthermore, Warsono, and Hariyanto (2017) emphasize that project-based learning can focus students on problems, which can motivate and encourage students to deal directly with the concepts and principles of knowledge as a hands-on learning experience. Thus, the development of electronic modules (e-modules) by applying project-based learning is needed to improve understanding of learning for vocational school students.

4 CONCLUSION

As many as 83.58% of students were less interested in Basic Graphic Design learning materials because the learning process that took place only guided the delivery of material or instructions given by the teacher. In practical learning activities using graphic design learning software, 92.54% of students experienced learning difficulties because there were no books, modules or student worksheets (LKS) that directly referred to the steps of graphic design. As many as 80.60% of students would like an electronic module (e-module) as a learning resource that can contain work steps and learning tutorial videos. Thus it is necessary to develop learning multimedia in the form of electronic modules (e-modules) based on project-based learning to create effective and efficient teaching materials for vocational education students.

ACKNOWLEDGMENT

Gratitude is directed to SMK Negeri 2 Sewon for giving permission to conduct the research and to Dr. Sunaryo Soenarto, M.Pd. who gave his guidance in the research process.

REFERENCES

Ambrose, G., & Harris, P. (2009). *The fundamentals of graphic design.* United Kingdom: AVA Publishing.

Bakri, F., Permana, A.H., & Siahaan, B.Z. (2016). Pengembangan modul digital fisika berbasis discovery learning pada pembelajaran fisika SMA [Development of digital physics modules based on discovery learning in high school physics learning]. *Prosiding Seminar Nasional Fisika dan Aplikasinya*, 228–229.

Daryanto. (2013). *Media pembelajaran: peranannya sangat penting dalam mencapai tujuan pembelajaran* [Learning media: their role is very important in achieving learning goals]. Yogyakarta, Indonesia: Gava Media.

Kustandi, C., & Sutjipto, B. (2011). *Media pembelajaran: Manual dan digital* [Learning media: Manual and digital]. Bogor, Indonesia: Ghalia Indonesia.

Maswan, & Muslimin, K. (2017). *Teknologi pendidikan: Penerapan pembelajaran yang sistematis* [Educational technology: The application of systematic learning]. Yogyakarta, Indonesia: Pustaka Pelajar.

Miarso, Y. (2009). *Menyemai benih teknologi pendidikan* [Seed the seeds of educational technology]. Jakarta, Indonesia: Kencana.

Mukarom, Z., & Rusdiana, A. (2017). *Komunikasi dan teknologi informasi pendidikan: Filosofi, konsep, dan aplikasi* [Communications and educational information technology: Philosophy, concepts, and applications]. Bandung, Indonesia: Pustaka Setia.

Prastowo, A. (2014). *Panduan kreatif membuat bahan ajar inovatif: Menciptakan metode pembelajaran yang menarik dan menyenangkan* [A creative guide to create innovative teaching materials: Creating interesting and fun learning methods]. Yogyakarta, Indonesia: Diva Press.

Putra, K.W.B., Wirawan, M.A., & Pradnyana, G. A. (2017). Pengembangan e-modul berbasis model pembelajaran discovery learning pada mata pelajaran sistem komputer untuk siswa kelas X multimedia SMK negeri 3 Singaraja [Development of e-modules based on discovery learning learning models on computer system subjects for multimedia class X students of State Vocational 3 Singaraja]. *Jurnal Innovation of Vocational Technology Education*, 1–41.

RoI. (1990). *Peraturan Pemerintah Nomor 29 Tahun 1990Tentang Pendidikan Menengah* [Government Regulation No. 29 of 1990 concerning Secondary Education]. Jakarta, Indonesia: Republic of Indonesia.

RoI. (2017a). *Surat Keputusan Direktur Jenderal Pendidikan Dasar dan Menengah Nomor: 330/D.D5/KEP/KR/2017, Kompetensi Inti dan Kompetensi Dasar Mata Pelajaran Muatan Nasional (A), Muatan Kewilayahan (B), Dasar Bidang Keahlian (C1), Dasar Program Keahlian (C2), dan Kompetensi Keahlian (C3).* [Decree of the director general of primary and secondary education Number: 330/D.D5/KEP/KR/2017,core competencies and basic competencies (KI-KD) of national content subjects (A), regional content (B), basic expertise (C1), basic expertise programs (C2), and skill competence (C3)]. Jakarta, Indonesia: Director General of Primary and Secondary Education, Republic of Indonesia.

RoI. (2017b). *Kompetensi inti dan kompetensi dasar SMK/MAK kompetensi keahlian multimedia* [The core competencies and basic competencies of SMK/MAK are multimedia skill competencies]. Jakarta, Indonesia: The Directorate of Vocational Middle School Development, Republic of Indonesia.

Serevina, V., Sunaryo., Astra, M., & Sari, I.J. (2018). Development of e-module based on problem based learning (PBL) on heat and temperature to improve student's science process skill. *The Turkish Online Journal of Educational Technology*, 17(3), 27–31.

Sugianto, D., Abdullah, A.G., Elvyanti, S., & Muladi, Y. (2013). Modul virtual: Multimedia flipbook dasar teknik digital [Virtual module: Basic flipbook multimedia digital technique]. *Jurnal Innovation of Vocational Technology Education*, 9(2), 101–116.

Sukiman. (2012). *Pengembangan media pembelajaran* [Development of learning media]. Yogyakarta, Indonesia: Pedagogia.

Surjono, H.D. (2017). *Multimedia pembelajaran interaktif: Konsep dan pengembangan* [Interactive multimedia learning: Concepts and development]. Yogyakarta, Indonesia: UNY Press.

Warsita, B. (2008). *Teknologi pembelajaran: Landasan dan aplikasinya* [Learning technology: The foundation and application]. Jakarta, Indonesia: Rineka Cipta.

Warsono, & Hariyanto. (2017). *Pembelajaran aktif* [Active learning]. Bandung, Indonesia: Remaja Rosdakarya.

Wijayanto, & Zuhri, M.S. (2014, September). Pengembangan e-modul berbasis flip book maker dengan model project based learning untuk mengembangkan kemampuan pemecahan masalah matematika [Development of flip book maker-based e-modules with project based learning models to develop mathematical problem solving abilities]. *Prosiding Mathematics and Sciences Forum*, 625–628.

Analysis of students' learning readiness in terms of their interest and motivation in achieving students' critical thinking skills

R. Putri & A. Ghufron
Universitas Negeri Yogyakarta, Yogyakarta, Indonesia

ABSTRACT: The purpose of this study was to determine students critical thinking skills as seen from their readiness to learn, in terms of interests and motivations, especially in the scope of science. This study aims to describe the following: (1) the learning readiness of high school students to achieve critical thinking skills in terms of their interests and motivations; (2) the factors that influence learning readiness of high school students. The research method used was descriptive qualitative. The results of this study are viewed from students' learning interests in achieving critical thinking skills. Students showed that they did not like the learning carried out (35.5% had interest in the low category) and only 30.8% were motivated in completing learning assignments (low category); this has a strong influence on the learning process and 27.6% of students' critical thinking skills were still low. The factors that influence high school students' readiness to learn originate from the students themselves, whether or not the learning is achieved, how to convey the learning, and whether the students' primary attention is devoted to activities outside of school.

1 GENERAL INTRODUCTION

1.1 *21st century*

The development of an increasingly competitive era is requiring 21st-century skills, which has an impact on the world of education. The challenges of the 21st century reflect the four skills competencies that must be mastered by students: critical thinking and problem solving, creativity and innovation, communication, and collaboration (Ministry Education and Culture, 2017). Responding to these challenges, especially in biology subjects, requires students to develop problem-solving abilities and critical thinking skills which include analysis, evaluation, explanation, inference, and interpretation, to master understanding or knowledge of biological material and meet the learning objectives of biology.

Learning objectives in the 21st century, are as follows: (1) develop the talents, interests, and potential of students so that they are characterized, competent, and literate; and (2) provide learning in each subject related to competence and context that encourages students to have thinking skills from the simple to the higher-order thinking process. To achieve these results requires a learning experience that varies from simple to complex applied in the learning process in the classroom (Ministry Education and Culture, 2017). In the learning objectives, the teacher must carry out learning and assessment relevant to the learning characteristics of the curriculum-2013, by creating a learner-centered process. The learning process is expected to be effective and efficient and focus on the importance of learning readiness of students in achieving predetermined learning goals.

Learning readiness can be seen from students' interest in learning and their motivation to achieve critical thinking skills, where learning are two things that cannot be separated. The educational process developed by the teacher must be able to create a comfortable atmosphere to develop the potential of students to the expected knowledge, skills, and attitude competencies. This will have an impact on learners' readiness in achieving critical thinking skills.

1.2 *Interest*

Students' interest in the learning process has a very important role in the world of education. It is one of the factors that allows students to be more focused and more enthusiastic so that they are not easily bored in the learning process and will not find it easy to forget their efforts to learn so will achieve learning goals.

Interest specifically refers to the learning activities that exist without any outside influences, where this interest will make students want to pursue the learning activities carried out. This opinion is supported by Slameto (2014), who stated that interest is a feeling of preference and a sense of interest in learning activities without being forced Interest can be interpreted in two ways: (1) interest as a cause, where the driving force makes individuals pay attention to certain things, situations or activities and not others; and (2) interest as a result, where effective experience is stimulated by something or because of participation in an activity.

1.3 *Motivation*

It is possible to assess students'motivation for achieving a critical thinking skill, where motivation is a complex part of students' psychology and behaviors that can influence how they choose to spend their time, how much time they take to complete an assignment, how they think, and how long the students survive on the assignment given. This is in accordance with intrinsic motivation that occurs when students carry out learning activities to meet their satisfaction related to learning activities themselves (Ryan & Deci, 2000).

Motivation is a process that supports and maintains behaviors that lead to a goal (Schunk, 2012). Motivation is described as the reason why someone chooses another action with great energy and frequency (Lin-siegler, Ahn, & Luna-Lucero, 2016). Motivation is a force that encourages a person to act or behave in a certain way (Rudhumbu, 2014). Furthermore, the learning motivation of students to achieve the goals of critical thinking skills will encourage the basic behavior in learning activities.Then they will try to maintain learning activities so they can achieve learning goals. Motivation is one of the determinants of learning success and has a significant impact on behavior and student learning processes (Alkis & Temimzel, 2018).

1.4 *Critical thinking*

Critical thinking is one of the skills that students should form. At the high school level, students are expected to have critical thinking skills through a scientific approach as self-development, which is learned in education units, so it is necessary to determine student learning readiness that can be viewed from their interests and motivations (Ministry Education and Culture, 2017).

Critical thinking skills are part of an intellectual model that is very important for students as a fundamental part of self-maturity. Therefore, it is necessary to determine the readiness of students to gain critical thinking skills, in terms of their interests and motivations. In accordance with Ennis (1993), critical thinking depends on two dispositions: attention to "being able to do it right", as far as possible and caring to present an honest position and clarity; the evaluation process used (applying criteria to assess possible answers), both implicitly and explicitly.

According to Gambril and Gibbs (2009), "critical thinking involves the use of standards such as clarity, accuracy, relevance, and completeness". It requires evaluating evidence, considering alternative views, and being genuinely fair-minded in presenting opposing views, which means that critical thinking includes work, accuracy, relevance, and completeness. Increasing critical thinking requires the ability to evaluate evidence, consider alternative perspectives, and think fairly in conveying different opinions appropriately.

Research conducted by Susilowati, Sajidan, and Ramli (2017) on students' critical thinking skills – interpretation, inference, explanation, analysis, and evaluation – showed that the average critical thinking skills of students is 51.60%, in the low category. Interpretation aspects scored 54.87% in the low category, analysis aspects scored 46.56% in the low category, and evaluation scored 54.58% in the low category. Students'critical thinking skills are still low, so the teacher is expected to be able to design a process of learning activities that can empower students' critical thinking skills.

This is also supported by research conducted by Lestari, Tindangen, and Akhmad (2016) in analyzing students' critical thinking skills – interpretation, inference, explanation, analysis, and evaluation – which found that the students' interpretation indicators was in the moderate category with the proportion of 53.10%, the explanatory ability indicator was in the moderate category with the proportion of 51.17%, the student evaluation indicators was in the moderate category with the proportion of 51.63%, the analysis indicators was in the poor category with the proportion of 34.55%, and the evaluation indicators are in the moderate category with the proportion of 48.51%. The results of the research show that students' critical thinking skills have not been optimally developed in the learning process and it is necessary to review students' learning readiness in terms of interest and motivation to achieve critical thinking skills. Fisher (2007) states that critical thinking is a method of thinking about any matter, substance or problem, where the thinker improves the quality of their thinking by handling skillfully the structures inherent in thought and applying intellectual standards, so that the only way to develop one's critical thinking ability is through thinking about one's own thoughts and consciously trying to improve them by referring to some good thinking models in that field.

The results of the research show that: (1) students' critical thinking skills have not been optimally developed in the learning process; (2) the learning process has not used the right learning strategies; and (3) profile of students' critical thinking skills is low, so that the teacher is expected to be able to design learning activities that can empower students' critical thinking skills. Based on the results, students' critical thinking skills must be developed or honed optimally in the learning process, so they are able to adapt to the demands of the 21st century and develop and measure their level of critical thinking skills.

Critical thinking is a "reasonable reflective thinking that is focused on deciding what to believe and do", which is reflective thinking that makes sense and is focused on deciding what to believe and do, by making a point that the components are not criteria, and assessment can be done mechanically. This is an important point about how critical thinking is related to teaching and learning (Eniss, 1993).

Facione (2011) stated that critical thinking is a complex thinking process consisting of analysis, evaluation, explanation, inference, interpretation, and self-regulation. To see the learning readiness of students to achieve critical thinking skills, the first task is to review the learning interests and motivation of the students themselves.

This research describes students' learning readiness in terms of their interest and motivation in achieving critical thinking skills in Senior High School 2 Bantul. The aspect of this research study refers to two components of the learning variable: (1) the initial conditions, which includes the objectives and characteristics of the study area, constraints in the implementation of the learning process, and the characteristics of students; (2) the learning methods, which includes the strategy for organizing the delivery and management of learning conducted by the teacher. These two variables are interrelated.

The analysis of each component is expected to show student learning readiness in terms of their interest and motivation in achieving critical thinking skills, and an in-depth description relating to the two variables that are factors that influence student learning readiness.

Based on the descriptions, the purpose of this study is to describe the analysis of and the factors that influence student learning readiness in terms of their interest and motivation in achieving critical thinking skills in Senior High School 2 Bantul.

2 METHOD

This study was conducted using a qualitative approach with the descriptive qualitative method. The data obtained were analyzed with a descriptive analysis approach by providing reviews and interpretations of the data obtained so that it becomes clear and meaningful.

The research procedure began with developing the research instruments – interviews and in-depth observations. The study was conducted on August 13 as a survey and continued from August 20 to August 24, 2018. The subjects were 50 students and two biology classes at Senior High School 2 Bantul. The object of this research is learning readiness in terms of students' interests and motivation to achieve critical thinking skills.

The study focused on students' learning readiness in terms of their interest and motivation in achieving critical thinking skills in the perspective of learning technology, by revealing two learning variables: initial conditions and learning methods (Degeng, 2013).

Primary and secondary data were collected through interviews, observations, and documentation. Primary data was obtained through interviews with biology teachers at Senior High School 2 Bantul related to student learning readiness and observations were made to collect data before and after the students undertook learning using learning models recommended in the 2013-curriculum. Secondary data were obtained from the sources in the form of archives, papers, books, student learning notes, and student learning outcomes. The documentation study was conducted by examining documents in the form of school profiles, as well as learning process activities that took place in the classroom.

The results obtained by Miles and Huberman were analyzed by three data reduction paths, data presentation, and conclusion drawing. The next step was the triangulation technique. Sugiyono (2011) explained that there are four types of triangulation examination techniques that utilize the use of sources, methods, investigators, and theories. With the triangulation technique, the data obtained will be more reliable and avoid bias.

3 RESULTS AND DISCUSSION

The research on the analysis of student learning readiness in terms of their interest and motivation in achieving critical skills thinking was based on two variable components: the initial conditions of learning and learning methods and revealing the factors that influence the student learning readiness. The results are summarized in Table 1.

Based on the data obtained, it is clearly seen that the average score for learning interest of the students was in the low category at 35.5%; motivation was in the low category at 30.8%; and critical thinking skills was in the low category at 27.6%. This means that students were not actively involved during the learning process. During the discussion activities, the students did not respond well, were chatting with their peers, and when the teacher explained the students did not record what the teacher was saying. Thus, it was difficult for the students to achieve critical thinking skills because it requires learning readiness. This is supported by Slameto (2014), who stated that interest is the basis of the learning process.

Table 1. The readiness of students' interest, students' learning motivation, and students' critical thinking skills.

Aspect	Number of students	Total score	Average	Category
Interest	142	400	35.5%	Low
Motivation	108	350	30.8%	Low
Critical thinking skills	69	250	27.6%	Low

This is supported by the results of interviews with two biology teachers who said that critical thinking skills that include interpretation, inference, explanation, analysis, and evaluation have not been optimally developed in the learning process. Learning activities in schools have not fully used strategies to develop critical thinking, so students are less motivated to follow the learning process. Many students who experienced difficulties in learning biology were less actively involved in the learning process.

Many students only copied the answers of friends, some collaborated with their peers, and others did not work at all. The students only received what the teacher said, and there was a lack of analytical skills in answering the questions given by the teacher during the learning process, so that students could not improve their critical thinking skills.

The teacher said that student learning readiness in terms of their interest and motivation toward critical thinking skills was still low. To achieve critical thinking skills, the key thing to improve was the students' interest and motivation, because if they liked the learning, they completed the assignments willingly, mastering critical thinking skills.

According to Mahapoonyanont (2012), the things that influence students' critical thinking skills are educational factors, the students themselves, child development and personal factors. Educational factors are related to learning strategies; the learner factors are the results of learning done, willingness to find out, read, and especially the self-motivation to carry out learning; child development factors and personal factors consist of personal status, attitudes, and child care.

Factors that influence high school student learning readiness originate from the students themselves, whether or not the learning is learned, how to deliver the learning, and whether the main attention of students is devoted to activities outside of school. The students' interest and motivation in the process of learning in school is still low, so the role of the teacher in the learning process is very important, to direct students in the learning process to achieve learning goals. Thus the students not only master the learning material but are also able to master critical thinking skills (Miarso, 2004).

For achieving critical thinking skills in the learning process, of course, we need a learning model that is appropriate to the critical thinking skills goals or predetermined learning goals. This is in accordance with Slavin (2005), who stated that it is necessary to choose a learning model oriented to the learning objectives and adjusted to the type of material and characteristics of students.

Teaching students how to think critically is a major problem in educational settings because critical thinking is key to participating effectively in a democratic society.The presence of decision-making skills has a direct impact on clinical judgment in the workplace.

Critical thinking is a way of thinking and a set of skills that drive an informed, conscious, systemic, considered and logical approach to deciding what to believe or do. Critical thinking leads to valid and proven arguments and conclusions.

The benefits of critical thinking include recognizing bias to guide self-development, contributing to study groups inside and outside the classroom, developing the best solutions to problems, gaining a better understanding of the arguments of others, giving good arguments, to create commitment to self-thinking, identifying important topics by staying focused on existing problems, and writing and speaking with relevant evidence (Feldman, 2010).

4 CONCLUSION

The average presentation of learning interest, motivation, and critical thinking skills of students of Senior High School 2 Bantul was found to be low. Students showed that they did not like the learning carried out (35.5% had interest in the low category) and only 30.8% were motivated in completing learning assignments (low category); this has a strong influence on the learning process and 27.6% of students' critical thinking skills were still low.

Factors that influence high school students' readiness to learn originate from the students themselves, whether or not the learning is learned by students, how to convey the learning, and the students' primary attention is devoted to activities outside of school.

REFERENCES

Alkis, N., & Temizel, T.T. (2018). The impact of motivation and personality on academic performance in online and blanded learning environments. *Journal of Educational Technology and Society*. 21(3), 35–47.

Degeng, N.S. (2013). *Ilmu pembelajaran: Klasifikasi variabel untuk pengembangan teori dan penelitian* [Learning science: Variable classification for theory and research development]. Bandung, Indonesia: Arasmedia.

Ennis, R.H. (1993). Critical thinking assesment. *Theory into practice, 32*(3), 179–186.

Facione, P.A. (2011). *Critical thinking: what it is and why it counts*. California, CA: Measured Reasons LLC and The Californias Academic Press.

Feldman, A.D. (2010). *Critical thinking*. USA: Crisp Publications.

Fisher, A. (2007). *Critical thinking. un introduction*. Cambridge: Cambridge University Press.

Gambril, E., & Gibbs, L. (2009). *Critical thinking for helping professionals: a skill-based workbook third edition*. New York, NY: Oxford University Press.

Ministry Education and Culture. (2017). *Panduan implementasi kecakapan abad 21 kurikulum 2013 di sekolah menengah Atasv* [Guide the implementation of 21st century skills in the 2013 curriculum in high school]. Jakarta, Indonesia: Ministry Education and Culture.

Lin-Siegler, X., Ahn, J.N., & Luna-Lucero, M. (2016). Even Einstein struggled: Effects of learning about great scientists struggles on high school student's motivation to learn science. *Journal of Educational Psychology, 108*(3), 314–328.

Lestari, A., Tindangen, M., & Akhmad. (2016). Analisis kemampuan berpikirkritis dengan model pembelajaran inkuiri pada pembelajaran Biologi kelas VII-A SMP Negeri 3 Long Kali tahun ajaran 2015/2016 [Analysis of critical thinking skills with inquiry learning models in Biology learning in class VII-A SMP Negeri 3 Long Kali, 2015/2016 academic year]. *Prosiding Seminar Nasional II Biologi, Sains, Lingkungan, dan Pembelajaran*, 355–373.

Mahapoonyanont, N. (2012). The causal model of some factors affecting critical thinking abilities. *Procedia-social and behavioral science* 171, 1255–1264.

Miarso, Y. (2004). *Menyemai benih teknologi pendidikan* [Seed the seeds of educational technology]. Jakarta, Indonesia: Preda Media.

Rudhumbu, N. (2014) Motivational strategies in the teaching of primary school mathematics in Zimbabwe. *International Journal of Educational Learning and Development UK, 2*(2), 76–103.

Ryan, R.M., & Deci, E.L. (2000). Intrinsic and extrinsic motivations: classic definitions and new directions. *Contemporary Educational Psychology*. 25(2), 54–67.

Schunk, D, H. (2012). *Learning theories* (Vol. 53). Boston, MA: Pearson Education, Inc.

Slameto. (2014). Developing critical thinking skills through school teacher training "training and development personnel" model and their determinants of success. *International Journal of Information and Education Techonology, 4*(2), 161–166.

Slavin, E.R. (2005). *Cooperative learning*. Bandung, Indonesia: Nusamedia.

Sugiyono. (2011). *Statistika pendidikan* [Educational statistics]. Bandung, Indonesia: Tarsito.

Susilowati, Sajidan, & Ramli, M. (2017, October). Analisis keterampilan berpikir kritis siswa madrasah aliyah negeri di KabupatenMagetan [Analysis of critical thinking skills in state madrasah aliyah students in Magetan Regency]. *Prosiding Seminar Nasional Pendidikan Sains (SNPS)*, 249–257.

Innovative Teaching and Learning Methods in Educational Systems – Retnowati et al. (Eds)
© 2020 Taylor & Francis Group, London, ISBN 978-1-03-224183-8

Study on a test scoring system for vocational secondary schools using Computerized Adaptive Testing (CAT)

F.P. Marsyaly & S. Hadi
Universitas Negeri Yogyakarta, Yogyakarta, Indonesia

ABSTRACT: This study aimed to determine: (1) the effectiveness of paper-based tests (PBT) and computer-based tests (CBT), (2) the best test assessment system, and (3) whether Computerized Adaptive Testing (CAT) is suitable for use in vocational secondary schools. This study is a literature review project focusing on a "Test Scoring System". The sources reviewed in this study come from four major sources: (1) academic books from libraries, (2) peer-reviewed educational research journal articles from digital databases, (3) government documents online, and (4) other online electronic resources. The result of this study is a recommendation to create a test scoring system based on item response theory: Computerized Adaptive Testing (CAT). This test scoring system is using item response theory to know student's ability.

1 INTRODUCTION

The development of technology and information had caused major shifts in several sectors, including the education sector. One result of the development of technology and information in the education sector lies in the assessment system for students' learning outcomes. The student learning outcome assessment system has moved from a paper-based test (PBT) assessment system to the computer-based test (CBT). The Minister of Education and Culture Regulation of the Republic of Indonesia Number 3 of 2017 concerning Assessment of Learning Outcomes by the Government and Assessment of Learning Outcomes by Education Units Article 9 Paragraph 1states that the implementation of the National Final Exam will be carried out vie a computer-based system. In line with the Minister of Education and Culture Regulation, Muhammad, as written by Widiyanto (2017), announced that in 2018 the Government would implement a Computer-Based National Final Exam for all 12th grade students at high school or vocational secondary school. The implementation of the Computer-Based National Final Exam wa s attempted to reach 100 percent. Therefore, the assessment of learning outcomes using computer assistance can be used for evaluating learning outcomes in class in an attempt to adjust to these technological and information developments.

However, the PBT assessment system is still widely used in the evaluation process of students in schools. The application of the PBT assessment system means that each student hasthe same amount of items and work time, regardless of the ability of each student. The essence of PBT assessment is that the test instrument given to each student is uniform, even though each student has different abilities. The assessment process means that students with high abilities will receive a few easy questions and there is a small chance of making errors, whereas students with low abilities will face some difficult questions with only a small chance of answering correctly. Therefore, the answers generated by students through PBT assessment have not maximally reflected the ability of each student.

In an effort to determine the ability of each student, the teacher can use a computerized adaptive testing (CAT) assessment system in the evaluation process of students' learning outcomes. Adaptive means that the items given to each student are adjusted to their level of ability with the CAT system (Lord, 1990). This allows each student to receive a different number of items, face different content, and complete the work at their own pace. The CAT appraisal system is also more effective and accurate than the PBT assessment model because there is almost no possibility of collaborating among students.

The application of CAT with the Triangle Decision Tree method or later can be abbreviated to CAT-TDT in the evaluation of school learning conducted by Winarno (2012). It is known that the possibility of a student asking other students the answers is very small, because the items seen by each student are not the same but based on the student's ability.

Therefore, an adaptive assessment model for learning competency skills in electric power installation engineering could be used to measure and assess students' understanding and mastery of existing competencies, and the achievement of learning objectives.

The objectives of this study are (1) to determine the effectiveness of PBT and CBT, (2) to find out the best test assessment system, and (3) to find out whether CAT is suitable for use in vocational secondary schools.

2   TEST SCORING SYSTEM

Tests or exams are used as an instrument for determining or the result of what participants learn. Thus, the teacher can form the next learning strategy. The test assessment methods used widely at present are paper-based tests (PBT), computer-based tests (CBT), and computerized adaptive testing (CAT).

PBTs are widely used for measuring someone's ability. The fundamental basis of this test is that teacher gives all the participants the same amount of questions. After all of the questions are attempted by the participants, the teacher then corrects all their answer and all of the participants can see their results from the test. In the same way, a computer-based test can give the same questions to all the participants. But, there are some differences, in that if a PBT can present the same questions for each participant, a CBT can present offer different questions to each participant. The result can still be seen immediately after completion.

CBTs are one of test methods managed by computers, from common to specific networks, or with technology peripherals connected to the internet with almost tests utilizing multiple choice questions (Bolboacă & Jäntschi, 2012). Furthermore, Cantillon, Irish, and Sales (2004) stated that there are two types of CBT. In the first method, test participants fill the answer or give a response on paper format, then is input to computer for correction. In the second method, CBT offers an assessment interface, and participants answers the questions on the computer, then receive feedback from the computer itself.

Brown, Race, and Bull (1999) cited by Jamil, Thariq, and Shami (2012) revealed that computer-based examinations can be used to promote more effective learning by testing various skills, knowledge, and understanding. Although a computer-aided test seems better than a paper test, the facts says differently. Lee and Hopkins (1985), as reported by Russell, Goldberg, and O'connor (2011), found that the average score from an arithmetic reasoning PBT was significantly higher than the average score from the same test via CBT.

Hadi (2015) explained that there were two approaches for analyzing the result of test items: the classical test, and item response theory. Allen and Yen (1979) explained that there was strong support for an assessment series from standard classical test theory, including basic CBT and PBT.

Hambleton (2004) explained that there were two basic disadvantages with classical test theory. The first was classic grain statistics, item difficulty level, and item discrimination based on the sample from certain participants. From this, we know that some items seem easy for participants with higher abilities and difficult for low ability participants. The classical test theory is not suitable to compare participants' performance - with a limited test score where the participants were given the same test or parallel test.

Bodmann and Robinson (2004) found that computer-based assessment was done faster by participants than paper-based assessment without any difference in scores. Piaw (2011) explained that the assessment system viq the computer is more "green" and is effective at reducing paper consumption. The result shows that this research significantly affects the assessment time and motivate the assessment for CBT mode. CBT mode is more reliable in internal and external validity terms, decreases the assessment time, and increases the participant's motivation for doing the test.

3 COMPUTERIZED ADAPTIVE TESTING

Computerized adaptive testing is as a modern assessment system with a vital component called Item Response Theory (IRT). IRT underlies the test items framework calibration, test participant's ability estimation, and item selection. Hol, Vorst, and Mellenbergh (2008) explained that computerized psychological assessment can modernize administration procedures and cannot be implemented in paper and pencil tests. Bagus (2012) explained that the CAT was more efficient than paper and pencil tests timewise because fewer items were presented. Conversely, a CAT system could be effective and adjust the items to the participant's ability.

From Wiess and Kingsbury's (1984) explanation of Wright (1992), the CAT model was an adaptive mastery test designed for determining the test participant's ability level, on top of a base level determined at the start. Hambleton (2004) said that CAT was a test managed by a computer, where items were managed and depended on the participant's achievement on the item that was given before. Test participants with a good level of achievement would receive harder items, and test participants with poorer levels of achievement would receive easier items. One reliability of the CAT method is explained in Economides and Roupas (2007): a CAT system's priority was security, reliability, and maintainability.

Schematic illustrations from the CAT system are shown in Figure 1. First, participants receive the first question at a moderate level. After that, the system will decide that the first answer is right or wrong and calculates the participant's ability. If participants get the wrong answer, the difficulty level of the second question will decrease, but if participants get the right answer, the difficulty level of the second question will increase. That method will loop until the system meets the stopping rule. At the end of the test, the level of each participant's ability will be available.

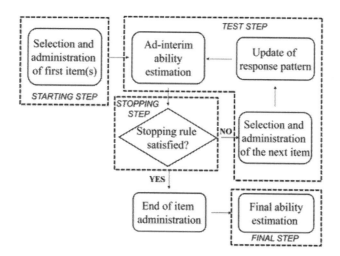

Figure 1.   CAT process schematic illustration (Magis, Yan & Von Davier, 2017).

4 COMPARISON OF THE STRENGTHS AND WEAKNESSES

4.1 *Advantages of the conventional test scoring system*

Magis, et al. (2017) – Linear Test

- Easily built
- Easily managed
- No extra effort needed to develop the test

  Cantillon, et al. (2004) – CBT
  Educational aspects

- Saves time for teachers by marking automatically and doing factor analysis
- Test participants get feedback after the end of the test
- The teacher can easily track the performance of each participant
- The test can be done or delivered one time in many sites
- Teacher receive will feedback and evaluation data
- Variety of media used (video, graphics, etc.)

  Administration aspects

- Possible objective marking without human bias
- Automatic and precise marking
- Fast to assemble from questions saved in the computer

4.2 *Disadvantages of the conventional test scoring system*

Magis, et al (2017) – Linear Test

- Long test length
- Not efficient for measurement
- Rigid test schedule for test participants
- Prone to cheating

  Cantillon, et al. (2004) – CBT
  Educational aspects

- Often used in objective test types, because essay tests are harder to correct
- Need particular skills in developing questions until they are stored
- Cannot distinguish participants with good ability for using the computer and good knowledge in the test material
- Hard to watch all participants because of side by side sitting positions
- Prone to plagiarism if participants can access the internet or their e-mail
- Need a reliable and secure sending system

  Administration aspects

- Expensive and need much time for adjustment.
- Must prepare a backup test if a failure occurs on hardware and network peripherals
- Staff must be trained properly before they can manage the test

4.3 *Advantages of the modern test scoring system*

Magis, et al. (2017)

- Short test size
- Efficient for measurement
- Flexible test schedule for test participants.
- Prevent Cheating

Economides & Roupas (2007)

- Secure
- Reliable
- Maintainable

Thompson (2011)

- Shorter test mode
- Same precise level for all participants
- More precise because many questions have been developed already
- Give different experiences for each test participantIncreases motivation
- Safe
- Frequent re-testing
- Better than CBT (non-adaptive computer-based training)

4.4 *Disadvantages of the modern test scoring system*

Magis, et al. (2017)

- Complicated implementation
- Built from strong model assumptions
- Need large set data calibration
- Need extra effort to develop the test
- Hard item exposure control
- Expensive computer management
- Robustness concerns

Economides & Roupas (2007)

- Limited feedback to participants
- Less attractive items presentation
- Does not offer support and advance functionality

Thompson (2011)

- No review
- Item exposure
- Recovery of poor starts
- Public relations
- Requires calibration
- Requirements

5 CONCLUSION

In this study, it was found that (1) the effectiveness of PBT and CBTs are the same. This is because there is no difference in the test results of each test participant. In addition, the concept is almost the same – it only looks different. (2) The best test scoring system is computerized adaptive testing (CAT), which uses modern item analysis and technology. The technology used is a computer and the analysis used is the item response theory (IRT). In addition, test participants with high abilities will be challenged to answer questions correctly, while test participants with low abilities will not face items that do not match their abilities. (3) The test scoring system suitable for vocational secondary schools is CAT because it can be used to measure the skills, knowledge, and abilities of test participants, while the learning process can be maximized.

REFERENCES

Allen, M., & Yen, W.M. (1979). *Introduction to measurement theory*. Monterey, California: Brooks.

Bagus, H.C. (2012). *The national exam administration by using computerized adaptive testing (CAT) model*. Jakarta Pusat, Indonesia:Puspendik Balitbang Kemdikbud.

Bodmann, S.M. & Robinson, D.H. (2004). Speed and performance differences among computer-based and paper-pencil tests. *Journal of Educational Computing Research, 31*(1), 51–60.

Bolboacă, S.D. & Jäntschi, L. (2012). Computer-based testing on physical chemistry topic: A case study. *International Journal of Education and Development using Information and Communication Technology (IJEDICT), 3*(1), 94–104.

Cantillon, P., Irish, B., & Sales, D. (2004). Using computers for assessment in medicine. *BMJ, 329*(7466), 606–609.

Economides, A.A., & Roupas, C. (2007). Evaluation of computer adaptive testing systems. *International Journal of Web Web-Based Learning and Teaching Technologies*.

Hadi, S. (2015). *Karakteristik instrumen* [Characteristics of instruments]. Yogyakarta, Indonesia: Yogyakarta State University.

Hambleton, R.K. (2004). Theory, methods, and practice in testing for the 21st century. *Psicothema, 16*(4), 696–701.

Hol, A.M., Vorst, H.A.M., & Mellenbergh. (2008). *Computerized adaptive testing of personality traits*. The Netherlands: Hogrefe & Huber Publishers.

Jamil, M., Thariq, H., & Shami P.A. (2012). Computer-based VS paper-based examinations: Perceptions of university teacher. *The Turkish Online Journal of Educational Technology (TOJET), 11*(4), 371–381.

Lord, F.M. (1990). *Application of item response theory to practical testing problem*. Hillsdale, NJ: Lawrence Erlbaum Associates.

Magis, D., Yan, D., & Von Davier, A.A. (2017). *Computerized adaptive and multistage testing with R*. Switzerland: Springer.

Piaw, C.Y. (2011). Comparisons between computer-based testing and paper-pencil testing: Testing effect, test scores, testing time, and testing motivation. *Proceedings of the Informatics Conference* (pp. 1–9).

RoI. (2018). *Peraturan Menteri Pendidikan dan Kebudayaan Republik Indonesia Nomor 3 Tahun 2017 tentang penilaian hasil belajar oleh satuan pendidikan* [Regulation of the Minister of Education and Culture of the Republic of Indonesia Number 3 of 2017 concerning Study of Learning Outcomes by the Government and Evaluation of Learning Outcomes]. Jakarta, Indonesia: Ministry of Education and Culture, Republik of Indonesia.

Russell, M., Goldberg, A., & O'connor. (2003). Computer-based testing and validity: a look back into the future. *Assessment in Education: Principles, Policy & Practice, 10*(3), 279–293.

Thompson, N.A. (2011). *Advantages of computerized adaptive testing (CAT)*. USA: Assessment System Worldwide.

Weiss, D.J. & Kingsbury, G. (1984). Application of computerized adaptive testing to educational problems. *Journal Educational Measurement, 21*(4), 361Computer-based testing and validity: a look back into the future.375.

Widiyanto, D. (2017, September 19). *2018, pemerintah target laksanakan UNBK* [2018, the government targets to implement UNBK]. Winarno. (2012, May). The results of computerized adaptive testing triangle decision tree (CAT-TDT) tests in improving honesty as a national character education value on learning evaluation in schools. *Paper presented at the National Seminar on Postgraduate Programs, Yogyakarta State University, Yogyakarta, Indonesia*.

Implementation of basic graphic design learning skills competence of multimedia in SMK Muhammadiyah Wonosari

L.F.A.N.F. Albana & Sujarwo
Yogyakarta State University, Indonesia

ABSTRACT: This study aims to describe (1) a learning plan of basic graphic design skills in multimedia vocational high schools, (2) the learning of basic graphic design skills, (3) the factors that affect implementing basic graphic design skills. The subject of this research consist of the head of Multimedia Expertise Competence, teachers and students. The research method used was descriptive qualitative. Data collection was done with observation, open questionnaire, interviews, and documentation study. The results revealed several findings. First, the learning plan developed by multimedia vocational teachers at a vocational high school in Yogyakarta, Indonesia, was designed according to curriculum-2013. Second, the implementation of learning conducted by the teacher in the classroom and laboratory includes preliminary activities, core activities, and closing activities. *Third*, the inhibiting factors experienced during the learning process for basic graphic design multimedia competency skills were the density of the material, a lack of time allocated for learning, the low level of student interest and self-regulated learning, no development of interesting teaching materials, and no self-regulation by the graphic design students. Supporting factors for learning basic graphic design multimedia skills include completeness of learning facilities and infrastructure supporting practical learning, and varied learning methods applied by the teacher.

1 INTRODUCTION

The process of delivering information or messages is developing rapidly in various ways, including verbal communication and visual communication. Graphic design is a form of visual communication that contains images to convey information or messages effectively. The process of delivering messages, from the planning, design, development, production, and communication stages, is inseparable from the role of technology in this. Current technological developments have a large influence on penetrating the boundaries of space and time.

The basic graphic design is one of the subjects that must be mastered by students of multimedia vocational high schools. Multimedia expertise skills are developing rapidly in line with an increase in digital technology, in the industrial revolution era 4.0. Basic graphic design is a subject for multimedia expertise competencies that can equip vocational students to face the ever-evolving world of work. This is in line with the objectives of vocational secondary education where it is key that students can develop their ability to do a particular task, in line with the statement in Government Regulation of RoI (1990). Vocational schools meet the needs of the community in channeling talents and abilities in a particular field. This is in accordance with one of the objectives described in RoI (2010), which is to equip students with the ability of science and technology and vocational skills of professions in accordance with the needs of the community.

Pertiwi and Ahmad (2017) concluded that there are several aspects and strategies of vocational schools in meeting the workforce needs in the industry: (1) attitude – graduates must have the confidence to complete a job; (2) knowledge – graduates must have procedural knowledge about industrial development methods.: (3) skill – graduates must have creativity in problem solving. There are two strategies that are appropriate for achieving competency in graduates: the external strategy to achieve graduate competencies based on expert opinions is the development of the teaching industry, while the internal strategy is the development of applied learning models.

One of the skills programs in great demand by students of junior high school graduates is Multimedia Skills Competency. Multimedia expertise competencies are implemented in vocational high schools to equip students with skills, knowledge, and competence, ensure they are friendly with a polite attitude, and they have faith and piety. This area of expertise in technology, information and communication is appealling and still dominates. There are many other things that become a factor in this expertise competency, starting from interesting subjects and leading to the creation of innovative and marketable products. Yogyakarta itself has 90 schools that offer multimedia expertise competencies (RoI, 2018, September 20). This is supported by Mukhadis, Putra, Nidhom, Dardiri, and Suswanto (2018), who revealed that the priorities in the western part of Indonesia are social and community services, agriculture, manufacturing or creative industries, and marketing. The creative industry and marketing are strong fields of work for vocational graduates, especially those with multimedia expertise competencies.

The specific objective of multimedia expertise competency is to equip students with skills, knowledge and attitudes in order to be competent in (1) operating digital illustration software and peripherals, digital imaging, and web design, (2) operating software and multimedia peripherals, presentation, 2D animation, and 3D animation, and (3) operating digital audio software and peripherals, digital videos, and visual effects as described in the revised curriculum-2013 spectrum. These goals are prepared with various considerations so that students will be able to face the increasingly developing challenges of the future.

Basic graphic design in the multimedia expertise competencies relate to the fields of information technology and communication expertise in computer engineering and informatography expertise programs. This subject emphasizes using tools, information and work procedures in accordance with the field and scope of work of basics graphic design. This skill competency is expected to produce superior graduates who are ready to work in the field of graphic design in the industrial/business world, and private institutions or companies (entrepreneurs).

Basic graphic design have core competencies that are specific to the third and fourth aspects of competence. Core competencies in the aspect of knowledge are understanding, implementing, analyzing, and evaluating factual, conceptual, basic, and metacognitive knowledge in accordance with the fields and scope of computer and informatics engineering. Determination of a course goal is of course adjusted to the curriculum as a guideline.

Changes in the curriculum in Indonesia greatly affect the implementation of multimedia competence in this vocational high school: the determination of subjects must follow the curriculum guidelines. Arvianto (2016) explained that all forms of curriculum operationalization will experience change. In essence, a change of curriculum is an effort to improve and refine the previous curriculum. The reforms made are a response to changing times, socio-cultural conditions of the community, and the rapid development of information technology (Artapati & Budiningsih, 2017). Many considerations become references to curriculum changes, such as challenges and future competencies, negative phenomena that are increasingly prevalent among education, and people's perceptions of the world of education.

Closely related to these considerations, there was a reshuffle of several areas in the curriculum. The impact of several revisions in the current curriculum makes things difficult for teachers when organizing learning. The current curriculum is a revised curriculum-2013 where improvements in substance are carried out on several aspects, including core competencies, basic competencies, the syllabus, evaluation of learning, and hours of learning (Alfa, 2017, May 19). Counseling and education and training are carried out by the government to ensure the implementation of the revised curriculum-2013 is successful. The journey to change the curriculum-2013 appears to not have included thorough planning. Socialization and education and training were held in three phases, in stages. This has resulted in delays in the implementation of the revised curriculum-2013 in several schools that participated in the three wave training. To date, there are still schools that have not implemented the revised curriculum-2013 in full. The government established a policy that the implementation of the curriculum adopted by vocational students from classes X-XII must remain the same. This causes schools to experience delays in education and training because they still have to teach curriculum-2013, to some students, especially students in class XII.

These changes have an impact on the basic subjects of graphic design for multimedia expertise competencies in vocational high schools. Curriculum changes will affect various components of education, such as principals, teachers, materials, facilities, infrastructure and various other educational devices. Basic graphic design learning will be successful if all components synergize in achieving the educational goals listed in the curriculum. Various challenges and obstacles regarding basic graphic design learning are interesting for a learning technologist.

Referring to the definition of educational technology described by AECT (2008), namely the study and ethical practices in facilitating and improving performance through the creation, use and management of appropriate technologicalprocesses and resourcesthe position of an educational technologist in the implementation of basic graphic design learning is in the studies and ethical practices. Educational technologists can conduct studies on the learning of basic graphic design and examine the extent to which practices occur in the field. The study can be used as a reference for a learning technologist to make critical formulations and solutions regarding managing and using processes. Factors that are supporting and inhibiting the implementation of basic graphic design learning can be determined through the study of the learning processes and resources applied.

Based on this, it is necessary to conduct research on the implementation of learning basic graphic design on multimedia vocational skills competencies. Through the observation process, it was found that the vocational high school in Gunungkidul that offers basic graphic design is Muhammadiyah Wonosari Vocational School. This is one of the 90 vocational schools in Yogyakarta that has multimedia expertise competencies and carries out basic graphic design courses. The reason for choosing Wonosari Muhammadiyah Vocational Schools as a research subject because this school is a vocational high school that applies basic graphic design learning in its curriculum. Learning systems supported by facilities and infrastructure, with the latest specifications and professional multimedia teaching staff in accordance with their respective fields of expertise, support the implementation of learning activities to the maximum.

What is the implementation of basic graphic design learning in Muhammadiyah Wonosari Vocational School? Based on this question, observations and interviews were conducted to gather the initial data. As a result, there were several problems experienced by the teachers. Basic graphic design learning has an impact on the teaching and learning process both from the aspects of planning and implementation. The following are pre-research results obtained from observations and interviews.

First, the teacher explained the lack of learning resources available in schools and that the utilization of facilities and infrastructure had not been carried out optimally. Moys, Collier, and Joyce (2018) revealed that in graphic communication learning student feedback showed that they wanted more support to develop practical learning. Second, the time allocated for basic graphic design learning is still less when it see from the student's learning burden. Third, the relation with student interaction shows that discipline is still relatively low. Fourth, the teacher explained that the motivation and interest of students toward learning is quite high but the independence of student learning is still relatively low.

Based on the description above, this study aims to (1) obtain information about the learning process of basic graphic design of multimedia skills competence from planning to implementation, and (2) obtain information about the inhibiting factors and supporting factors for the implementation of basic graphic design multimedia skills competencies at vocational high schools. The study was then carried out in Muhammadiyah Wonosari Vocational School.

2 METHOD

The approach used in this study is a qualitative approach, while the type of research suitable to describe the findings produced by this study is qualitative descriptive type.

The results of the research data were analyzed with a qualitative approach then presented in a descriptive narrative. Data analysis was carried out continuously from the beginning of collection until the end of data collection. Data collection in the study was obtained through observation, both of participants and non-participants, analysis of documents, and in-depth interviews.

This research was conducted in Muhammadiyah Wonosari Vocational School in March to April 2018. Subjects in this study were all those involved with the implementation of basic graphic design learning in Wonosari Muhammadiyah Vocational School: the head of multimedia expertise competence, teachers of multimedia subjects producing graphic tools of multimedia, totaling six people, and 28 students. The object of this research was the implementation of basic graphic design learning in Muhammadiyah Wonosari Vocational School. The implementation of learning used as an object was planning and implementing activities. The inhibiting factors and supporting learning processes were also part of the object of this study.

In a study it is necessary to have a data as a reinforcement of the final results of the research. The data was collected using the following techniques and instruments.

First, this study was conducted by observation in data collection techniques. Observations were carried out with passive participation in learning, so they were not directly involved in the implementation. Observations took place during the learning process, including the observation of preliminary activities, core activities, and closing activities. Possible obstacles and efforts by the teacher to overcome the obstacles encountered during the learning process were also included in the observation. In addition, observations were also made on learning planning, implementation of learning designed by the teacher, and complemented by observation guidelines for student activities in learning.

Second, this study also performed interview techniques in data collection in the field. The interview technique used in this study was semi-structured interviews. This type of interview is an in-depth interview, where the implementation is freer when compared to structured interviews. Before conducting interview activities, interview guidelines were prepared, with the aim that the process remains focused on the main objective. The interviewees selected by the researchers in this study were principals, K3 Multimedia, teachers producing multimedia, and students.

Third, documentation studies were used to support the use of observation and interview methods in descriptive research. The documentation study was carried out to obtain data regarding learning planning and implementation of student learning. Documenting of learning activities were carried out in the classroom and in the laboratory. Documents in the form of school profiles, syllabus subjects and learning implementation plans (RPP) were also collected to supplement the data in this study.

Data validity testing was done by observing persistence, data triangulation and auditing or checking data. In connection with this, the steps taken in testing the validity of the data were as follows.

First, the perseverance of observation intended to find the characteristics and elements in situations that were relevant to the problem and then focus on these things in detail. Next, looking for consistent interpretations in various ways in the analysis process by finding what consistency can be calculated with what cannot be done so that the resulting data is valid, accurate, and accountable

Second, data triangulation, which checked the validity of data by utilizing sources, techniques and theories related to research. Source triangulation allows researchers to re-check and complete information with other sources. Source triangulation in this study was carried out between the principal, Multimedia K3, teachers producing multimedia, and students. Triangulation of the process technique was done by testing the credibility of the data, by checking the data on the same source but with different techniques. Data obtained by interview were then checked with data obtained from observation and documentation. The triangulation technique used in this study was between observation, interviews and documentation. Triangulation with theory was used to sharpen research analysis by examining the degree of data confidence.

In this study, data analysis was carried out continuously from the beginning of data collection. The process consists of three interrelated stages: data reduction, data presentation, and conclusions. The data analysis process was carried out from the beginning to the end of the research data collection.

All data that has been collected through in-depth interviews, observations in both participants and non-participants, and document analysis, were reduced in the following ways:

First, by selecting, determining focus, simplifying and transforming the data that appears on the field notes. Data reduction was done in the form of summary writing, sharpening, coding, focusing, disposal, and compiling data so that conclusions were taken, proven, and accounted for. The data obtained were also classified into sub-data. The classification in this study consisted of planning learning and implementing learning by using authentic assessment, and obstacles and obstacles faced.

Second, data analysis was done both on learning planning, learning implementation, and obstacles and obstacles in the implementation of learning. Data on the implementation of learning was analyzed based on the Regulation of the Minister of Education and Culture of the Republic of Indonesia Number 65 of 2013 (RoI, 2013). After the data analysis was complete, the data was presented in a narrative form that described learning planning, implementation of learning, and obstacles and constraints in implementation. Further discussion of the data description was carried out. The discussion aims to find an overview and interpret what was contained in the planning of learning and the implementation of learning. Based on the data description and discussion, the conclusion should answer the questions that were in the formulation of the research problem.

3 RESULTS AND DISCUSSION

Research on the implementation of basic graphic design learning in Muhammadiyah Wonosari Vocational School was based on three variables components, planning, implementation, and obstacles encountered, as well as efforts made by the teacher to overcome obstacles. The data below is a collection obtained from observations, interviews, documentation studies, and document reviews. The data below was reviewed through research principles such as perseverance, regular observation, data triangulation, and auditing. Data analysis was the next step, by selecting, determining focus, simplifying and transforming data based on records in the field. The studies analyzed in this study are learning planning, learning implementation, and obstacles and obstacles in the implementation of learning. The eesults of the implementation of basic graphic design learning data in Muhammadiyah Wonosari Vocational School are below.

3.1 Lesson plan

The data obtained in the planning of learning are as follows.

Learning planning based on the curriculum-2013 initially made it difficult for teachers to formulate learning implementation plans. This is because teachers are not optimally adapted to curriculum changes. Education and training on this continues to be offered by the government to make the learning process successful for the curriculum-2013. The development of the KTSP resulted in changes in several aspects of K13 and in the study of basic graphic design in the curriculum-2013, including changes to subjects, study hours, syllabus and several other aspects that greatly influence the preparation of learning implementation plans.

To support the maximum implementation of curriculum-2013, Muhammadiyah Wonosari Vocational School runs an in-school training program, which is held every new school year. This program was carried out by inviting supervisors from the Department of Education and Culture of Gunungkidul Regency to provide workshops on the preparation of teacher workbooks, which included books 1–3.

With the implementation of the program, the ability of teachers to compile teacher workbooks including RPP is increasing. RPP on basic graphic design is prepared in accordance with the direction of the workshop. The results of the preparation of inter-teacher RPP on multimedia expertise competencies in general are almost the same. RPP is designed in detail and structured in accordance with the guidelines for drafting the RPP, which refers to the syllabus of the revised curriculum-2013. The RPP prepared by teachers of Wonsoari Muhammadiyah Vocational School uses a scientific approach in accordance with the rules of curriculum-2013.

These goals can be achieved with the right organizing strategies. Degeng (2013) explained that the organizing strategy refers to ways to make sequencing and synthesize facts, concepts, procedures and related principles. This organizing strategy cannot be separated from the characteristics of the content structure of the field of study, which has important implications for efforts to make sequences and synthesis between a field of study. The structure of this field of study includes learning structures or learning hierarchies, procedural structures, conceptual structures, and theoretical structures.

Particularly in the learning of basic graphic design itself, the organizing strategy is planned carefully according to needs. The hierarchy of learning, procedural structures, conceptual structures, and theoretical structures are structured in order to be able to achieve the objectives of the multimedia expertise competencies outlined in the curriculum, starting from the sequence of simple material to more complex material, explaining the relationship between the concept and procedural theory.

3.2 *Learning implementation*

Observations from March to April 2018 revealed that learning was going well even though there were some obstacles. Supporting learning facilities at Muhammadiyah Wonosari Vocational School are adequate. These facilities include LCD projector, laptop/computer, internet access, learning media, library room, computer lab, and other supporting facilities.

Based on observations in class X Multimedia, which became the object of research, it was found that the implementation of learning consisted of preliminary activities, core activities, and closing activities. The following is a descriptive description of the implementation of learning using curriculum-2013 with project-based learning methods.

Preliminary. The learning begins with prayer. Then students receive direction related to the learning of Basic Graphic Design, which is continued with an explanation of the learning objectives. After that, the first stage of the project-based learning (PjBL) syntax was carried out, giving students basic questions about the learning material.

Core activities. The implementation of this core activity involves stages two to six in the PjBL syntax. These stages include tstage two of preparing project planning, stage three of arranging a schedule, stage four of monitoring students, stage five of project progress and stage six of evaluation of experience evaluation results. The application of learning models allows students to enhance collaboration, and solving these problems provides an overview of the real scenarios that students will face in the world of work (Bell, 2010).

Closing-down. The observation revealed that summarizing the material together between the teacher and the students became part of the learning closing. It was then followed by a joint prayer at the end of the learning session.

3.3 *Inhibitor in learning*

3.3.1 *Lesson plan*

Problems experienced by teachers in preparing learning plans include making RPPs. The teacher experienced several obstacles in determining the steps of scientific learning in accordance with the curriculum-2013 and the characteristics and needs of students, especially in learning Basic Graphic Design. In addition, another obstacle is the allocation of learning time, as often the planning in the lesson plan is not implemented during the learning process.

3.3.2 *Learning implementation*

Inhibiting factors experienced during the learning of the basic graphic design vocational skills competence multimedia is the density of material, but the allocation of learning time is still minimal. Students' learning interest and independence are low, and this is reflected in students' behavior during practical learning, where they are always waiting for the teacher's step-by-step direction. In addition, interesting teaching materials have not yet been developed that specifically discuss graphic design as a guide for student learning during practical activities in learning to increase students' interest and independence.

Supporting factors of learning basic graphic design multimedia competency skills include completeness of learning facilities and infrastructure supporting practical learning, and varied learning methods applied by the teacher.

4 CONCLUSION

Based on the elaboration of the results of the research and discussion above, this study concludes some of the followings.

First, the basic graphic design learning plan conducted through the preparation of RPP by teachers in Muhammadiyah Wonosari Vocational School has been well prepared based on the rules of curriculum-2013.

Second, the implementation of basic graphic design learning includes introduction, core activities, and concludes by applying project-based learning methods with syntax: (1) providing basic questions, (2) preparing project planning, (3) arranging schedules, (4) monitoring students, (5) project progress, and (6) evaluation of experience evaluation results.

Third, the inhibiting factors experienced during the learning process of basic graphic design multimedia competency skills, among others, are the density of the material, lack of time allocated for learning, the low student interest and self-regulated learning, and no development of interesting teaching materials or self-regulated learning by students. Supporting factors of learning basic graphic design multimedia competency skills include completeness of learning facilities and infrastructure supporting practical learning, and varied learning methods applied by the teacher.

REFERENCES

AECT Definition and Terminology Committee. (2008). Definition. In A. Januszewski & M. Molenda (Eds.). *Educational technology: A definition with commentary*. New York, NY: Lawrence Erlbaum.

Alfa, Y. (2017). *Revisi Kurikulum 2013* [Curriculum-2013 Revision]. Retrieved from https://www.kompasiana.com/ysfalfan/591dfeb81bafbd2c7f4a3ee0/revisi-kurikulum-2013

Artapati. L.W. & Budiningsih. C. (2017). Pelaksanaan pembelajaran Kurikulum 2013 di SD Negeri Serayu Yogyakarta [Implementation of Curriculum-2013 in SD NegeriSerayu Yogyakarta]. *Jurnal Inovasi Teknologi Pendidikan* 186–200.

Arvianto, F. (2016). *Perkembangan kurikulum Indonesia* [Development of the Indonesian curriculum]. Jawa Tengah, Indonesia: CV Kekata Group. ISBN 978-602-6413-39-0.

Bell, S. (2010). Project-based learning for the 21st century: skills for the future. The Clearing House: *A Journal of Educational Strategies*, Issues and Ideas, *83*(2), 39–43.

Degeng, N.S. (2013). *Learning sciences*. Bandung: Kalam Hidup.

Moys,. J.L., Collier,. J., & Joyce,. D. (2018). By design: Engaging graphic communication students in curriculum development (a video case study). *Journal of Educational Innovation, Partnership and Change*, *1*, 1–4. doi: 10.21100/jeipc.v4i1.752

Mukhadis, A., Putra, A.B.N.R., Nidhom, A.M., Dardiri, A., & Suswanto, H. (2018). The relevance of vocational high school program with regional potency priority in Indonesia. *Journal of Physics: Conference Series*, *1028*, 1–8.

Pertiwi, N. & Ahmad, I.A. (2017). Construction industry needs for vocational high school graduates. Advances in Social Science, *Education and Humanities Research (ASSEHR)*, *127*, 157–160. doi: 10.2991/icaaip-17.2018.35

RoI. (1990). *Peraturan Pemerintah Nomor 29 Tahun 1990 tentang Pendidikan Menengah* [Government Regulation Number 29 of 1990 concerning Secondary Education]. Jakarta, Indonesia: Republic of Indonesia.

RoI. (2010). *Peraturan Pemerintah nomor 17 Tahun 2010 tentang Pengelolaan dan Penyelenggaraan Pendidikan* [Government Regulation Number 17 of 2010 concerning Management and Implementation of Education]. Jakarta, Indonesia: Republic of Indonesia.

RoI. (2013). *Peraturan Menteri Pendidikan dan KebudayaanNomor 65 Tahun 2013 Tentang Standar Proses* [Regulation of the Minister of Education and Culture of the Republic of Indonesia Number 65 of 2013 concerning the Standards of Process of Primary and Secondary Education and based on existing theories]. Jakarta, Indonesia: Republic of Indonesia.

RoI. (2018, September 20). *Data pokok SMK* [Main Data of VHS]. Retrieved from http://datapokok.ditpsmk.net/

Professional Teacher

Developing video-based learning resources for music teachers in Singapore

A. Bautista
National Institute of Education

S-L. Chua
Singapore Teacher Academy for the Arts, Ministry of Education, Malan Road, Singapore

J. Wong & C. Tan
Nanyang Technological University, Nanyang Walk, Singapore

ABSTRACT: In recent decades there has been an exponential increase in the use of class-room video as a tool to improve teacher skills, in all subject areas, at all grade levels, and all over the world. This chapter provides an overview of the project 'Towards Responsive Professional Development for Singapore Music Teachers. Phase 2: Developing Video-Based Learning Resources'. The Singapore Teachers' Academy for the Arts (STAR) and the National Institute of Education (NIE) have partnered to develop a repository of classroom videos intended to foster the professional development of Singapore's school music teachers. First, the team coded 211 video-recorded full lessons, conducted mainly by regular music teachers. Subsequently, we produced a collection of video clips that illustrate ways in which Singapore's music syllabus can be taught and/or how student-centric principles can be enacted in the class-room. Finally, we collected a wealth of empirical data (using surveys, interviews, and focus group discussions) to examine music teachers' perceptions of the usefulness of the video-based learning resources developed. The videos are currently being used in online and blended professional development (PD) courses delivered by STAR.

1   INTRODUCTION

In recent decades there has been an exponential increase in the use of classroom video as a tool to improve teacher skills in all content areas (for a recent review, see Gaudin & Chaliès, 2015). Videos of classroom practice allow teachers "to enter the world of the classroom without having to be in the position of teaching in-the-moment" (Sherin, 2004, p. 13). Having the opportunity to watch oneself and/or others teach and analyzing instructional practices have great potential to improve teaching effectiveness and student learning (Desimone & Garet, 2015). Teacher educators and professional development (PD) providers have used classroom video for a variety of purposes, for example to illustrate good instructional practices, to foster exchange of pedagogical strategies, to engage educators in sustained thinking about teaching, as well as for assessment and evaluation (Marsh & Mitchell, 2014; Masats & Dooly, 2011).

In music education, there is also a long tradition of incorporating classroom video into teacher education programs. During the 19070s and 80s, many countries implemented system-wide programs where music teachers were asked to analyze their own video clips for self-assessment purposes, using predetermined checklists of classroom events and behaviors (West, 2013). In the 1990s, scholars started to emphasize the multiple advantages of video technology and how music teachers could improve by watching and analyzing classroom videotapes. For example, Grashel (1991) argued that because videotapes provide a visual and aural record of classroom events, specific segments can be reviewed multiple times and hence music teachers may reflect more deeply on music teaching and learning. In addition, many teachers can benefit from the very same video, watching it together or in different locations. Berg and Smith (1996) also discussed how "video club" meetings could serve as a platform for music teachers to help each other become more effective in the classroom.

2 CONTEXT FOR THE DEVELOPMENT PROJECT

The development project described in this article is part of a long-term research program, in partnership between the National Institute of Education (NIE) and the Singapore Teachers' Academy for the Arts (STAR). The ultimate aim is to strengthen the level of preparation and development of Singapore's music teachers, while contributing to the field of music teacher PD research (both nationally and internationally). The project is a continuation from prior studies in which we found that school music teachers in Singapore – both music specialists and generalists – place value on value life and video-mediated peer observation as forms of professional learning. Through interviews, surveys, and focus group discussions, we learned that one of music teachers' preferred ways of learning involves observing how other experienced music educators (e.g., fellow colleagues, STAR facilitators, music pedagogues) implement music activities in actual classrooms. Music teachers value peer observation precisely because it inspires lesson design, allows for a better understanding and application of teaching strategies, allows for the anticipation of students' reactions, and contributes to building their confidence. In addition, music teachers are highly open to video-mediated peer observation because it is convenient, easily accessible, and may provide contextualized examples of good classroom practices (Bautista, Toh, & Wong, 2018; Bautista, Wong, & Cabedo-Mas, 2018). However, given numerous hindrances (including lack of time, different priorities, difficulties in being released to participate in music-related PD), not all music teachers in Singapore have regular access to these types of peer-observation opportunities (Bautista & Wong, 2017). It is therefore essential to continue to explore strategies to reach out to all music teachers, both specialists and generalists, providing them with opportunities to better themselves as music educators.

3 PURPOSE OF DEVELOPMENT PROJECT

The project 'Towards Responsive Professional Development for Singapore Music Teachers. Phase 2: Developing Video-Based Learning Resources' (AFD 03/15 AB) was funded by the Ministry of Education (MOE) in 2016. The main purpose was to create video-based learning resources for music teachers from primary and secondary schools. While working with music teachers over the years, STAR had collected several hundred video recordings on aspects related to music teaching and learning. Most videos were lessons conducted by music teachers in Singapore classrooms. Others were videos from master teachers demonstrating innovative approaches, as well as videos of PD workshops focusing on a wide variety of music skills and pedagogies. The development project had five specific objectives:

- **Objective 1:** Generate a coding framework that allows for the organization and management of STAR's repository of videos;
- **Objective 2:** Code the video recordings according to this framework;
- **Objective 3:** Identify videos containing high-quality teaching and generate a collection of short video clips that illustrate ways in which the music syllabus can be taught and/or how different pedagogical moves can be enacted in the music classroom;
- **Objective 4:** Evaluate the usefulness of the video clips as perceived by music teachers – both specialists and generalists;
- **Objective 5:** Refine the video clips and make them available for PD purposes, as part of STAR's PD courses and programs.

In the following, we describe how the team proceeded to meet these five objectives and achieve the deliverables outlined for this project.

3.1 *Objective 1: Design of coding framework*

The framework draws on MOE's Primary and Lower Secondary Music Programme (MOE, 2014) and STAR's Evaluation Framework for the PAM research project (MOE, 2016). This unique framework was collaboratively designed by an interdisciplinary team comprising members from STAR (experts on music teacher PD), from NIE (music education researchers), and from Arts Education Branch (curriculum designers). It was pilot-tested with two groups of music teachers (three primary and three secondary). The team had six three-hour long workshops with these teachers, during which we discussed and refined the analytical codes. The coding framework was able to provide a second-by-second description of the teacher moves during the entire music lesson. The team presented the framework in a number of national and international conferences, receiving acceptance among the academic community.

3.2 *Objective 2: Coding and archiving of the repository of full lessons*

The 211 full lessons shared by STAR were coded according to the above-mentioned framework. There were 59 primary lessons and 152 secondary lessons. These were mainly lessons conducted by primary and secondary music teachers and videos of PD workshops conducted by subject/topic experts, usually STAR master teachers. Examples include videos collected within the scope of a Kodaly workshop (primary), information & communication technology (ICT) trials (secondary), and informal and non-formal (secondary) lessons.

The MaxQDA qualitative analysis software was utilized to code, organize, and archive the repository of full lessons shared by STAR. As shown in Figure 1, MaxQDA allows researchers to enter multiple time-based codes, depending on the behaviors and/or events observed at each specific time point. The layout of MaxQDA allows for a clear visualization of how codes evolve as the lesson unfolds.

Figure 1.   Illustrative screenshot of coding layout in MaxQDA.

3.3 Objective 3: Creation of video clips

The team created 153 video clips that illustrate examples of teaching practices in the Singapore music classroom, which could be used for professional development purposes. The duration of the video clips ranged from 23 seconds (the shortest) to 21 minutes and 42 seconds (the longest), with an average duration of 4 minutes. Most clips were 3–5 minutes long. The video clips were edited with Adobe Premiere Pro and Apple Final Cut Pro X.

We created an Excel database in which each video clip was described and tagged according to several parameters: a) title; b) description (50–75 words, approximately); c) duration; d) grade level of students; e) music-specific pedagogies (Kodaly, Dalcroze, etc.); f) type of episode (tuning in, management, closure); g) classroom format (e.g., whole class, group work, individual); h) syllabus content covered (e.g., notation, improvisation, singing);and i) pedagogical processes (e.g., critical thinking, reflection).

With reference to the learning objectives specified in the Primary and Lower Secondary Music curriculum (MOE, 2014), the 153 video clips created were tagged as follows: Performing (73%, 112 clips); Understanding Musical Concepts (52%, 80 clips); Listening & Responding (39%, 60 clips); Appreciation of Music (36%, 56 clips); and Improvisation/Creation (26%, 40 clips). Regarding student-centric pedagogical principles, the highest number of video clips (34%, 52 clips) illustrate Critical Thinking and Reflection in Music, followed by Musical Collaboration (32%, 50 clips), and Musically Supportive Moves (29%, 45 clips). The proportion of video clips illustrating Musical Creativity, Providing Choice and Empowerment, Questioning, and Fluency and Momentum ranged from 20% to 26%. The collection of videos had a relatively lower number of exemplars related to Assessment and Evaluation, Contextualizing Lessons, and Differentiated Instruction (15% or lower).

3.4 Objective 4: Evaluation of teachers' perceptions and views

3.4.1 Participants, data collection tools, and design

The team collected a wealth of empirical data to examine music teachers' perceptions and views regarding the usefulness of the video resources created. In total, 154 music teachers (106 music specialists and 48 generalists) participated in the evaluation component of the project. Participants ranged from 24 to 58 years of age ($M = 35.5$, $SD = 7.9$), had 0.5–38 years of general teaching experience ($M = 10.3$, $SD = 7.3$), and 0–28 years of music teaching experience ($M = 8.7$, $SD = 6.6$). There were 133 female teachers and 21 male teachers. More specifically, we had six survey sessions with 111 teachers (80 specialists and 31 non-specialists), individual interviews with 21 teachers (12 specialists and nine non-specialists), and six focus group discussions with 22 teachers (14 specialists and eight non-specialists).

In all data collection sessions, teachers were shown one or several video clips created during the development phase of the study. The various data collection instruments utilized during the evaluation phase (namely paper-and-pencil surveys, an online survey, interview protocols, and focus group discussion protocols) included questions pertaining to the following topics: applicability and usefulness of the video clips; extent to which the features of the videos influenced teachers' learning; potential use of the videos within PD settings; and teachers' overall satisfaction with the videos (e.g., general opinions about their quality, aspects that teachers found interesting, suggestions for improvement). We were also interested in exploring potential differences in the perceptions and views of music specialists and generalists, as this would be important for the design of future video-based PD initiatives. Surveys were primarily based on Likert-scale items and were distributed by the end of the PD session (in the case of the online survey, the video clip was embedded in the Qualtrics platform). Completion time ranged from 5 to 10 minutes. The interviews and focus group discussions combined open-ended verbal questions with short paper-and-pencil questionnaires containing Likert-scale items. Interviews ranged from 40 to 50 minutes, and focus group discussions from 70 to 90 minutes.

3.4.2 *Findings*

Surveys. Teachers gave very positive responses regarding the applicability and usefulness of the video clips. Most agreed that the content of the videos was directly applicable or transferrable in their own classrooms. They acknowledged that other music teachers in Singapore would benefit from the video clips and would find them useful in improving their own practice. Regarding the features of the video clips and how these influenced their learning, teachers positively valued having the videos annotated and accompanied with the corresponding lesson plans. Most teachers enjoyed having the opportunity to discuss the video clips with other fellow colleagues during the STAR sessions and expressed interest in watching the full lesson in the future. Finally, we asked teachers about their overall satisfaction with the videos clips shown during the courses. Most music teachers agreed or strongly agreed with the view that having video clips integrated as part of the PD sessions made their learning more meaningful and authentic. Teachers also stated that having videos made it easier for them to put what they had learned during the course into practice, and that, overall, the video clips enhanced their satisfaction with the PD workshop. We did not find differences between responses of music specialists and generalists.

Individual interviews. Among other activities, the participants were asked to watch two video clips during the interviews, one chosen by the research team and another one of their own choosing. After watching both clips, we asked teachers to rate the various descriptors presented in the Excel database according to several dimensions (including accuracy, attractiveness, and overall quality). Most respondents rated the title, the description, the syllabus content covered, and the pedagogical processes as being accurate or very accurate. The video clip title and description were rated as attractive or very attractive by most teachers. The team asked follow-up questions to the participants when their responses were neutral or negative, to better understand their thinking and attend to their concerns during the refinement phase. We asked teachers to rate the overall quality of both video clips watched and their descriptors, on a scale from 1 (poor) to 10 (excellent). On average, teachers gave a score of 6.1 (min = 2, max = 10; SD = 2.4). The most common response was 7.

Another topic explored within the interviews was how music teachers use reason when searching for video clips. In particular, we compared teachers' reasoning according to two variables: music education specialization (specialists *vs.* non-specialists) and educational level in which the teacher teaches (primary *vs.* secondary). The participants were shown the Excel video clip repository described above. We first asked them to brainstorm keywords they would use to search for video clips, according to their interests and needs. Once two or three video clips were shortlisted, the teachers were asked to read through the various columns of the Excel repository to select the video clip that was most interesting and/or relevant to them. Finally, we asked several follow-up questions to explore the reasoning behind teachers' final choices. Findings showed that the description was the most important parameter for all teachers across the board, regardless of their profile, whereas other parameters such as video duration, type of grouping, and music-specific pedagogies were rarely considered. Syllabus content and grade level were more important for primary teachers, while the information about pedagogical processes was more important for secondary teachers. With regard to specialization level, we found that grade level and type of classroom setting were important parameters for non-music specialists, whereas music specialists prioritized pedagogical processes. The conclusion is that music teachers' reasoning while searching for video clips in a repository might differ substantially depending on key demographic characteristics, particularly their level of content-specific training and the levels in which they teach. The main implication is that the very same video clip might need to be presented differently depending on the audience and the purpose. Tailoring video clips for different types of teachers and providing scaffolds to guide their learning would maximize the impact of video-medicate learning resources.

Focus group discussions. In one of the activities, our goal was to compare the aspects that music teachers with different specializations notice when analyzing video clips. We found that music specialists and generalists focused on different types of aspect, and described them with different levels of sophistication. While the music specialists focused on aspects that

155

interrelated various elements and/or processes (typically pertaining to curriculum objectives, pedagogical moves, quality of teacher-student interactions, lesson materials and activities), the generalists focused on aspects where elements were described in isolation (primarily musical concepts, lesson sections, or classroom management actions). Additionally, music specialists described what they noticed using a high level of specificity, referring to specific events captured in the video clips, whereas generalists tended to provide broad descriptions, without reference to concrete evidence. The conclusion is that teachers may adopt completely different lenses when watching and analyzing classroom video clips, which might coincide (or not) with the intent with which video clips were developed by PD researchers. The implication is that providers of video-based PD should be mindful of these differences to better scaffold and guide teachers' learning through videos of practice.

3.5 *Objective 5: Refinement of video-based learning resources*

Based on teachers' perceptions and views, as well as the feedback provided by STAR on several occasions throughout the project, the team made further improvements to the video-based learning resources created. The project's research assistants incorporated the feedback to fine-tune the parameters of each of the video clips developed, and to re-cut, render, and finalize the video clips repository. The finalization of the video clips included adding transition (text-on-black) screens, shortening redundant sections, improving the lighting or audio quality of the video clip (when possible).

4 DISCUSSION

Videos of practice can be effective as a learning artifact for teachers who want to implement new teaching strategies, as it allows alternative pedagogical moves to be demonstrated in their natural environments (Gaudin & Chaliès, 2015; Marsh & Mitchell, 2014; Masats & Dooly, 2011). The availability and easy accessibility of video-mediated learning resources, typically not present in pure face-to-face PD courses, is one reason why many PD programs are moving to online or blended (i.e., hybrid) models, in an attempt to combine the strengths of personal interactions and online technologies (Dede, Jass Ketelhut, Whitehouse, Breit, & McCloskey, 2009).

The use of videos of practice is particularly powerful and necessary in music teacher education because, as described in the literature, music educators across the world experience feelings of isolation and lack of collegial support due to the low number of music teachers in most schools (Bautista, Yau, & Wong, 2017). For music teachers, being able to observe the lessons conducted by other teachers may lead to new ideas for their own lessons, especially if there is a chance for collective reflection, analysis, and discussion post-observation, for example with the mediation of online platforms (Sindberg, 2016). Video can also aid music teachers in understanding student musical learning and thinking, allowing teachers to optimize and differentiate their lessons based on students' needs, interests, and preferences (Zhukov, 2007). This is especially so in the music education environment, where video can better help teachers see how students practice and perform, and the difficulties they face when making music – be it singing, playing, composing, improvising, etc. As argued by West (2013), it is clear that video technology "involves more human senses than does written narrative, and thus increases the level of personal engagement" (p. 12).

The deliverables of the development project described here are important contributions to the practice of music teacher PD within the Singapore context, particularly at STAR. The first deliverable is the design of an analytical framework specific to Singapore's music education curriculum (MOE, 2014; MOE, 2016). This framework has allowed us to code the content and processes present in STAR's existing video recordings and will continue to be useful for the coding of new lessons. The second deliverable is the thematic codification of STAR's large database of 211 video recordings. The third deliverable is a collection of over 150 short video clips that illustrate how music content can be taught and/or how pedagogical moves can be executed. All these video-based learning resources are available to the fraternity of Singapore music teachers as part of STAR's blended and online PD programs, hence have the potential to benefit many in-service music teachers across the country. In addition, NIE is currently looking into utilizing the video clips within their courses and programs with pre-service music teachers. Another important contribution of this project is the data related to teachers' perceptions and views about the video-based learning resources created. The various surveys, individual interviews and focus group discussions have allowed us to gather information on the applicability and usefulness of the video clips, the extent to which the features of the videos influenced teachers' learning, teachers' opinions about potential usage of the videos within PD settings, and their overall satisfaction with the videos. This information contributes to the theory of music teacher PD, both nationally and internationally.

5 FUTURE DIRECTIONS

Based on the video learning resources we created in this development project, our team is currently working on a follow-up MOE-commissioned research study entitled 'Towards Responsive Professional Development for Singapore Music Teachers. Phase 3: Investigating Efficacy of Video-Based Professional Development' (AFR 01/18 AB). This study aims to examine how video-based PD could be effective in fostering the music pedagogical content knowledge (PCK) of primary and secondary music teachers, including specialists and generalists. The project is currently piloting and investigating the impact of three video-based PD prototypes (delivered by STAR) to identify ways to further enhance their effectiveness in future iterations. These three PD prototypes will be an online course for generalist primary music teachers and two blended-learning PD programs for beginning primary and secondary music teachers.

ACKNOWLEDGMENTS

This study was funded by the Education Research Funding Programme, National Institute of Education (NIE), Nanyang Technological University, Singapore, project no. AFD 03/15 AB. The views expressed in this paper are the author's and do not necessarily represent the views of the host institution.

REFERENCES

Bautista, A., Toh, G.-Z., & Wong, J. (2018). Primary school music teachers' professional development motivations, needs, and preferences: Does specialization make a difference? *Musicae Scientiae, 22*(2), 196–223. doi:10.1177/1029864916678654

Bautista, A., & Wong, J. (2017). Music teachers' perceptions of the features of most and least helpful professional development. *Arts Education Policy Review*, 1-14. doi:10.1080/10632913.2017.1328379

Bautista, A., Wong, J., & Cabedo-Mas, A. (2018). Music teachers' perspectives on live and video-mediated peer observation as a form of professional development. *Journal of Music Teacher Education.* doi:10.1177/1057083718819504

Bautista, A., Yau, X., & Wong, J. (2017). High-quality music teacher professional development: A review of the literature. *Music Education Research, 19*(4), 455–469. doi:10.1080/14613808.2016.1249357

Berg, M. H., & Smith, J. P. (1996). Using videotapes to improve teaching. *Music Educators Journal, 82*(4), 31–37. doi:10.2307/3398914

Dede, C., Jass Ketelhut, D., Whitehouse, P., Breit, L., & McCloskey, E. M. (2009). A research agenda for online teacher professional development. *Journal of Teacher Education, 60*(1), 8–19. doi:10.1177/0022487108327554

Gaudin, C., & Chaliès, S. (2015). Video viewing in teacher education and professional development: A literature review. *Educational Research Review, 16*, 41–67. doi:10.1016/j.edurev.2015.06.001

Grashel, J. (1991). Teaching basic conducting skills through video. *Music Educators Journal*, *77*(6), 36–37. doi:10.2307/3398211

Marsh, B., & Mitchell, N. (2014). The role of video in teacher professional development. *Teacher Development*, *18*(3), 403–417. doi:10.1080/13664530.2014.938106

Masats, D., & Dooly, M. (2011). Rethinking the use of video in teacher education: A holistic approach. *Teaching and Teacher Education*, *27*(7), 1151–1162. doi:10.1016/j.tate.2011.04.004

MOE. (2014). *Music teaching and learning syllabus. Primary & Lower Secondary*. Singapore: Ministry of Education. Retrieved from: http://www.moe.gov.sg/education/syllabuses/arts-education/files/2015-general-music-programme-syllabus.pdf.

Sherin, M. G. (2004). New perspectives on the role of video in teacher education. *Advances in research on teaching*, *10*, 1–27. doi:10.1016/S1479-3687(03)10001-6

Sindberg, L. K. (2016). Elements of a successful professional learning community for music teachers using comprehensive musicianship through performance. *Journal Of Research in Music Education*, *64*(2), 202–219. doi:10.1177/0022429416648945

West, C. (2013). Developing reflective practitioners: Using video-cases in music teacher education. *Journal of Music Teacher Education*, *22*(2), 11–19. doi:10.1177/1057083712437041

Zhukov, K. (2007). Student learning styles in advanced instrumental music lessons. *Music Education Research*, *9*(1), 111–127. doi:10.1080/14613800601127585

Innovative Teaching and Learning Methods in Educational Systems – Retnowati et al. (Eds)
© 2020 Taylor & Francis Group, London, ISBN 978-1-03-224183-8

Teacher–student communication style and bullying behavior: Sociometry evaluation

I. Sholekhah, S. Indartono & D.W. Guntoro
Universitas Negeri Yogyakarta, Yogyakarta, Indonesia

ABSTRACT: Teacher–student communication style in the classroom determines the classroom climate as a whole. Furthermore, social relations and peer reactions also significantly influence the classroom climate for students. This article briefly highlights the influence of teacher–student communication styles and bullying behavior using a sociometry. The aim was to determine the choices, communication, interaction patterns and structure of relationships among students in indicating bullying behavior in the classroom and in school. The survey was conducted with junior high school students by distributing sociometric questionnaires. The results of the study are illustrated in a sociogram chart that shows students' social relations, and the existence of "acceptance" and "rejection" of students in the environment. Social relations with other students are grouped into several categories: popular, rejected, neglected, and controversial. Teachers must be aware of the refusal of a student by the peer environment so as not to interfere with the learning process. The results contribute to a variety of evaluation techniques that can help teachers identify and help students who have difficulty in adjusting to their environment. The limitations and further research will be discussed.

1 INTRODUCTION

A school is a place where interactions between students and the environment are integrated into a social relationship. In the implementation of the learning process at school, there are two parties involved: teachers and students. According to Neacsu in Roxana (2013), the involvement of teachers and students in learning in schools has their own characteristics and goals, each also has a different social role and status. The school environment must provide a positive climate so that the social relationships that occur between children are better for their social development.

The role of teachers in schools is very important in determining the success of education. The teacher's main tasks are to educate, teach, guide, direct, assess, and evaluate students in the learning process in the classroom. In the communication process in the class, the teacher uses a particular communication style as a way of conveying learning messages. This communication style consists of individual characteristics that are reflected in the actions of communication, which are special ways to receive or read messages, interpret messages, respond, and provide feedback (Roxana, 2013). Teacher–student communication style in the classroom determines the classroom climate as a whole.

Student interactions in the classroom also provide various communication possibilities that influence the learning process. In addition to student relations with teachers, their relations with classmates also play a very important role. How can they build communication so that there is quality interaction in the learning process? Teachers must be aware of student relations with other students to avoid negative behaviors such as bullying. Rebecca & Dabney (2015) suggested that loneliness related to friendship causes stronger symptoms of depression than loneliness from parents. This provides a very strong relationship that students' friendships will greatly affect their learning achievement.

The various concepts regarding the role of teacher–student communication and interactions among students are very different from the reality. In fact, schools can be a daunting place for students. There may be violence by teachers toward students, incompatibility among friends, and rejection of the school environment, which is very influential on the personality and social relations of students. In Indonesia, the occurence of violence toward children at the education level remains high (see Figure 1).

Table 1 shows that violence or bullying is no longer unusual in Indonesia. Bullying is the act of using power to hurt someone or a group of people verbally, physically or psychologically so that the victim feels traumatized and depressed. In fact, this bullying behavior often occurs in schools and is conducted by teenagers (see Table 1). The impact caused by bullying includes an increased risk of various health problems and mental problems. Bullying victims will be very easily depressed, nervous, feel isolated and insecure while in the school, as well as have decreased enthusiasm for learning and academic achievement that may be carried into adulthood.

This research was conducted on students in junior high schools to determine the role of teacher–student communication and the emergence of bullying behavior in the class or school. A middle school was chosen as the subject of research because the students are aged 10-12 years, and experience a period of transition between childhood and adulthood. In this phase, adolescents are considered to have reached level of independences and are searching for their identity (more logical thinking and high idealism), and more time is spent outdoors. To foster their maturity, teenagers need help and assistance from those more experienced: parents, teachers, and the community.

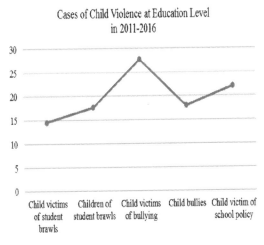

Figure 1.   Cases of violence against children at the education level, 2011–2016.
Source: KPAI 2016.

Table 1.   Types and criteria of students.

Type of students	Criteria
Popular	High level of acceptance
Rejected	High level of rejection
Controversial	Both acceptance and rejection
Neglected	Students name appears infrequently or not at all

Source: Leung & Silberling (2006).

160

This article briefly discusses the use of an evaluation tool called sociometry to uncover the influence of teacher–student communication styles on bullying behavior. Sociometry has proven useful in predicting the average assessment of teachers at various levels, from elementary schools to high schools, the armed forces and industry (McCandless & Helen, 1957). This research on the communication style used teachers in the classroom aimed to discover how learning interactions provided by the teacher affected the student's output. Teacher–student communication will shape students' personality as part of the learning process. In addition, the sociometric evaluation also provides indications of bullying on social relations among students in the class.

This research provides a solution to the problems that occur when identifying the need for interventions when there are indications of bullying behavior in the classroom or in school. It combines the results of the sociometric evaluation with learning achievement, both for students who are indicated as victims of bullying or the perpetrators. The results of the study contribute to a variety of evaluation techniques that can help teachers identify and help students who have difficulty in adjusting to their environment.

2   LITERATURE REVIEW

2.1   *Teacher–student communication style*

The model used to describe a teacher's communication style was first introduced by Timothy Leary in 1957 (Levy & Wubbels, 1992). This model is mapped and studied in the Model for Interpersonal Teacher Behavior (MITB), which states there are two dimensions in teacher–student relations: influence (dominance-submission) and proximity (opposition-cooperation). These dimensions are represented as two axes which underlie eight types of teacher behavior: leading, helpful/friendly, understanding, student responsibility and freedom, uncertain, dissatisfied, admonishing and strict (Wubbels & Brekelmans, 2005).

Urea (2013) explained that teacher–student communication styles are divided into four types: not assertive, aggressive, manipulative, and assertive. An unassertive style is indicated by the tendency to hide or escape, and lack of courage in dealing with people. An aggressive style is characterized by a tendency to always be present, and to be a winner at any cost even if it causes damage or sadness to others. A manipulative style is characterized as a preference for acting backstage, and with a tendency to look for what is hidden behind other people's statements. Whereas an assertive style is characterized by self-expression, honesty, and a direct approach, accompanied by the ability to give direct opinions without endangering others, and pursuing interests without violating the needs of others.

The various studies have shown that positive teacher–student relations are a component of school ties and a school climate associated with fewer bullying behaviors (Ma, 2002; Roland & Galloway, 2004; Cunningham, 2007; Murray-Harvey & Slee, 2010; Raskauskas et al., 2010; Richard et al., 2012). Adolescents often engage in naughty and criminal behavior because economic and social conditions limit the ability of the community (e.g. parents and schools) to manage or supervise their behavior (Espelage & Swearer, 2009). Conversely, adolescents tend to be less involved in misbehavior when they have a safe and positive bond with important people in their lives, such as teachers (Hirschi, 1969). This reinforces that positive relationships with the people closest to students will create positive behaviors, and in this case the positive relationship comes from teacher–student relations and relations among students.

2.2   *Bullying behavior*

Bullying is a problem in a dangerous society (Marci et al, 2013). Bullying is a form of aggressive behavior described as a situation when a student is exposed repeatedly and over time to negative actions on the part of one or more students' (Olweus, in Fekkes et al., 2005). This negative action occurs due to a power imbalance (asymmetric power relationship), where the person who becomes a victim of negative actions has difficulty in defending themself and is powerless against abuse (Olweus, 1994). Bullying behavior can be 'physical' (such as hitting, pushing, kicking), 'verbal' (such as calling names, provoking, threatening, spreading slander), or it can be in the form of other habits such as social exclusion (Fekkes et al., 2005).

Most bullies use bullying to gain or maintain dominance and tend to decrease empathy for their victims (Beale in Smokowski et al., 2005). Victims of physical intimidation tend to be small and weak compared to bullies and cannot protect themselves from acts of violence (McNamara & McNamara, 1997). A study found that victims of bullying showed poorer social and emotional adjustment, had greater difficulty in making friends, had fewer relationships with peers and feet lonely (Nansel et al., 2001). This bullying behavior must be seen as an indicator of the risk for various mental disorders in adolescents (Kaltiala-Heino et al., 2000).

2.3 Sociometry evaluation

Sociometry is a procedure for summarizing and composing – and to a certain extent can quantify – the opinions of students about the acceptance of their peers and the relationship between them (Arifin, 2013). According to Moreno (Remer, 2018) sociometry is a measurement of socius, an interpersonal aspect of human relations. Sociometric techniques are considered suitable for assessing the closest friends of each child at the appropriate time in a group (Boyd & Helen, 1957). The aim of sociometry is to identify highly classified networks in the classroom, the overall social structure of the class, and the role of children in networks or structures (Cairns, 1983).

Sociometric evaluation is a type of non-test evaluation that can be used to determine students' social abilities in social relationships. The results of research by Bryn and Gilmour (1994) show that the sociometry method has considerable significance for clinical practice. First, it can provide a greater understanding of the process of behavior and emotional disturbances. Second, sociometry techniques can give doctors and other specialists the ability to measure contemporary impacts that disturb children and adolescents in their peer groups. Rejection by peers is a major risk in children for their adjustment to the social context, such as school (Zettergren, 2003).

In this study, sociometric evaluation is used as a tool to indicate the bullying behavior that occurs in students. The results are then compared with the teacher–student communication method in the class, especially in how teachers communicate and treat their students. The effect is also seen in student learning outcomes that are still lacking, through the treatment.

3 METHOD

An evaluation technique called sociometry was used to collect data about teacher–student communication styles and bullying behavior in the classroom. Sociometry is widely used by psychologists to find out about a person's social relationships with others (Leung & Silberling, 2006). This sociometric evaluation was designed via a questionnaire to reveal the social relations of students in the class, and distributed to junior high school students in Jember Regency. Ambulu Junior High School 3 Ambulu Jember was place of research, with a sample of 30 students. The sociometry questionnaire was designed with the following questions: choose the friend who is most liked and disliked in the study group, choose friends to be leaders or class leaders, choose a friend of mine, and choose friends to ask for help if there are learning difficulties.

Data analysis was conducted by quantitative and qualitative sociometric analysis. Quantitative analysis was done by calculating the number of voters for each individual subject to sociometry. In this analysis, there are three individual positions in the group: Election Status (Choice Status: CS), Rejection Status (Rejection Status: RS), and Election and Rejection Status (CRS).

To determine the three statuses, the formula according to Susilo Rahardjo (2013) was used:

1 *Choice Status (CS)*
Number of selector A: CS A = N × p Information:
A = Code of students who are seeking status in groups
N = Number of students in a group
P = Number of choices for each student determined in the sociometry questionnaire
2 *Rejection Status (RS)*
Number of rejectors B: RS B = x – 1 N × t
Information:
B = Code of students who sought status in groups
N = Number of students in a group
t = Number of rejections for each student determined in the sociometry questionnaire
3 *Choice and Rejection Status (CRS)*
Number of voters C - Number of rejectors C: RS C = N × q
Information:
C = Code of students who sought status in groups
N = Number of students in a group
q = Number of choices/rejections for each student determined in the sociometry questionnaire

4 RESULT AND DISCUSSION

Social relations show how the environment accepts a person so that it can give judgment about someone who can be "accepted" or "rejected" by the environment (McCandless & Marshall, 1957). Social choice is made with several criteria, which can be very subjective and based on someone's likes or dislikes at a first impression. However, these social choice criteria can be more objective when knowing that someone has or does not have certain skills needed in a group (Leung and Silberling, 2006). This choice can help described the social relations that occur in a group. The results are depicted in a group social network indicated by a sociogram.

The data on the sociogram show a table or matrix of the choices made by several students in the class. Students' social relation as described in the sociogram identify four types of students according to Leung & Silberling (2006).

The results of the study are illustrated in a *sociogram* chart (Figure 2) of students' social relations.

The sociogram chart shows several choices in the class. The results show students' social relation status with their classmates, such as popular, rejected, controversial, and neglected. Students with a popular status are in the yellow and purple circles on the sociogram figure. Then a gray and green circle shows students with controversial status, both by acceptance and

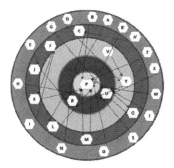

Figure 2. Sociogram of student social relations.

rejection. The outermost circle on the sociogram shows students who are neglected, for example students who have never appeared in the sociometry. Neglected status for these students is the same as for students who get a high rejection in their community in class.

Popular status is given to those who have good social attitudes and relationships with their classmates. Popular students usually have a positive attitude such as smart, diligent, patient, good, and not mean with money. Controversial students are humorous, friendly, fun, helpful and not arrogant. Neglected/rejected students have a tendency to be unfavorable in social relations with their classmates because they have negative attitudes such as lazy, arrogant, naughty, and ignorant.

Popular and controversial students have satisfying learning outcomes in class, with good and satisfying learning achievements. This happens because the school environment and their friends class provide support for learning. Neglected and rejected students who had a tendency to be negative have unsatisfactory learning outcomes. This is not because the learning climate caused by classmates is bad, but because neglected and rejected students tend to have low learning motivations, which causes their learning outcomes to be unsatisfactory.

It is likely that student bullying behavior can be seen using non-sociometric type tests. Some answers given by students indicated bullying behavior that may occur in students in the class. In this study, bullying behavior should be known from the neglected and rejected status of students in the class. However, it is not proven that students with neglected and rejected status are bullies or victims.

The researcher's observations on teacher behavior in learning indicate that the teachers used a forceful communication style. They tended to act in accordance with applicable regulations, which provide rewards for student achievements and punish students who make mistakes. This type of communication style makes students learn more actively in class, be polite and respect teachers, and be fair to students.

Teachers should provide more supervision for neglected and rejected students. Attention must also be given to students who are given the status of "rejection" by friends in their class. The goal is to make the students' learning outcomes better. Efforts that can be made include providing guidance, advice, and in-depth learning coaching. Students who get the highest "popularity" score in the class, can also be monitored to check they continue to behave according to the rules in school, so as not to become arrogant, and that they give assistance to other friends.

The overall results of the study indicate that sociometric evaluation tools can be used to measure students' social relationships in the classroom. The research findings are discussed with several theories such as social relations theory, social rejection theory, and social isolation theory. Social relation theory or social interaction is any relationship between two or more individuals. Carolyn & Daniel (1987) conducted a study to investigate the provisions of a person's social relations in adapting to stress. The result shows that there is a link between social support and good psychological and physical health in the interaction between individuals.

Students' social relationships are reflected in their interactions with peers in class and interaction with teachers in learning activities. The relationship between the two (both student–teacher and student–teacher relations) has an influence on student achievement and personality in the classroom. In this study, for example, social relations between students have the consequence of social "acceptance" that greatly affects students' learning outcomes. Children with high social acceptance status tend to be very adaptable to the environment and obtain satisfying learning achievements. In this, the student–teacher relationship is equally important. The teacher has a very large role in the success of students in providing social support.

Although humans are social beings, some social rejection is a part of life that cannot be avoided. Rejection by a group of people can cause extremely negative effects, especially when producing social isolation (Williams et al, 2005). Based on social rejection theory, the experience of rejection can cause a variety of adverse psychological consequences, such as loneliness, low self-esteem, aggression, and depression (McDougall et al, 2001). Other consequences caused by social rejection are feelings of insecurity and high sensitivity to rejection in the future (Richman, Laura & Leary, 2009).

A social rejection that occurs in neglected/rejected students must be a point of concern for a teacher. Indications of bullying behavior that have not appeared in neglected/rejected students also need to be anticipated and these students need assistance from the teacher so that they can improve their learning achievement in class. In addition, mentoring and teacher attention should ensure that the impact of social rejection is not too large in how it affects the students.

A serious impact of social rejection is social isolation. Social isolation theory describes isolation as a state of complete or near-complete lack of contact between an individual and a society. It is different from loneliness, which reflects a temporary lack of contact with other humans. Existing research shows that students' social development is an important factor in creating create social, emotional, and academic success (Rebecca & Dabney, 2015). Pamela et al. (2009), explained that socially isolated children tend to have lower educational outcomes, become part of disadvantaged social classes and are more likely to experience psychological distress in adulthood. By accepting social assistance, children can more easily cope with high levels of stress (Rebecca, Meena & Mel, 2014).

The results of this study contribute practically and managerially. The practical implications of research are mainly felt by teachers. The results of this study are used as input for teachers and applicant teachers to highlight the relationship of teacher–student communication and interaction between students in the class. The social relationship must always be a teacher's focus in connection with the learning that has been done so that student learning achievement increases, the learning climate is created positively, and there are no learning disadvantages in the classroom or at school.

The managerial contribution of this research is for schools and the government. For schools, the results of this study should provide information relating to the condition of students, the relationship between teacher–student communication, and the level of violence or bullying that occurs in the school. The government could use the results of the research as a basis or consideration for making policis related to decreasing violence and bullying that occurs to adolescents at school.

5 CONCLUSION

The purpose of the article using sociometry evaluation in the interaction of teacher–student communication styles and bullying behavior has been accomplished. Things that can be understood through the sociometry evaluation are the revised choices, communication, interaction patterns and the relationship structures among students in indicating the relationship of intimidation in the classroom and at school. Teacher–student communication style is extremely influential on student learning outcomes. The way teachers act and communicate with students in the classroom provides a positive learning climate. A teacher's communication style with a firm attitude has a positive effect on the learning process and is very influential on student output.

Likewise, bullying behavior, an interaction between students in the class, also influences the learning outcomes. Students who belong to the "popular" category are those who have positive qualities, such as smart, assertive, courageous, responsible, humorous, and not arrogant. Students who belong to the category of "rejected" by the class environment tend to have negative qualities, such as being naughty, lazy, arrogant, etc. Further investigation by the researchers shows that there is no indication of bullying in students' behavior from the social relations that occur in the classroom. However, teachers must always focus on their students in order to obtain maximum learning outcomes.

5.1 *Limitation and future research*

This research is limited to only small class groups with a total of 30 students. Subsequent research can be conducted in large class groups with a sample of more students so that the results can be generalized. The results of the sociometry questionnaire can also be used to create student study groups based on intelligence, expertise, and learning achievement to become heterogeneous learning groups.

REFERENCES

Arifin, Z. (2013). *Evaluasi pembelajaran: Prinsip, teknik, prosedur* [Learning evaluation: Principles, techniques, and procedure]. Bandung, Indonesia: Remaja Rosdakarya.

Cairns, R.B. (1983). Sociometry, psychometry, and social structure: A commentary on six recent studies of popular, rejected, and neglected children. *Merrill-Palmer Quarterly, 29*(4), 429–438.

Citrona, C.E. & Russell, D.W. (1987). The provisions of social relationships and adaptation to stress. *Personal Relationships, 1,* 37–67.

Cunningham, N.J. (2007). Level of bonding to school and perception of the scgool environment by bullies, victim, and bully victims. *Journal of Early Adolescence, 27,* 457–478.

Espelage, D., & Swearer, M.S. (2009). Contributions of three social theories to understanding bullying perpetration and victimization among school-aged youth. In M.J. Harris (Ed.), *Bullying, rejection, and peer victimization: A social cognitive neuroscience perspective* (pp. 151–170). New York, NY, US: Springer.

Fekkes, M., Pijpers, F.I.M., & Verloove-Vanhorick, S.P. (2005). Bullying: What does what, when and where? Involvement of children, teachers and parents in bullying behavior. *Health Education Research, 20*(1), 81–91.

Forgas, J., Williams, K.D., & Von Hippel, W. (2005). The social outcast: Ostracism, social exclusion, rejection, and bullying. East Sussex, UK: Psychology Press.

Hertz, M.F., Donato, I., & Wright. J. (2013). Bullying and suicide: A public health approach. *Journal of Adolescent Health, 53*(1), S1–S3.

Hirschi, T. (1969). *Causes of Delinquency. Berkley.* CA: University of California Press.

Lacey, R.E., Kumari, M. & Bartley, M. (2014). Social isolation in childhood and adult inflammation: Evidence from the national child development study. *Psychoneuroendocrinology. 50*(2014), 85–94.

Leung, B.P., & Silberling, J. (2006). Using sociograms to identify social status in the classroom. *The California School Psychologist, 11*(1), 57–61.

Levy, J., & Wubbels, T. (1992). Student and teacher characteristics and perceptions of teacher communication style. *Journal of Classroom Interaction, 27*(1), 23–29.

London, R. & Ingram, D. (2015). The health consequences of social isolation: It hurts more than you think. San Rafael, CA: Beyond Differences.

Ma, X. (2002). Bullying in middle school: Individual and school characteristics of victims and offenders. *School Effectiveness and School Improvement: An International Journal of Research, Policy and Practice. 13,* 63–89.

McCandless, B.R. & Marshall, H.R. (1957). A picture sociometric technique for preschool children and its relation to teacher judgments of friendship. *Child Development, 28*(2), 139–147.

McDougall P., Hymel S., Vaillancourt T. & Mercer L. (2001). The consequences of childhood rejection. In M.R. Leary (Ed), *Interpersonal Rejection* (pp. 231–247). New York, NY: Oxford University Press.

McNamara, B., & McNamara, F. (1997). Keys to dealing with bullies. Hauppauge, NY: Barron's.

Murray-Harvey, R., & Slee, P.T. (2010). School and home relationships and their impact on school bullying. *School Psychology International,* 31, 271–295.

Nansel, T.R., Overpeck, M., Pilla, R.S., Ruan, W.J., Simons-Morton, B., & Scheidt, P. (2001). Bullying Behaviors Among US Youth: Prevalence and Association With Psychosocial Adjustment. *JAMA, 285,* 2094–2110.

Olweus, D. (1994). Bullying at school long-term outcomes for the victims and an effective school-based intervention program. *Plenum Series in Social/Clinical Psychology. Agressive Behavior: Current Perspecives* (pp. 97–130). New York, NY: Plenum Press.

Qualter, P., Brown, S.L, Penny, M. & Rotenberg, K.J. (2009). Childhood loneliness as a predictor of adolescent depressive symptoms: An 8-year longitudinal study. *European Child & Adolescent Psychiatry, 19*(6), 493–501.

Rahardjo, S., & Gudnanto. (2013). *Pemahaman individu teknik nontes (Edisi revisi)* [Non-test technique of individual understanding (Revised edition)]. Jakarta, Indonesia: Kencana Prenadamedia Group.

Raskauskas, J.L., Gregory, J., Harvey, S.T., Rifshana, F., & Evans, I.M. (2010). Bullying among primary school children in New Zealand: Relationships with prosocial behaviour and classroom climate. *Educational Research, 52*, 1–13.

Remer, R. (1995). Strong sociometry: A definition. *Journal of Group Psychotherapy, Psychodrama & Sociometry, 48*(2), p69.

Richard, J.F., Schneider, B.H., & Mallet, P. (2012). Revisiting the whole-school approach to bullying: Really looking at the whole school. *School Psychology International, 33*, 263–284.

Richman, L.S. & Leary, M.R. (2009). Reactions to discrimination, stigmatization, ostracism, and other forms of interpersonal rejection. *Psychological Review, 116*(2), 365–383.

Riittakerttu, K., Rimpela, M., Paivi R., & Arja R. (2000). Bullying at school-an indicator of adolescents at risk for mental disorders. *Journal of Adolescence, 23*, 661–674.

Roland, E., & Galloway, D. (2004). Professional cultures in schools with high and low rates of bullying. *School Effectiveness and School Improvement: An International Journal of Research, Policy and Practice, 15*, 241–260.

Smokowski, P.R., & Kopasz, K.H. (2005). Bullying in school: An overview of types, effects, family characteristics, and intervention strategies. *Children & Schools, 27*(2), 101–110.

Urea, R. (2013). The impact of teachers' communication styles on pupils' self-safety throughout the learning process. *Procedia - Social and Behavioral Sciences. 93*(2015), 164–168.

Williams, B.T., & Gilmour, J.D. (1994). Annotation: Sociometry and peer relationships. *Journal of Child Psychology and Psychiatry, 35*(6), 997–1013.

Wubbels, T., & Brekelmans, M. (2005). Two decades of research on teacher-student relationships in class. *International Journal of Educational Research. 43*(1–2), 6–24.

Zettergeren, P. (2003). School adjustment in adolescence for previously rejected, average and popular children. *British Journal of Educational Psychology, 73*(2), 207–221.

Innovative Teaching and Learning Methods in Educational Systems – Retnowati et al. (Eds)
© 2020 Taylor & Francis Group, London, ISBN 978-1-03-224183-8

Analysis of students' mistakes in solving algebra word problems using the Newman Procedure

R. Keumalasari & Turmudi
Universitas Pendidikan Indonesia, Bandung, Indonesia

ABSTRACT: The process of learning mathematics in school does not always run smoothly because the ability of each student is different. One of the problems in learning mathematics is that students have difficulty understanding the material, which causes them to make mistakes. Analyzing students' mistake may reveal a faulty problem-solving process and provide information on the understanding of, and their attitudes toward, mathematical problems. One of the analytical procedures that has been developed for this is the Newman Procedure. Newman maintained that when a person attempted to answer a standard written mathematics word problem they had to be able to pass over a number of successive hurdles: (1) reading (or decoding), (2) comprehension, (3) transformation, (4) process skills, and (5) encoding. Based on the research results of 12 students from Class VII at a junior high school in Bandung, the percentage of mistakes students make in solving algebra word problems are as follows: reading (or decoding) 18.30%; 15.49%, transformation 18.30%; process skills 22.53%; and encoding 25.35%.

1 INTRODUCTION

The process of learning mathematics in school does not always" run smoothly because the ability of each student is different. One of the problems in learning mathematics is that students have difficulty in understanding the material, which causes them to make mistakes. Student mistakes the "illustrate" individual difficulties and they show that the student has failed to understand or grasp certain concepts, techniques, problems, etc. in a "scientific" or "adult" manner (Radatz, 1980).

In the development of learning mathematics in schools, emerging approaches that link between materials taught in schools and the students' daily activities are manifested through the use of word problems. Mathematics problems presented in the form of words that describe everyday problems require skills to interpret the material into the language of mathematics. The purpose of solving word problems is not just to get results in the form of answers; it is important that the student knows and understands the thinking process or steps to get the answer (Wahyuddin, 2016). Therefore a mathematical word problem exam is unique and more challenging than the ordinary mathematics task (Sajadi, Amiripour, & Malkhalifeh, 2013).

One of the difficulties experienced by students in mathematics learning is word problems. At all levels of mathematics, one of the most difficult topics to teach is how to solve word problems because there are varied and complex tasks involved at different stages of the process, and the procedure must be separated into several steps (Travis, 1981). In addition, one of the mathematical topics related to daily life is algebra. Many problems in everyday life that can be solved simply by the language of symbols in algebra, make algebra an important skill.

Therefore, the researcher analyzed students' mistakes in solving algebra word problems. Analyzing students' mistake may reveal the faulty problem-solving process and provide information on the understanding of, and the attitudes, toward mathematical problems. This analysis of students' mistakes aims to provide a deep understanding of the learning process. Students' mistake is an important tool used to diagnose the difficulties experienced by students who require direct help (Jabeen, 2015; Susanto, Farid, & Ulum, 2017).

One of the analytical procedures that has been developed is the Newman Procedure. Newman maintained that when a person attempted to answer a standard written mathematics word problem then the person had to be able to pass over a number of successive hurdles: (1) reading (or decoding), (2) comprehension, (3) transformation, (4) process skills, and (5) encoding (White, 2010). Further details of the five phases of the procedure according to the NEA are as follows. The reading phase (or decoding) is the stage where students read and understand the question. Students may make mistakes because they cannot read and understand the terms in question. Such mistakes are characterized by the inability of students to write what is known in the question. The comprehension phase is where the student decides to understand the problem. Mistakes in this phase are characterized by the inability of students to write what was asked for or makes errors, in part, on what is known.

The next phase is transformation. The students should be able to write a mathematical model according to the questions asked. When students cannot use the right strategy in word problems or make a mistake in using what is known in the strategy, then the students have problems with this phase. In the process skill phase, the students can solve word problems that have been modeled with the rules, procedures, or the appropriate algorithm. The last phase of the Newman Procedure is encoding. Students should be able to write down the answers correctly in this phase. In the process skills phase, students may have conducted a series of procedures and the exact algorithm, but are not able or are less careful in rewriting what was required in the word problem. These types of mistakes will be observed in this study. If the last phase has been passed by the students, then they have solved the word problem. In this research, student mistakes are analyzed to fulfill the purpose of the research can be achieved, which is to describe the mistakes made by students and determine the factors that cause student mistakes when completing algebraic word problems.

2 METHOD

This study aims to analyze students' mistakes in solving algebra word problems based on Newman procedures, to make it easier for teachers to identify student mistakes and so students can understand the material more easily. A qualitative method was chosen as research method. Qualitative research is characterized by an interpretative paradigm, which emphasizes subjective experiences and the meanings they have for an individual (Starman, 2013). Furthermore, qualitative researchers seek to understand a phenomenon by focusing on the total picture rather than breaking it down into variables (Ary, Jacobs, & Sorensen, 2010).

The research was conducted at SMP Kartika Bandung. The subjects were 12 students of Class VII C in the academic year 2017/2018. The selection of research subjects was conducted by purposive sampling (Fraenkel, Wallen, & Hellen, 2012). From the result of analyzing the students' work, several mistake were chosen that represented Newman's five types of error: reading (or decoding), comprehension, transformation, process skills, and encoding. Data collection techniques used in this study included test methods to obtain data that will be processed and analyzed, interviews to determine the cause of errors, and documentation to obtain data documentation research.

The validity of the instruments used in this study was content validity. Validity testing is done by examining or reviewing the items that have been determined by the validator. Data validity is done by the triangulation method, by comparing data from student test results, interview result data, and documentation so that the data obtained through each method can be used as a complement to strengthen the data of the research results. The data analysis technique in this research was three activities: (1) data reduction by correcting the results of students' work, which is then analyzed to find the types of mistakes made when solving algebra word problems, and determine which students will be the subjects for the interviews; (2) data presentation, which is the process of collecting data from organized research results to find any mistake made by students in solving algebra word problems and the causes, (3) data verification and conclusion, which are the activities that can answer the problem in research.

3 RESULT AND DISCUSSION

This research instrument uses algebra word problems. The test question consisted of two questions that feature algebraic problems in everyday life situations. The mistakes made by students can be seen in Table 1.

Table 1 shows that the percentage of mistakes made in solving algebra word problems in the respective phases are as follows: reading (or decoding) 18.30%; comprehension 15.49%; transformation 18.30%; process skills 22.53%; and encoding 25.35%. The lowest percentage was for mistakes in understanding the problem (comprehension). This indicates that most students have understood what was being asked in the question. However, in the reading (or decoding) phase, there are still many students who did not accurately write what was known in the question. In addition, the highest percentage of mistakes were in the encoding phase, indicating that the previous phases are very influential on the last phase. If students have made mistakes in the previous phase this will influence the encoding phase. Also, students are not used to writing the final answer to the problem, which made it difficult for the researchers to trace the mistake that occurred in this final process. There are still many students who are not able to represent the information asked in the question as a whole.

The mistakes made in the reading phase (or decoding) totaled 18.30%. Based on the analysis of test results and interview results, some students made mistakes in interpreting the meaning of the problem and determining what is known. Although students could read the problem fluently and did not experience significant difficulties, there were still students who could not interpret the sentences appropriately.

From the students' answers, it appears that they made mistakes in understanding what was asked in the question. Some students do not seem to understand the sentences in questions such as, "The width of the park is 2 meters shorter than its length, while its length is 3 times longer than the longest pool diagonal, when the length of the diagonal of each pool is 6 meters and 8 meters".

In Figure 1 below, the student interpreted the sentence by writing l = 2 m and p = 3 m. Based on the results of the interview, it can be seen that the student did not understand the mathematical sentence contained in the word problem. Figure 2 shows that another student understood the sentence by writing width = 2 meters and length = 24. Based on the interview results, the student showed understanding with the phrase "the length is 3 times longer than

Table 1. Type of student mistake based on the Newman Procedure.

Type of student mistake	Question		Total	Percentage
	1	2		
Reading (or decoding)	4	9	13	18.30%
Comprehension	3	8	11	15.49%
Transformation	4	9	13	18.30%
Process skills	6	10	16	22.53%
Encoding	7	11	18	25.35%

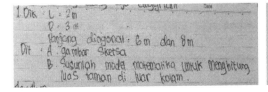

Figures 1 and 2. Examples of students mistake (reading).

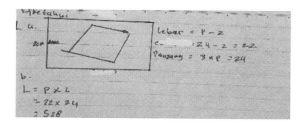

Figures 3. Examples of students' mistake (comprehension).

the longest pool diagonal" because the length of the diagonal of the second pool is known. However, he claimed to have made a mistake in writing a mathematical sentence to determine the width because of not reading the problem carefully. Based on the analysis of test results and interview results, it can be concluded that the mistakes made in the reading phasewere caused bystudents not reading and understanding the questions thoroughly, and students were lacking creativity in identifying real problems in the mathematical model.

In the comprehension phase, the students generally failed to understand the problem as a whole and were less careful in sorting information. The percentage for student mistakes made in this phase was 15.49%. Although this was the lowest compared to other types of mistake, the mistake in understanding what was being asked in the question results in whether or not the correct final answer will be obtained.

Based on the results of the interview, students admitted to making mistakes in interpreting the questions contained in algebra word problems. When interviewed further, it transpired thatthe students thought "park area outside the pool" in the problem meant the area of the rectangle. Therefore, the students only used the formula of the rectangle area and assumed that was all they had to do. The analysis of the test results and interview results showed that the mistakes were made in the comprehension phase because students did not understand the problem thoroughly.

Furthermore, the mistakes made in the transformation phase totaled 18.30%. Some students made mistakes in designing a mathematical model based on a given word problem so in the next step they got the wrong result.

Figures 4 and 5 show that students made mistakes in transforming what was known into the form of mathematical models. This meant they got the wrong answer in the next step. Based on the results of the interview, students made mistakes in determining the length. They understood what was meant in the problem, but could not write it using equations. From this mistake, students also made mistakes in the next step, and as a result could not find the right solution. One student wrote "the park width is 2 meters shorter than its length" with "l = 2 - p" and the other students interpreted it as "6 - 2 = 4".

Based on the analysis of test results and student interview results it can be concluded that the students' mistakes in the transformation phase are due to a lack of ability in

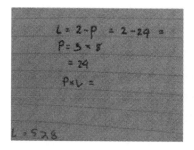
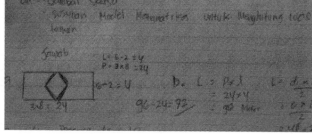

Figure 4 and 5.   Examples of students' mistake (transformation).

transforming the words into the mathematical model and also the students did not understand how to form the equations for algebraic problems known in the word problem.

In the process skills phase, 22.53% of mistakes were made. Based on the analysis of test results and student interviews, the mistakes made in solving the algebra story problem include moving the numbers and variables of different sections and doing the algebra calculation which adds two variables. Figure 6 shows that one of the students made mistakes in the calculations on the fourth line. Although the previous line was correct, a mistake in the process skills phase resulted in an incorrect final answer. Ananalysis of the test results and interviews shows that process skills mistakes were made due to students' lack of accuracy in doing algebraic operations and rushing the calculations.

In the encoding phase, 25.35% of mistakes were made. This was where the highest percentage of errors were made by students. Figure 7 shows that the student was able to solve the problem up to the process skills stage but could not complete the stage of encoding. This mistake was very unfortunate because the student had successfully reached the stage of data processing but failed to get the final solution. This error occured because of clerical errors rather than the student the issues in question.

Based on the results of the interview, students admitted they forgot and did not carefully read the commands contained in the problem, so they thought the answer was completed until the search for the value of the variable, but they did not substitute the variable into the requested equation. In addition, many students fare not used to writing conclusions. This shows that encoding mistakes made by students were due to a lack of ability and accuracy in understanding the command and the students were also not familiar with writing conclusions.

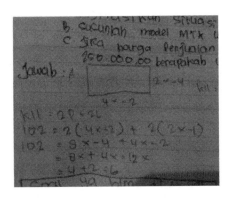

Figure 6.   Examples of students mistakes (process skills).

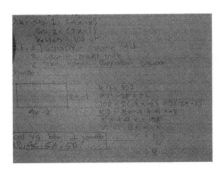

Figure 7.   Examples of students mistake (encoding).

4 CONCLUSION

The results of the research show students' mistakes in solving algebra word problems based on Newman's Procedure. Such mistakes include reading (or decoding), comprehension, transformation, process skills, and encoding. The most common mistake students make is encoding, at 25.35%. Students make this mistake because they do not carefully read the commands contained in the problem, so think the answer is completed bysearching for a variable value only and do not substitute the variable into the requested equation.

The second most common mistake made by students is with the process skills, at 22.53%. This mistake occurs when the student moves the numbers and variables of different sections and makes errors in algebraic calculations. Transformation mistakes accounted for 18.30% of the total. Students make a mistake in transforming what is known into the form of a mathematical model. Reading (or decoding) mistakes were 18.30% of the total, made when students incorrectly interpreted the meaning of the problem. The comprehension mistakes were the least common at 15.49%, where the student has not understood the algebra word problem so made a mistake in interpreting the questions.

The results of the above research can provide guidance for teachers and other researchers. Teachers should be able to find out the mistakes made by students in word problems and can provide effective teaching based on these mistakes. Other researchers may conduct further research on the development of students mistake analysis tools for algebra or other topics.

ACKNOWLEDGMENTS

We thank the students and teacher for their participation and help in this research.

REFERENCES

Ary, D., Jacobs, L.C., & Sorensen, C.K. (2010). *Introduction to research in education.* Canada: Wadsworth Cengage Learning.

Fraenkel, J.R., Wallen, N.E., & Hellen, H.H. (2012). *How to design and evaluate research in education* (8th edn.). New York, NY: McGraw Hill Companies Incorporate.

Jabeen, A. (2015). The role of error analysis in teaching and learning of second and foreign language. *Education and Linguistics Research, 1*(2), 52–61. doi: 10.5296/elr.v1i1.8189

Radatz, H. (1980). Student's errors in the mathematical learning process: a survey. *For the Learning of Mathematics, 1*(1), 16–20.

Sajadi, M., Amiripour, P., & Malkhalifeh, M.R. (2013). The examining mathematical word problems solving ability under efficient representation. *Aspect Mathematics Education Trends and Research,* 1–11. doi: 10.5899/2013/metr-00007

Starman, A.B. (2013). The case study as a type of qualitative research. *Journal Of Contemporary Educational Studies. 1*(1), 28–43.

Susanto, D.A., Farid, A., & Ulum, B. (2017). Error analysis of students working about word problem of linear program with NEA procedure. *Journal of Physics: Conference Series, 855,* 1–8. doi: 10.1088/1742-6596/855/1/012043

Travis, B.P. (1981). *Error Analysis in Solving Algebra Word Problems.* Retrieved from https://files.eric.ed.gov/fulltext/ED209095.pdf

Wahyuddin. (2016). Analisis kemampuan menyelesaikan soal cerita Matematika ditinjau dari kemampuan verbal [Analytical ability to solve math word problems in terms of verbal ability]. *Tadris Matematika, 9*(2), 148–160. doi: 10.20414/betajtm.v9i2.9

White, A.L. (2010). Numeracy, literacy and Newman's error analysis. *Journal of Science and Mathematics Education in Southeast Asia, 33*(2), 129–148.

Lesson study as an alternative for teacher creativity development in reflecting and improving the quality of learning of the Indonesian language

R.W. Eriyanti
Universitas Muhammadiyah, Malang, Indonesia

ABSTRACT: Regulation of the Minister of National Education Number 16 of 2007 mandates that every teacher should develop their profession sustainably by reflecting on learning and improving the quality of their learning through classroom action research. However, in reality teachers have difficulty in implementing this because of a lack of understanding of classroom action research methodology and experience in conducting it. The purpose of this activity is to improve teachers' understanding and experience in improving and reflecting on the quality of learning. A lesson study was conducted involving Indonesian teachers. With this lesson study, teachers' creativity was found to improve when reflecting on learning, they designed better learning, and learning increased. In addition, teachers become more confident in carrying out learning.

1 INTRODUCTION

Teachers are professional educators and special skills are needed to perform professional duties (RoI., 2005). The Ministry of Education in Indonesia mandates that a teacher must meet the minimum academic qualification requirement of Strata 1 (S-1) or Diploma 4 (D-4) and have professional, pedagogical, social, and personal competence. The four competencies are integrated into teacher performance (RoI, 2007).

Pedagogic competence is an important aspect for teachers, which includes the ability to understand the characteristics of students in terms of physical, moral, spiritual, social, cultural, emotional, and intellectual aspects. This is necessary because the teachers face various characteristics each day. Moreover, they should understand the theory of learning and the principles of educational learning. In addition, teachers need to understand and be able to develop a curriculum of subjects that are taught; and they should organize learning that educates and uses information and communication technology. Teachers should facilitate the development of potential students to actualize their potentials and should be able to communicate effectively, empathetically, and courteously with students. They must also conduct assessment and evaluation of learning processes and outcomes, utilize the results of the assessment and evaluation for the interests of learning, and take reflective action to improve the quality of learning (Hakim, 2015).

Personality competence, including acting in accordance with Indonesian religious, legal, social, and national norms is typical for Indonesian teachers. They should present themselves as honest, noble, role models for students and society, and steady, stable, wise, and authoritative. Displaying a strong work ethic, high responsibility, a sense of pride and self-confidence and upholding the professional code of ethics are typical characteristics of teachers because they act as a role models for the students (Abduh & Zainudin, 2016).

Social competence for teachers this includes being inclusive, objective, and non-discriminatory of gender, religion, race, physical condition, family background, and socioeconomic status. Teachers may communicate effectively, empathically, and courteously with fellow educators, education personnel, parents, and the community. They are expected to adapt these characteristics depending on where they work throughout the territory of the Republic of Indonesia, which has vast socio-cultural diversity, and communicate within their professional communities and with other professions orally and in writing or other forms. This is necessary for teachers because they also have to establish direct communication with students and parents, and other citizens.

Professional competence is an essential aspect for teachers. This includes understanding the materials, structures, concepts, and scientific thinking patterns that support the subjects they teach. They should also develop creative learning materials in a sustainable manner by taking reflective action and use information and communication technology to communicate and develop their skills. Professional competence of teachers in the field of science is in accordance with the special education level for science teachers.

Teacher's duties and responsibilities are not only limited to the development of the students' physical aspects, but the psychological, spiritual, personal, and intellectual aspects are also important. All aspects must be developed completely so that each student can develop his personality completely.

The developments of science and technology put pressure on human behavior to meet the needs and demands of life. In the field of education, this raises new awareness of teachers, who prepare students and future young people that are able to respond to the progress of science and technology and to community needs. In relation to this, teachers need to improve their professional abilities so as to improve the quality of their learning in order to produce graduates with skills for the future.

It has been stipulated in the Regulation of the Minister of National Education, particularly on the aspect of pedagogical and professional competence, that teachers should take reflective action to improve the quality of learning and develop professionally in a sustainable manner. This implies that in carrying out learning tasks, the teacher must reflect on an ongoing basis in order to make sustainable improvements in the quality of their learning. Based on the results of reflection, improvements in learning in accordance with needs are then proposed for teachers.

In reality, teachers have difficulty in implementing lesson study in class. This can be seen in this study from the lesson and class action research designs prepared by the teacher. In the teacher-designed learning plan, there was a reflection activity in closing activities (Ono, Chikamori, & Rogan, 2013). However, there was no teacher creativity to implement the reflection of learning. The reflection activity only contains activities to make a summary or conclusion of materials learned by students. Thus, the learning improvement activities designed by the teacher are not based on the results of reflection.

Thus, it is necessary to assist teachers in improving the quality of their learning in the form of lesson study. Lesson study was chosen as a way to increase the creativity of teachers in improving the quality of learning, through studying with colleagues and when teachers feel open to expressing their impressions during learning.

2 METHOD

The development of teachers' creativity in Indonesian language learning was conducted through lesson study activities on a school basis. The lesson study participants were all teachers in two schools, up to 105 participants, who taught various subjects and prior to the lesson study, groups were formed based on similar subjects taught. Teacher creativity development was the focus for teachers of Indonesian. In the group, the teachers decided who would model the lesson as the teacher and who would be observers, and what the focus of learning would be.

Lesson study was carried out in four stages: learning planning, learning implementation, observation, and reflection on learning (Lewis, Perry, & Murata, 2006). The learning planning phase was implemented collaboratively. The purpose of activities learning design activities is to produce a design that helps students study effectively and generates active participation by students in the learning. The activities began with the model teacher creating a learning plan/learning design based on the focus decided earlier. Then, the learning plan was discussed to get input for refinement from other teachers. The learning plan was then simulated before it was applied in the class. At this stage, observation procedures and instruments required in the observation were also established.

The learning implementation phase was based on the learning plan designed collaboratively. One teacher modeled the lesson and others performed observations using the instruments prepared earlier. Observation activities during the lesson focused on the student activities. The purpose of the observer was to study the ongoing learning process.

The observation and reflection stage is intended to identify the advantages and disadvantages of learning implementation. The teacher who acts as the model initiates discussion through the delivery of impressions during the lesson. Discussion activities continued with the delivery of input from observers. In addition, the lesson designed in the learning process was also delivered.

3   RESULTS AND DISCUSSION

The purpose of this lesson study is to develop the creativity of teachers in implementing Indonesian language learning, especially in designing learning activities, implementing innovative learning, evaluating and reflecting on learning. The results of this activity show that Indonesian teachers are now more creative in designing, implementing, evaluating, and reflecting on Indonesian language learning. The results of this activity are set out in Table 1.

Table 1 shows that through lesson study, teachers' creativity in Indonesian language learning can be developed. Teachers were able to design learning effectively, creatively, and innovatively. This can be achieved by teachers through reflective learning activities in a collaborative way. Through reflection activities, teachers knew the weaknesses of learning that was carried out and were able to diagnose the causes in order to improve the next learning effort appropriately.

Sunardi and Sujadi (2017) stated that reflection activities need to be done by the teacher after the completion of learning process. Through reflection, teachers can identify the learning characteristics of each student, identify what empowers them, and diagnose the cause of the weaknesses. Thus, learning improvement activities can be designed for the next learning. The event is only implemented through lesson study.

Through lesson study, teachers could share information about learning activities that were implemented. Teacher observers could provide more positive information about the advantages and disadvantages of learning implemented by the model teachers. On the other hand, the observers could also learn together from the model teachers. Model teachers could convey impressions during perceived learning.

Table 1 also shows that through lesson study, language teachers improved their learning implementation. Teachers built contexts more creatively, could chose the teaching materials in accordance with the competence and characteristics of learners, creatively selected and used strategies and learning media that motivated students, and the lesson was more effective.

After the lesson study was implemented, the teacher carried out the learning reflection that had been implemented. Teachers could precisely identify the way students learned, evaluate the learning that has been done, and determine follow-up learning. Teachers were also increasingly aware of the shortcomings and advantages of learning.

This is in accordance with the results of lesson study conducted by Lutfi, et al. (2008) that during the first year, students felt the lesson moreopen than usual and more enjoyable., The lesson study participated teachers stated that they were motivated to offer better learning and

Table 1. Development of teacher creativity in text-based Indonesian lessons.

No.	Developed aspect	Before activity	After activity
1.	Designing text-based Indonesian Lessons	- Do not understand the purpose of learning Indonesian - Difficult to distinguish competence of knowledge and competence of Indonesian language skill - Difficult to elaborate basic competencies into achievement indicators of competence - Less creative to build the context of Indonesian language learning - Dependant in determiningthe teaching materials on textbooks - Less creative in choosing a strategy and designing learning scenarios. - Difficulties in developing assessment instruments - Difficulties in developing assessment instruments	- Understand the purpose of learning Indonesian at school - Able to distinguish the competence of knowledge and skills of Indonesian language - Able to describe basic competencies as indicators of achievement of competence - Creative build context of learning Indonesian - Able to develop creative teaching materials and relate them to reality - Creative in choosing learning strategies in accordance with the competence and characteristics of learners. -Able to develop assessment instruments
2.	Implementing problem-solving learning Bahasa Indonesia	- Less involving students in the useof instructional media - Less able to create a creative and innovative learning environment - Less able to activate in learning the Indonesian language - Less creative in the use of questions to guide students in learning Indonesian - Creative in involving students in the utilization of instructional media	- Able to create a creative and innovative Indonesian learning environment - Able to actively engage in learning the Indonesian language - Creative in the use of questions to guide students in using the Indonesian language
3..	Evaluating and reflecting on Indonesian lessons	- Difficulties in carrying out process assessments because the instruments are not yet fully available - Cannot determine KKM yet Have not been able to distinguish the application of assessment approach (PAP and PAN) - Not yet creative determine the follow-up evaluation results - Have not been able to learn reflecting techniques	- Able to carry out process assessments effectively and efficiently - Can determine the KKM - Can differentiate the application of assessment approach (PAP and PAN) - Creative in determining follow-up evaluation results - Able to apply techniques to reflect on learning

follow the lead of their peers who had implemented the learning. A school principal's view of school-based lesson study impacts the school's popularity in the surrounding community.

4 CONCLUSION

Teachers are required to develop their professionals in a sustainable manner. Lesson study is one alternative strategy that can be implemented to develop teachers' creativity in improving the learning implemented. Through lesson study, teachers can share knowledge and experience of learning. Thus, teachers can reflect on the lessons learned and follow-up in the form of improved learning.

REFERENCES

Abduh, M., & Zainudin, A. (2016). The role of Indonesian teachers' competencies in developingchild-friendly school. *Proceedings of International Conference on Child-Friendly Education* (pp. 164–172). Surakarta, Indonesia.

Hakim, A. (2015). Contribution of competence teacher (pedagogical, personality, professional competence and social) on the performance of learning. *The International Journal of Engineering and Science*, *4*(2), 1–12.

Lewis, C., Perry, R., & Murata, A. (2006). How should research contribute to instructional improvement? The case of lesson study. *SAGE Journal*, *35*(3), 3–14.

Lutfi, A., Mitarlis, Muchlis, Yonata, B., & Novita, D. (2008). *Aksi*, *9*(1), 22–30.

Ono, Y., Chikamori, K., & Rogan, J. M. (2013) How reflective are lesson study discussion sessions? Developing an instrument to analyze collective reflection. *International Journal of Education*, *5*(3), 52–67.

RoI. (2005). *Undang-undangNo.14 Tahun 2005 tentang guru dan dosen* [Law number 14 year 2005 concerning teachers and lecturers]. Jakarta, Indonesia: Ministry of Education, Republic of Indonesia.

RoI. (2007). *Undang-undangNo.16 Tahun 2007 tentang standar kualifikasi akademik dan kompetensi guru* [Law number 16year 2007 concerning academic qualification standard and teachers' competencies]. Jakarta, Indonesia: Ministry of Education, Republic of Indonesia.

Sunardi, & Sujadi, I. (2017). *Sumber belajar calon peserta program PLPG penilaian dan evaluasi pembelajaran* [Learning resources for candidates in the PLPG program for learning assessment and evaluation]. Jakarta, Indonesia: Ministry of Education.

Innovative Teaching and Learning Methods in Educational Systems – Retnowati et al. (Eds)
© 2020 Taylor & Francis Group, London, ISBN 978-1-03-224183-8

Lesson study to improve teacher creativity in solving problems of mathematics learning

Y.M. Cholily
Universitas Muhammadiyah Malang, Malang, Indonesia

ABSTRACT: Mathematics is considered to be as a difficult lesson by students. Mathematics learning materials are abstract and formulated by language. Teachers' understanding of the importance of the mastery of language by students in learning mathematics tends to be limited, so the focus of learning is still on the problem-solving procedure. This Lesson study activity was conducted to improve the teacher's creativity in problem-solving learning in mathematics. Activities were carried out collaboratively, from determining problems, designing problem-solving learning, implementing problem-solving learning, evaluating, and reflecting on learning. With lesson study, teachers are more creative in designing, implementing, evaluating and reflecting on problem-solving learning in mathematics.

1 INTRODUCTION

Mathematics has a very important role in daily life, the development of science, and the development of thinking ability (Shaikh, 2013). In everyday life, mathematics is necessary for problem-solving in the fields of economics, industry, health, education, etc. (Ernest, 2015; Hodaňová & Nocar, 2016). Mathematics is also required for other sciences, such as economics, natural sciences, health sciences and medicine, physics, agricultural sciences, statistics, etc. (Hodaňová & Nocar, 2016). Rapid developments in the field of information and communication technology are based on the development of mathematics in the field of number theory, algebra, analysis, probability theory, and discrete mathematics (Ernest, 2015). Mathematics is also needed to develop thinking and reasoning abilities.

As mathematics is very important in various fields of life, it must mastered by all students in Indonesia. Through learning mathematics, students can develop logical, analytical, critical, and creative thinking skills, and systematics (Hodaňová & Nocar, 2016). Mathematics skills are required for students to master other lessons.

Mathematics is a very important lesson in class (Tan & Lim, 2010), and the purpose of learning mathematics in schools is to give students mathematical skills (mathematical literacy), not as math as a science. Mathematical learning in these subjects is intended to provide the knowledge and practical skills necessary for solving mathematical problems and applying mathematics to solving simple real-life problems (Ernest, 2015; Kaminski, Sloutsky, & Heckler, 2008).

Skills that must be achieved by students in learning mathematics in schools include aspects of knowledge. These skills include using the mathematical pattern as a conjecture in problem-solving, and making generalizations based on existing phenomena or data; communicating ideas, reasoning and constructing mathematical evidence using full sentences, symbols, tables, diagrams, or other media to clarify circumstances or problems; and having an appreciation of the usefulness of mathematics in life, curiosity, attention, and interest in learning mathematics, and a tenacious attitude and confidence in problem-solving. The skills or abilities are intimately linked, one reinforcing as well as relying on the other. Although not explicitly stated, the ability to communicate is required in a variety of skills, for example, to explain ideas on conceptual understanding, presenting formulas and problem-solving, or making arguments on reasoning. Thus it is clear that the ability to solve problems when learning mathematics in school is very important for students.

The characteristics of mathematics learning are that it is an abstract object study, the nature of truth is logical, and mathematics uses symbols in its application. This results in most students assuming that mathematics lessons are difficult to understand (Anthony & Walshaw, 2009; Dodeen, Abdelfattah, & Alshumrani, 2014; Park, 2006; Shaikh, 2013; Tan & Lim, 2010). The student may assume that mathematics lessons are difficult because their teachers are anxious about teaching the subject (Whyte & Anthony, 2012). Difficulties experienced by students in mathematics are also caused by a lack of understanding by teachers toward the direction of learning. Some teachers do not understand the purpose of learning mathematics in schools so the implementation of learning is more directed to the mastery of mathematics as a science. Teachers do not help students to link mathematics learning materials with real-life everyday issues. As a result, learning mathematics in schools is less meaningful for students.

Teachers as professional educators should develop their skills on a continual basis. The development of science and technology (IPTEK) at an extraordinary rate put pressure on human behavior to be able to meet the needs and demands of life. In the field of education, this raises a new awareness to revitalize the performance of teachers and education personnel in order to prepare students and young people to respond to the progress of science and technology, and the needs and demands of society. In connection, teachers are required to always improve their professionalism so as to improve the quality of learning. In this way, it is expected that schools will produce graduates who are able to face the future.

The demand for 21st-century skills requires teachers who are able to adapt to their environmental needs, understand their discipline in various contexts and respond quickly to the development needs of learners and society. Teachers are required to be critical, creative, and innovative in order to adapt to the rapidly evolving developments and needs of life. Professional teachers are capable of offering students interactive, inspirational, fun, challenging, motivational, active, creative, innovative and self-motivated learning in accordance with their talents, interests, and development needs (RoI, 2005).

It has been established in the regulations by the Minister of National Education, particularly for the aspects of pedagogical and professional competence, that teachers should take reflective action to improve their quality of learning and develop professionalism in a sustainable manner. This implies that teachers should reflect on the lessons learned in order to improve their skills. Therefore, teachers should develop their competencies in a sustainable way.

Teacher professionalism can be improved through formal and informal education. Nelson Mandela stated that education is a powerful weapon to change the world. Through education, the potential and competence of teachers (Challen, Siinalaanelaid, & Kukkes, 2017) and students can be developed.

One activity that can be done to increase the creativity of teachers in learning is lesson study. Lesson study is a model of professional education through collaborative learning and continuous learning based on the principles of collegiality and mutual learning to build learning a community (Susilo & Chotimah, 2010). Through the lesson study, teachers meet collaboratively to study the curriculum, define problems and learning objectives, design learning based on established goals, implement and observe learning, and reflect on discussing, reviewing and designing the learning improvements that have been implemented (Lewis, Perry, & Murata, 2006). This improves the effectiveness and efficiency of implemented learning.

A lesson study program to develop continuous professionalism of educators has been established since 2006 in the JICA Program under the name "Strengthening in Teacher Training of Mathematics and Science Education at Junior Secondary Level" (SISTTEMS). The results of this activity have shown that teachers' skills can be improved through lesson study and the ability of lecturers and teachers to improve their professionalism can be developed (Susilo & Chotimah, 2010).

Based on these problems, a school-based lesson study was conducted to develop teachers' creativity in implementing problem-solving learning in the field of mathematics, especially in designing learning activities, implementing innovative learning, evaluating and reflecting on problem-solving learning in mathematics.

2 METHOD

Teachers' creativity in problem-solving learning in mathematics can be developed through lesson study activities. Before the implementation of lesson study activities groups were formed (Lewis, 2006) based on teaching similar subjects. Lesson study participants consisted of teachers of mathematics, Indonesian, physics, biology, history, religion, physical and health, English, and school counselors.

In each group, the teachers decided who would play the roles of teacher and observers and determined the problems for the learning focus. Based on this group discussion each lesson study was conducted.

Lesson study is carried out in three stages: learning planning (plan), implementation of learning (do), observation and reflection of learning (see). The learning planning phase is implemented collaboratively in the teacher's group by subject. The purpose of activities to design learning is to produce a learning design that will teach students effectively and generate active participation by students in learning. The activities begin by the person in the 'teacher' role drafting a lesson plans based on the problem previously decided by the group. Then, the lesson plan is discussed to get input for refinement from other teachers. Furthermore, the lesson plan is role-played and observed before it is applied in the classroom and established observation procedures and instruments are required.

The learning implementation phase is based on the lesson plan designed collaboratively. The person in the 'teacher' role models the lesson while the others make observations using instruments that have been prepared. Observation activities focused on student activities during the lesson. The purpose of the observer is to study the ongoing learning process.

The observation and reflection stage is intended to identify the advantages and disadvantages of learning implementation. The teacher who acts as the model initiates the discussion through the delivery of impressions during the lesson. Discussion activities continue with input from observers. In addition, the lesson is delivered from the learning process.

3 RESULTS AND DISCUSSION

The purpose of this lesson study is to develop teachers' creativity in implementing problem-solving learning in the field of mathematics, especially in designing learning activities, implementing innovative learning, evaluating, and reflecting on learning. The results of this activity indicate that teachers are more creative in the design, implementation, evaluation, and reflection of problem-solving learning in mathematics. The results of this activity are set out in Table1.

Table 1 shows that learning in solving mathematics problems can be developed through lesson study. Teachers were able to design learning effectively, creatively, and innovatively. Teachers understood the purpose of learning mathematics; they could analyze basic competence, describe indicators of basic competence, formulate learning objectives, develop creative teaching materials and relate them to reality; and could design learning scenarios to teach students creatively and innovatively, and develop assessment instruments.

This can be achieved by teachers through reflective learning activities conducted in a collaborative way. Through reflection activities, teachers knew the weaknesses of learning and were able to diagnose the causes of these learning weaknesses so could design activities to improve the next learning efforts appropriately.

Table1. The development of master's creativity in mathematical problem-solving learning.

No.	Developed aspect	Before activity	After activity
1.	Designing learning problem-solving in mathematics	- Cannot understand the purpose of mathematics learning yet - Difficulties in understanding the basic competence of Mathematics - Difficulties in elaborating basic competencies into achievement indicators of competence - Difficulties in formulating learning objectives - Less creative linking materials with daily life - Less creative choosing strategy and designing innovative learning scenarios - Difficulties in developing assessment instruments	- Understands the purpose of learning mathematics - Able to analyze basic competence - Able to describe basic competencies as indicators of achievement of competence - Able to formulate learning objectives - Able to develop creative teaching materials and relate them to reality - Able to design learning scenarios that can teach students creatively and innovatively - Able to develop assessment instruments
2.	Implementing mathematical problem-solving learning	- Less involvement of students in the utilization of instructional media - Less able to create a creative and innovative problem learning environment - Less able to involve students in solving math problems - Less creative in the use of questions to guide students in problem-solving - Less able to involve students in the use of instructional media	- Able to create a creative and innovative learning environment - Able to make students in solving math problems - Creative in the use of questions to guide students in problem- solving
3.	Evaluating and reflecting learning mathematical problem-solving	- Difficulties in conducting process assessments because the instrument is not yet fully available - Cannot determine KKM yet - Has not been able to distinguish the application of assessment approach (PAP and PAN) - Not yet creatively determining the follow-up evaluation results - Has not been able to learn reflecting techniques - Able to carry out process assessments effectively and efficiently	- Can determine the KKM - Can differentiate the application of assessment approach (PAP and PAN) - Creativity determines follow-up evaluation results - Ability to apply reflective learning techniques

Sunardi and Sujadi (2017) stated that reflection activities should be done by the teacher after the completion of continuous learning. Through reflection, teachers can identify the learning characteristics of each student, identify the empowerment carried out, and diagnose the cause of any weaknesses. Therefore learning improvement activities can be designed. Those activities are implemented through lesson study.

Through lesson study, teachers can share information about learning activities that were implemented. Observer teachers could provide positive information about the advantages and disadvantages of learning implemented by the role-play teachers and conversely, they can also learn together from the lesson given by the role-play teacher. Whereas the role-play teacher could give feedback related to the learning process.

Table 1 shows that through lesson study, teachers improved their learning implementation. Teachers were more creative in building contexts, choosing teaching materials in accordance with the competence and characteristics of learners, selecting and utilizing strategies and learning media, so their lesson was more effective.

After the lesson study was implemented, the creative teacher carried out the learning reflection related to the activities had been done. Teachers could precisely identify how students learned and were introspective about how the learning was done, so it could determine follow-up learning. Teachers were also increasingly aware of the shortcomings and advantages of learning.

4 CONCLUSION

Teachers are required to develop their professional skills in a continuous manner. Lesson study is one alternative strategy that can be implemented to develop teachers' creativity in improving learning. Through lesson study, teachers can share knowledge and experience of learning. Thus, they can reflect on the lessons learned and follow-up in the form of improved learning.

REFERENCES

Anthony, G., & Walshaw, M. (2009). Characteristics of effective teaching of mathematics: A view from the West. *Journal of Mathematics Education, 2*(2), 147–164. doi: 10.21831/JPE.V1I2.2633

Challen, V., Siinalaanelaid, Z., & Kukkes, T. (2017). A qualitative study of perceptions of professionalism amongst radiography students. *Radiography, 23*(1), 23–29.

Dodeen, H., Abdelfattah, F., & Alshumrani, S. (2014). Test-taking skills of secondary students: the relationship with motivation, attitudes, anxiety, and attitudes towards tests. *South African Journal of Education, 34*(2), 1–18. doi: 10.15700/201412071153

Ernest, P. (2015). The social outcomes of learning mathematics: Standard, unintended or visionary? the standard aims of school mathematics unintended outcomes of school mathematics. *International Journal of Education in Mathematics, Science and Technology, 3*(3), 187–192. doi: 10.18404/ijemst.29471

Hodaňová, J., & Nocar, D. (2016). Mathematics importance in our life. Conference: 10th annual International Technology, Education and Development Conference (INTED2016), At Valencia, Volume: INTED2016 Proceedings (pp. 3086–30863092). doi: 10.21125/inted.2016.0172

Kaminski, J.A., Sloutsky, V.M., & Heckler, A.F. (2008). The advantage of abstract examples in learning math. *Science, 320*(5875), 454–455.

Lewis, C., Perry, R., & Murata, A. (2006). How should research contribute to instructional improvement? the case of lesson study. *Educational Researcher, 35*(3), 3–14.

Park, H.S. (2006). Development of a mathematics, science, and technology education integrated program for a Maglev. *Eurasia Journal of Mathematics Science and Technology Education, 2*(3), 1–14.

RoI (2005). *Peraturan pemerintah Republik Indonesia Nomor 19 Tahun 2005 tentang standar nasional Pendidikan* [Republic of Indonesia Government Regulation Number 19 of 2005 concerning national education standards]. Jakarta, Indonesia: Republic of Indonesia.

Shaikh, S.N. (2013). Mathematics anxiety factors and their influence on performance in mathematics in selected international schools in Bangkok. *Journal of Education and Vocational Research, 4*(3), 77–85.

Sunardi., & Sujadi, I. (2017). *Sumber belajar calon peserta PLPG* [Learning resources for PLPG candidates]. (Sadjidan, Ed.). Jakarta: Kementerian Pendidikan dan kebudayaan Dirjen Guru dan Tenaga Kependidikan.

Susilo, H., & Chotimah, H. (2010). *Lesson study berbasis MGMP sebagai sarana pengembangan keprofesionalan guru* [Lesson study based on MGMP as a means of developing teacher proficiency]. Malang: Surya Pena Gemilang.

Tan, S.F., & Lim, C.S. (2010). Effective mathematics lesson from the lenses of primary pupils: Preliminary analysis. *Social and Behavioral Sciences, 8*(5), 242–247.

Whyte, J., & Anthony, G. (2012). Maths anxiety: The fear factor in the mathematics classroom. *New Zealand Journal of Teachers' Work, 9*(1), 6–15.

Mapping the innovation of Professional Learning Communities (PLC) in primary schools: A review

P.S. Cholifah & H.I. Oktaviani
Universitas Negeri Malang, Malang, Indonesia

ABSTRACT: Professional Learning Communities (PLC) for teachers are part of a Continuing Professional Development (CPD) scheme. There have been many innovations published in PLC studies articles during its development. This article aims to provide an overview of the types of PLCs developed, the participation types, and a complete picture of PLC gathered from the perceptions of PLC members, as part of mapping the innovations related to this theme. This study describes a systematic review of several studies in the primary school context related to PLC. In total, 22 studies from 2009 to 2018 were identified that met the selection criterion. Result findings indicated several key points. First, three different types of PLC delivery were identified: face-to-face (F2F) sessions, online synchronous sessions, and blended learning sessions. Second, there were several types of PLC participation sessions including, scaffolding, coaching and mentoring, reflective dialogue, leadership, and strong relationships. The future limitations of this study will be discussed in this paper.

1 INTRODUCTION

Commitment to improving professionalism is an absolute necessity in all fields, including primary education. As one of the key professions in the field of education, it is fundamental that teachers continually improve their professional skills. This is based on the general idea that teachers are lifelong learners who must constantly develop and improve their professional competencies.

One thing that has been done in increasing professionalism is related to the production of knowledge that involves collaboration through professional learning communities (PLC). Collaboration offers a simultaneous process in supporting individual and organizational capacity building, which assumes the existence of focus dissemination activities, the obligation to learn, and a disciplined approach to achieving common goals (ASCD, 2000). Furthermore, collaborative learning communities can inspire and encourage teachers to commit to professional development as a priority in their work (Lassonde & Israel, 2010).

PLC, which is part of a continuing professional development (CPD) scheme, has been widely used and studied since it was first mentioned in publications, and there have been various PLC innovations. This systematic review aims to provide complete insight into the urgency of the PLC that involved the teachers to become a professional figure. This article gives an overview of the types of PLCs developed and a complete picture of PLC identification from the perspectives of PLC members.

2 LITERATURE REVIEW

2.1 *Professional teacher*

The main goal of the systematic process in the professional development of teachers is to improve the quality of learning undertaken. Guskey (2003) emphasized that professional development aims to improve student learning outcomes and processes. Several studies have described significant results regarding the implementation of teacher professional development and student achievement (Powers, et al., 2015; Hill, Bicer & Capraro, 2017).

Several models have been developed both in the context of teacher professional development (TPD) in general and of sustainable professional development (CPD). Lieberman (1995) classified three forms of TPD: through direct teaching (such as courses, workshops, conferences, etc.); studying in schools (such as mentoring, coaching, action research, and peer-teaching); and learning outside school (such as network, school collaboration, partnership, etc.). In the context of the development of sustainable professionalism, Kennedy (2005) initiated CPD models to improve teacher professionalism, including training, cascades, coaching/mentoring, community of practice, action research and transformation.

2.2   *Professional learning community*

The main emphasis in this paper, related to the forms and models of professional development, prioritizes networking and collaboration specifically carried out through community practice. Wenger (1998) stated that community practice is a broader development of the social system of learning.

The assumptions that arise regarding adult learning refer to the development of a PLC. ASCD (2000) reveals five fundamental things related to PLCs that are within the scope of adult learning: (1) inquiry and placement of assumptions that depend on the learning process; (2) the learning is a continuous process; (3) learning is carried out by learners related to meaningful issues; (4) learning is a natural experimental process; and (5) learning must be fulfilled with abundant, diverse, and easily accessible sources. Research also indicates that teachers involved in professional development, especially when they have similar vision and goals, tend to be more able to improve their skills (DuFour, 2004).

Hord (2009) outlined six essential points of a PLC: (1) shared values or visions among PLC members at a school; (2) supportive leadership across the community; (3) supportive conditions, such as time, system, place, or resources; (4) supportive relational components among the PLC members; (5) collective learning to elevate the students' achievement; and (6) sharing practices among PLC members in gaining feedback and support. Those essential features could be generated by several activities, such as observation of peers, experimentation, developing new methods, and collaborative in conducting action research. Sampoong (2015) described the experiences of PLC in undertaking collaborative works, such as planning, discussing, and reflecting small group work informally every day. Furthermore, those practices can be generated in a PLC among teachers within a school, several schools, or by schools and advisors such as universities or teacher centers (du Plessis &Muzaffar, 2010).

3   METHOD OF THE RESEARCH

A meta-synthesis is useful for identifying patterns in a topic, especially when it comes to selecting or developing a program or policy (Erwin, Brotherson, & Summers, 2011). The mechanism of this study type promotes a better understanding of the topics, by reviewing existing studies to build a connection, especially when identifying gaps in the published research. The meta-synthesis process used in this study followed the processes below.

Related studies were identified through database searches via ERIC and Google Scholar, with the study focuses of peer-reviewed, open access, and written in English. This process involved selecting papers published between 2008 and 2018 using the following key terms: "primary school", "elementary school", "professional learning community", "professional learning communities", and "PLC". We excluded articles from a journal or conference proceeding that consisted of a systematic review or book chapter.

As many as 213 articles from the first screening phase related to the main term used. After reading the title, keywords, publication type, and abstract of each, 34 items were identified and read comprehensively by both authors. After the screening process 22 articles were selected for the last step of analyzing the data.

4 RESULTS

4.1 *Data presentation*

Before answering all the research questions, the first part of this results analysis was the method and the participants. In total, 11 (50%) studies used qualitative research, 3 (14%) used a mixed-method approach, and 8 (36%) used quantitative research. For quantitative research, most of the studies identified used a survey and correctional design, while case study design was used in most of the studies with a qualitative approach. Regarding participants, the studies mainly involved primary teachers, principals, and only three included staff members and pupils in their research.

4.2 *Type of PLC*

The type of PLC used in this review was based on these activities occurring during all the PLC sessions. For this purpose, we categorized the studies into three different categories: face-to-face, online, and blended PLC sessions. Face-to-face sessions include regular meetings among PLC members for training, coaching, and mentoring, or discussions using video-type analysis (Cranston, 2009; Webb, Vulliamy, Sarja, Hämäläinen, & Poikonen, 2009; Leclerc, Moreau, Dumouchel, Sallafranque-St-Louis, 2012; Sleegers, den Brok, Verbiest, Moolenaar, & Daly, 2013; Chaseling, Boyd, Robson, & Brown, 2014; DeMatthews, 2014; Owen, 2014; Asanok & Chookhampaeng, 2015; Fahara, Quintanilla & Bulnes, 2015; Sampoong, et al., 2015; Smith, 2015; Gee & Whaley, 2016; Shanks, 2016; Steeg, 2016). An online synchronous model was used in five sessions with a pre-session online meeting and four sessions of online learning experiences (Francis & Jacobsen, 2013). Moreover, a blended model was developed through the use of a blog for web sessions and individual coaching and PLC sessions taught by two trainers (Mieke, Peeters, & Decelle, 2017).

4.3 *PLC and members' participation*

There are several types of member participations during PLC sessions: (1) using scaffolding and obtaining students' need (Steeg, 2016; Ratts, et al., 2015); (2) coaching and mentoring (Asanok & Chookhampaeng, 2015); (3) building reflective dialogue (Intanam, Wongwanich, & Lawthong, 2012; Cranston, 2009; Francis & Jacobsen, 2013); (4) supporting leadership within schools (Gray & Summers, 2016; Leclerc, et al., 2012; DeMatthews, 2014; Kalkan, 2016; Song & Choi, 2017); (5) building strong relationships, meaning collaborations and attitudes to working together (Shanks, 2016; Webb, et al., 2009; Owen, 2014; Smith, 2015).

4.4 *Perceptions of PLC*

Based on the criteria, the perceptions of PLC showed they demonstrated roles of trust, efficacy, or respect (Gray & Summers, 2016). The teachers showed positive perceptions of the PLC models applied (Intanam, et al., 2012; Kalkan, 2016; Ratts, et al., 2015; Sleegers, et al., 2013; Smith, 2015), positive relationships between level of professional behaviors or as an innovative professional development (Cansoy & Parlar, 2017, Smith, 2015; Ratts, et al., 2015); and positive correlations between PLC and school effectiveness (Song & Choi, 2017; Ratts, et al., 2015). Although there were positive perceptions in most of the study results, teachers saw PLC as difficult and challenging, yet valuable (DeMatthews, 2014).

5 DISCUSSION

The main issues in TPD are related to ease of access, flexibility, and quality. The face-to-face model is the most common method used, especially from the aspect of the need to face-to-face or discussion. Several studies have shown a paradigm shift from the implementation of traditional professional to technology-integrated learning communities. Zygouris-Coe & Swan (2010) stated that ICT integration has the potential to change TPD and support the collaboration of PLCs, allowing teachers from various schools, regions, and even countries to learn together, share successes and barriers, and transfer learning.

Fahara, et al. (2015) suggested a scheme for an online PLC. A more specific implementation of an online mode is the use of blended learning in a PLC program proposed by Salazar et al. (2010), which is related to flexibility and variation of learning resources. The use of blended learning allows teachers to expand the content of the program and deliver a variety of access options, either due to place, environment, or varying levels of technological mastery, but there is still an opportunity to conduct a face-to-face session.

The large-scale PLC is reflected through (1) the quality of the online professional development community, (2) improvements in learning, (3) professional collaboration (Hord, 2009), (4) knowledge and professional skills, (5) a model for continuous improvement, and (6) the impact it has on teachers' knowledge and skills in learning (Zygouris-Coe & Swan, 2010).

Related to participation in the PLC, Hord (2009) stated that supportive conditions, such as time, system, place, or resources, relational components, and sharing practices, are essential to the process of participation. Ratts et al. (2015) stated that participation of PLC members – peer observation, providing feedback on instructional practices, actively collaborating with other PLC members to judge students' work quality, and reviewing student works to improve instructional analysis – had a positive correlation with quality improvements in their teaching. Furthermore, positive and active collaboration among PLC members helped the teachers to see insulation and isolation as a restrictive component of their professional lives as a teacher (Smith, 2015). In this case, the systems put in place during PLC will also help the process of ensuring PLC is done regularly, such as a place for meetings, meetings at a regular time, availability of resources, and sharing practices using action research or lesson study (Gee & Whaley, 2016) to change practices among PLC members. The ethos for each perception parts consists of trust, respect, collegiality, and structural relationship (Chaseling et al., 2014; Cranston, 2009). Based on the criteria, the perceptions teachers have of a PLC are crucial to being actively engaged in their development process.

6 CONCLUSION

This study provides a systematic review of several studies in the primary school context related to the PLC. In total, 22 studies from 2009 to 2018 were identified to meet the criterion used in order to outline the innovations in PLC research. The findings indicated several key points. Firstly, there were three different types of PLC delivery identified: face-to-face, online synchronous and blended learning sessions. Second, PLC participation sessions included scaffolding, coaching and mentoring, reflective dialogue, leadership, and strong relationships.

Although there are many publications on this main theme for PLC, selection of papers based on the inclusion criteria must affect the result. Furthermore, this study did not use a qualitative analysis program to optimize e the review results, especially in mapping the trends. Based on that limitation, future research should be done using a qualitative analysis program.

REFERENCES

Asanok, M. & Chookhampaeng, C. (2015). Coaching and mentoring model based on teachers' professional development for enhancing their teaching competency in schools (Thailand) using videotape. *Educational Research and Reviews, 11*(4), 134–140. doi: 10.5897/ERR2015.2357.

ASCD. (2000). *Educators as learners: creating a professional learning community in your school.* Alexandria, VA: Association for Supervision and Curriculum Development (ASCD).

Cansoy, R. & Parlar, H. (2017). Examining the relationships between the level of schools for being professional learning communities and teacher professionalism. *Malaysian Online Journal of Educational Sciences, 5*(3), 15–27. Retrieved from https://files.eric.ed.gov/fulltext/EJ1150436.pdf.

Chaseling, M., Boyd, W.E., Robson, K., & Brown, L. (2014). Whatever it takes! Developing professional learning communities in primary school mathematics education. *Creative Education, 5,* 864–876, doi: 10.4236/ce.2014.511100.

Cranston, J. (2009). Holding the reins of the professional learning community: Eight themes from research on principals' perceptions of professional learning communities. *Canadian Journal of Educational Administration and Policy, 90.* Retrieved from https://files.eric.ed.gov/fulltext/EJ842519.pdf.

DeMatthews, D. (2014). Principal and teacher collaboration: An exploration of distributed leadership in professional learning communities. *International Journal of Educational Leadership and Management, 2* (2), 176–206. doi: 10.4471/ijelm.2014.16.

du Plessis, J. & Muzaffar, I. (2010). Professional learning communities in the teachers' college: A resource for teacher educators. *USAID.* Retrieved from https://files.eric.ed.gov/fulltext/ED524465.pdf.

DuFour, R. (2004). What is a professional learning community? *Educational Leadership, 61*(8), 6–11.

Erwin, E., Brotherson, M., & Summers, J. (2011). Understanding qualitative metasynthesis. *Journal of Early Intervention, 33*(3), 186–200. doi: 10.1177/1053815111425493.

Fahara, M.F., Quintanilla, M.G., & Bulnes, M. G.R. (2015). Building a professional learning community: A way of teacher participation in Mexican Public Elementary Schools. *IJELM – International Journal of Educational Leadership and Management, 3*(2), 114–142. doi: 10.17583/ijelm.2015.1338.

Francis, K. & Jacobsen, M. (2013). Synchronous online collaborative professional development for elementary mathematics teachers. *The International Review of Research in Open and Distance Learning, 14* (3), 319–343. doi: 10.19173/irrodl.v14i3.1460.

Gee, D. & Whaley, J. (2016). Learning together: Practice-centred professional development to enhance mathematics instruction. *Mathematics Teacher Education and Development, 18*(1), 87–99. Retrieved from https://files.eric.ed.gov/fulltext/EJ1103502.pdf.

Gray, J.A. & Summers, R. (2016). Enabling school structures, trust, and collective efficacy in private international schools. *International Journal of Education Policy & Leadership, 11*(3), 1–15. doi: 10.22230/ijepl.2016v11n3a651.

Guskey, T.R. (2003). What makes professional development effective? *The Phi Delta Kappan, 84*(10), 748–750. doi: 10.1177/003172170308401007.

Hill, K.K., Bicer, A., & Capraro, R.M. (2017). Effect of teachers' professional development from MathForward™ on students' math achievement. *International Journal of Research in Education and Science (IJRES), 3*(1), 67–74. Retrieved from http://files.eric.ed.gov/fulltext/EJ1126679.pdf.

Hord, S.M. (2009). Professional learning communities. *Journal of the National Staff Development Council, 30*(1), 40–43. Retrieved from http://www.earlychildhoodleadership.com.

Intanam, N., Wongwanich, S., & Lawthong, N. (2012). Development of a model for building professional learning communities in school: Teachers' perspectives in Thai educational context. *Journal of Case Studies in Education, 3,* 1–11. Retrieved from https://files.eric.ed.gov/fulltext/EJ1109718.pdf.

Kalkan, F. (2016). Relationship between professional learning community, bureaucratic structure and organizational trust in primary education schools. *Educational Sciences: Theory & Practice, 16*(5), 1619–1637, doi: 10.12738/estp.2016.5.0022.

Kennedy, A. (2005). Models of continuing professional development: a framework for analysis. *Journal of In-service Education, 31*(2), 235–250. doi: 10.1080/13674580500200277.

Lassonde, C.A. & Israel, S.E. (2010). *Teacher collaboration for professional learning: facilitating study, research, and inquiry communities.* San Francisco, CA: Jossey-Bass.

Leclerc, M., Moreau, A.C., Dumouchel, C., & Sallafranque-St-Louis, F. (2012). Factors that promote progression in schools functioning as professional learning community. *International Journal of Education Policy and Leadership, 7*(7), 1–14. Retrieved from www.ijepl.org.

Lieberman, A. (1995). Practices that support teacher development: Transforming conceptions of professional learning. *Phi Delta Kappan, 76*(8), 591–596. Retrieved from http://www.jstor.org/stable/20405409.

Mieke, G., Peeters, S., & Decelle, A. (2017). Mind the gap! The impact of professional learning communities focussed on the primary-secondary school transition. Conference proceedings ATEE 2017 Annual Conference, 418–434.

Owen, S. (2014). Teacher professional learning communities: Going beyond contrived collegiality toward challenging debate and collegial learning and professional growth. *Australian Journal of Adult Learning, 54*(2), 54–77. Retrieved from https://files.eric.ed.gov/fulltext/EJ1033925.pdf.

Powers, K., Shin, S., Hagans, K.S., & Cordova, M. (2015). The impact of a teacher professional development program on student engagement. *International Journal of School & Educational Psychology, 3*(4), 231–240. doi: 10.1080/21683603.2015.1064840.

Ratts, R.F., Pate, J.L., Archibald, J.G., Andrews, S.P., Ballard, C.C., & Lowney, K.S. (2015). The influence of professional learning communities on student achievement in elementary schools. *Journal of Education & Social Policy, 2*(4), 51–61. Retrieved from http://jespnet.com/journals/Vol_2_No_4_October_2015/5.pdf.

Sampoong, S., Erawan, P., & Dharm-tad-sa-na-non, S. (2015). The development of professional learning community in primary schools. *Educational Research and Reviews, 10*(21), 2789–2796, doi: 10.5897/ERR2015.2343.

Salazar, D., Aguirre-Muñoz, A., Fox, K., & Nuanez-Lucas, L. (2010). On-line professional learning communities: Increasing teacher learning and productivity in isolated rural communities. *Systemics, Cybernetics, and Informatics, 8*(4), 1–7. Retrieved from http://www.iiisci.org/journal/CV$/sci/pdfs/GE220YX.pdf.

Shanks, J. (2016). Implementing action research and professional learning communities in a professional development school setting to tupport teacher candidate learning. *School-University Partnerships, 9*(1), 45–53. Retrieved from https://files.eric.ed.gov/fulltext/EJ1107086.pdf.

Sleegers, P., den Brok, P., Verbiest, E., Moolenaar, N. M., & Daly, A. J. (2013). Toward conceptual clarity: A multidimensional, multilevel model of professional learning communities in Dutch elementary schools. *The Elementary School Journal, 114*(1), 118–137. Retrieved from http://www.jstor.org/stable/10.1086/671063.

Smith, G. (2015). Using an innovative model of professional development in primary science to develop small Irish rural schools as professional learning communities. *Global Journal of Educational Studies, 1*(1), 78–94. doi: 10.5296/gjes.v1i1.7620.

Song, K. & Choi, J. (2017). Structural analysis of factors that influence professional learning communities in Korean elementary schools. *International Electronic Journal of Elementary Education, 10*(1), 1–9. doi: 10.26822/iejee.2017131882.

Steeg, S.M. (2016). A case study of teacher reflection: Examining teacher participation in a video-based professional learning community. *Journal of Language & Literacy Education, 12*(1), 122–141. Retrieved from https://files.eric.ed.gov/fulltext/EJ1101024.pdf.

Webb, R., Vulliamy, G., Sarja, A., Hämäläinen, S., & Poikonen, P. (2009). Professional learning communities and teacher well-being? A comparative analysis of primary schools in England and Finland. *Oxford Review of Education, 35*(3), 405–422, doi: 10.1080/03054980902935008.

Wenger, E. (1998). *Communities of practice; learning, meaning, and identity.* New York, NY: Cambridge University Press.

Zygouris-Coe, V.I. & Swan, B. (2010). Challenges of online teacher professional development communities a statewide case study in the United States. *Online Learning Communities and Teacher Professional Development: Methods for Improved Education Delivery,* 114–134. doi: 10.4018/978-1-60566-780-5.ch007.

Implementation of snowball drilling learning model on discrete mathematics to improve student's independence and learning outcomes

L. Novamizanti
Telkom University, Bandung, Indonesia

ABSTRACT: This research aims to increase the independence and learing outcomes for combinatorial and tree applications in discrete mathematics by class action research, which was conducted using the cooperative learning model of snowball drilling. The research subject was 38 students of Telkom University's class TT-40-G4 (Telecommunications Engineering class in the 2017/2018 odd semester). The data of learning outcomes were obtained from the test method, while the data of student learning independence were obtained from the observation method. Data analysis used a quantitative descriptive analysis method. The pre-test results showed an average of 36.91, of which only 23% of students reach the minimum learning completeness standard (SKBM). After the first cycle, the average of the test increased to 72.36, with 74% of students achieving SKBM, and in second cycle the average of the test increased to 75.83, with 91% of students achieving SKBM. Based on the interview results, it can be concluded that students were pleased with the snowball drilling model.

1 INTRODUCTION

The quality of human resources can be improved by elevating the quality of education. One method to improve the quality of education is by upgrading the learning process. Efforts to improve learning quality have also been done by Telkom University through the Center for Teaching and Learning Excellence. The improvement was done via two aspects: the applied learning method and the learning support facilities. There are two main concepts to consider when creating an effective mathematics teaching and learning process: 1) to understand that mathematics is a fun process: 2) the learning process should center the student (Polly, McGee, Wang, Van de Walle, Karp & Williams, 2007). According to Ningsih (2014), mathematics is one of the lessons given at all levels of school and is closely related to the problems of everyday life.

Discrete Mathematics is a compulsory subject of the Telecommunication Engineering Major, which can be taken by the students in the second or third semester. In this 2017/2018 odd semester, there are 14 parallel classes. Many students in those classes had a GPA under 3.0, and there were also students retaking the class. Discrete Mathematics class provides students with knowledge about the concepts of set, function, and relation, combinatorial theory, and solutions of some cases related to its concept. This subject also provides students with the ability to complete graph and tree applications.

Based on the previous semester, the Discrete Mathematics class was conducted through a conventional learning method in the classroom teaching activities. The lecturer stands in front of the class to explain the materials without involving the students. Almost all of the students are quiet and look like understand when the lecturer was explaining the materials. Some students lack attention during the class because of this one-way teaching method. When the lecturer asked the students to explain the materials, many of them remained quiet and did not understand the explanation. When they were asked to do some exercises in front of the class, only some of them attempted this. There were also students who did not want to do the exercises in front of the class. When they were forced to do so, they looked at their friends' answers, without really understanding.

Not all students were independent in learning. Some did the exercises without being told by the lecturer, but there were also those who did not. This showed that the students still have poor learning independence. Therefore, this needs a varied learning model that can stimulate student independence in the learning process. The purpose is to make the students more active and improve their interest and learning outcomes.

The solution we have chosen is cooperative learning, where the use of such techniques can improve the academic performance of students in mathematics, and have a positive impact on teachers and students (Chan & Idris, 2017). Many studies conclude that students are happier in the classroom when they can discuss and be responsible for their own learning (Mohammadjani & Tonkaboni, 2015).

Cooperative learning involves instructional strategies in small groups so that students work together to achieve a common goal (Vijayakumari & D'Souza, 2013; Alabekee, Samuel, & Osaat, 2015). Johnson & Johnson (1999) define cooperative learning as an instructional use of small groups so students work together to maximize each other's learning. Paulsen & Chambers (2004) define cooperative learning as a teaching strategy in which students are actively and intentionally work together in small groups to improve their own learning and their teammate.

The cooperative learning method we used to improve this class's performance was snowball drilling learning. This method is done by giving a question to one of the students, and if the student can work on a given problem, they then point to other students to work on the next one. However, if the student cannot work on the question given, then they are given the next problem until they succeed. The snowball drilling learning method is more likely to be successful because this method forces the students to be ready any time they are called to the front of class to do the exercises. In practice, the snowball drilling method requires all students to work on the questions properly, because they must be ready to present their work at any time, with n0 warning. The snowball drilling method also aims to make the evaluation phase run smoothly, increasing the motivation and enthusiasm of students for working on the problem (Purnamasari, 2014).

Research by Ningtyas shows that the average learning outcomes of the snowball drilling method along with STAD model were more effective than the snowball throwing model (Ningtyas, 2012). This is supported by other research by Sumargiyani, which shows that in advanced calculus applying this method could improve students' independence in learning (Sumargiyani, 2014).

This research was conducted at PGMIPA (mathematics and natural science teacher education) FKIP (Faculty of Teacher Training and Education), Ahmad Dahlan University, Mathematics Education Major of the 2012/2013 academic year. The results of another study showed that the application of the cooperative learning method of snowball drilling can increase the activity of learning and mastering the learning outcomes for students of class XI in the Starter System topic (Murfi & Yuswono, 2018). From the relevant research above, there are positive outputs from the implementation of the snowball drilling cooperative learning model – it can improve students independence and learning outcomes.

2 LITERATURE REVIEW

The method is a series of learning systems have a very important role in determining the outcome of the process. Gintings (2008) stated that learning method can be interpreted as a pattern or type method in utilizing various basic principles of education and other related techniques and resources that occur during the student learning process. In the other words, learning method is a delivery technique that is mastered by a lecturer to present the subject matter to students in the classroom, either individually or in groups, so that the subject materials can be absorbed, understood, and utilized by the students (Ahmadi & Prasetya, 2005).

According to Slavin, cooperative learning is a form of learning undertaken by dividing the students into a small group of four to six members with a heterogeneous structure that can make them more keen to learn (Isjoni, 2010). Snowball drilling is a cooperative learning model developed to strengthen the knowledge gained by students from the learning material. The steps of snowball drilling cooperative learning model are as follows.

a. Describing learning goals and preparing the students to be ready to learn.
b. Presenting materials to students.
c. Dividing students into several small groups.
d. Providing an explanation to students about the team formation procedures.
e. Preparing questions on paper and sharing them in every group.
f. Giving students 10–15 minutes to solve the questions.
g. Asking the students to form a paper ball from the questions that have been given.
h. Giving students instructions to pass the paper ball from one student to another in each group.
i. Giving the stop sign, so the student who holds the paper ball chooses a question to answer on the board.
j. Repeating steps h and i until all the questions have been answered or the teacher feels it has gone long enough.
k. Checking the students' answers.
l. Rewarding the students' efforts.

Independence comes from a word that means standing alone. According to Desmita (2011), independence has several meanings:

a. Having a condition where a person has a competitive will for the good of themselves
b. Being able to make decisions and use initiatives to solve problems
c. Having confidence in finishing all duties
d. Being responsible for what they do

Independence in learning activities is driven more by students' own will, own choices, and their responsibilities (Tirtarahardja & La Sulo, 2005). From the description, we can conclude that the independence of learning is a student willing to learn, which comes from students themselves. To achieve the learning goals, students can control their learning, their own decisions, and responsibilities, and are also motivated individually or not dependent on others.

3   RESEARCH METHOD

This study is classroom action research, conducted in the author's own class and assisted by research partners as observers. From this study, it is expected that the use of snowball drilling learning methods can improve the teaching and learning process of students in the Discrete Mathematics class at Telkom University, so that students become more independent and increase the learning outcome. The study was conducted in the odd semester of the 2017/2018 academic year, from September 17 – December 30, 2017. The research location is in the Telecommunication Engineering Major of Telkom University in the Discrete Mathematics class. This research was undertaken in two steps: cycle-I that was held before the mid-semester, and cycle-II was held after the mid-semester. The research sample is Telkom University TT-40-G4

Telecommunication Engineering class's students, who took the Discrete Mathematics class, totaling 38 students.

The design determined in this research is classroom action research, a model developed by Kemmis, McTaggart, and Nixon (2013). Cycle-I consists of four components: planning, acting, observing, and reflecting. Data collection techniques and instruments used in this study are as follows.

a. Observation sheet, to record behavior, event, and all things that are considered important for students. It could also be used to describe the learning process in detail. The observation sheet is a private questionnaire to determine the students' independence during the learning process.
b. Documentation, which loads data about things or variables in the form of field notes, transcript, books, achievements, agendas, and others.
c. The questionnaire, which is used to ask respondents/students to choose a sentence or description that is closest to their opinion, feeling, judgment, or position.
d. Test questions which are a series of questions or exercises and other tools used to measure skills, intelligence, abilities, or talents that are owned by individuals or groups. The test can be used to measure basic abilities and achievements or learning achievements. This test is taken by students individually after learning a topic, at the end of learning in cycle-I and cycle-II.

The results of observations on student learning independence were analyzed according to the following steps.

a. Calculate the number of students who have to learn independence in the learning process.
b. Calculate the percentage score of the student learning independence according to the formula:

$$O = \frac{\sum n}{N} \times 100\% \tag{1}$$

where O is the percentage of observation sheet, n is the number of items that are checked, and N is the number of all data.

The qualification percentage was then divided into five criteria: very good ($80 < O \leq 100$), good $60 < O \leq 80$, enough $80 < O \leq 100$, bad $20 < O \leq 40$, and very bad $0 < O \leq 20$. The descriptive analysis techniques were applied to determine the mean of the data in the Discrete Mathematics learning outcomes.

Indicators as a successful implementation measure of classroom action are as follows.

a. Students' independence in the Discrete Mathematics class on the combinatorial and tree materials, with the application of the snowball drilling learning model increases with the criteria at least in the category of "good".
b. 75% of students in the TT-40-05 class of Telecommunication Engineering Major experienced individual learning completeness of ≥ 50.01 in Discrete Mathematics on the combinatorial and tree materials, with the application of the snowball drilling learning model depending on others.

4 RESULT AND DISCUSSION

The observation of the students' learning independence was conducted during the learning activities of combinatorial and tree materials with the snowball drilling learning model. Subsequently, we analyze the data obtained to find out whether there was an increase in learning independence in each cycle. Table 1 shows the results of observations of learning independence in each cycle. The type of study conducted is action research.

Table 1. Students learning independence observation sheet results in qualifications.

No	Indicator	Cycle-1	Cycle-2
1	Competitiveness willing	74%	92%
2	Confident and optimistic	78%	89%
3	Responsible	78%	89%
4	Initiate dealing with problems	76%	84%
5	Think critically, creatively, and innovatively	65%	78%
6	Trying to work hard, diligently, and discipline	71%	89%

Based on Table 1, students' independence in all the following activities in the Discrete Mathematics teaching and learning process increased in cycle -II compared to cycle-I. Students tried hard to compete with other groups. When the lecturer told the students to take turns to come forward, there are no problems, doubts, or fears, because if they cannot answer the questions their friends in the group can help. The observation of this researcher was strengthened by the results of several students. The responsibility of students toward snowball drilling cooperative learning is very good because every student takes full responsibility for solving all the questions on the task of trying to do the questions in the front of the class, even though not all of their answers are correct. After the researchers encouraged the students to work on the given questions, they did not despair. Using their groups to discuss this together, it transpired that the results of the indicators critical thinking, creativeness, and innovativeness of students in the second cycle had increased by 13%, from 65% to 78% which is in good category.

Figure 1 shows a comparison of the results of the pre-test (the test before the implementation of the snowball drilling method) and post-test (the test after the implementation of the snowball drilling method). It can be seen that after each cycle, the students' test scores were increased. At the end of the cycle, the lowest score was 42, the highest score was 100, and the average student test score was 75.83.

Figure 2 shows a comparison of the results of learning completeness in each cycle. At the end of cycle-II, the completeness of learning had increased from 23% at pre-test to 74% at cycle-II post-test, which means that there was a 68% increase in the completeness of learning of the snowball drilling type of learning model used. At the end of the cycle, 91% completeness was obtained, which means that the results reached the institution's target that 87.5% of students pass the target score of 50.01.

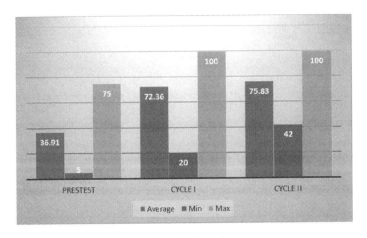

Figure 1. Comparison of pre-test, cycle-I, and cycle-II results.

194

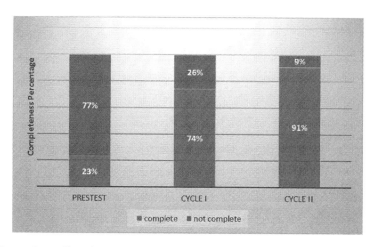

Figure 2. Comparison of learning outcomes completeness in each cycle.

5 CONCLUSION

Based on the results of this research, it can be concluded that students' independence in the Discrete Mathematics class, on combinatorial and tree materials, through the application of cooperative learning methods of type snowball drilling increased, with a score of 87% at the end of the cycle with very good criteria. Student learning completeness had also increased. At the beginning of the test, the number of students who pass the minimum score was 23%, while in the second cycle the number is increased to 91%. With this method, the student learning outcomes increased, with the average scores of 72.36 and of 75.83 in cycle-I and cycle-II.

REFERENCES

Abrami, P.C. & Chambers, B. (2004). Teacher Motivation to implement an educational innovation: Factors differentiating users and non-users of cooperative learning. *Educational Psychology, 24*, 201–216.

Ahmadi, A. & Prasetya, J.T. (2005). *Strategi belajar mengajar* [Teaching strategies]. Bandung, Indonesia: Pustaka Setia.

Allamakee, E.C., Samuel, A., & Osaat, S.D. (2015). Effect of Cooperative Learning Strategy on Students Learning Experience and Achievements in Mathematics. *International Journal of Education Learning and Development, 3*(4), 67–75.

Chan, L. L. & Idris, N. (2017). Cooperative Learning in Mathematics Education. *International Journal of Academic Research in Business and Social Sciences, 7*(3), 539–553.

Desmita. (2011). *Psikologi perkembangan* [developmental psychology]. Bandung, Indonesia: Remaja Rosdakarya.

Ginting, A. (2008). *Esensi praktis belajar dan pembelajaran* [The practical essence of learning]. Bandung, Indonesia: Humaniora.

Isjoni. (2010). *Cooperative learning: Efektivitas pembelajaran kelompok* [Cooperative learning: Group learning effectiveness]. Bandung, Indonesia: Alfabeta.

Johnson, D. W, & Johnson R. T. (1999). Learning together and alone: Cooperative competitive, and individualistic learning (5th ed.). Englewood Cliffs, NJ: Prentice-Hall.

Kemmis, S., McTaggart, R., & Nixon, R. (2013). *The action research planner: Doing critical participatory action research*. Singapore: Springer Verlag.

Mohammadjani, F., & Tonkaboni, F. (2015). A Comparison between the Effect of Cooperative Learning Teaching Method and Lecture Teaching Method on Students' Learning and Satisfaction Level. *International Education Studies, 8*(9), 107–112.

Murfi, A. F. & Yuswono, L. C. (2018). Penerapan Metode Pembelajaran Kooperatif Snowball Drilling untuk Menigkatkan Keaktifan Aktifitas Positif dan Ketuntasan Hasil Belajar Peserta Didik

[Implementation of the Snowball Drilling Cooperative Learning Method to Improve the Active Activities of Positive Students and the Mastery of Student Learning Outcomes]. *Jurnal Pendidikan Teknik Otomotif, XXIII*(2), 131–138.

Ningsih, E. F. 2014. Eksperimentasi Model Pembelajaran Kooperatif Tipe Team Assisted Individualization (TAI) dengan Metode Snowball Drilling Terhadap Prestasi Belajar Matematika Ditinjau dari Kemandirian Belajar [Experimentation Model of Team Assisted Individualization (TAI) Cooperative Learning Model with the Snowball Drilling Method Against Mathematics Learning Achievement Judging from Learning Independence]. *Jurnal Elektronik Pembelajaran Matematika,2*(7), 758–770.

Ningtyas, F. W. (2012). *Keefektifan model STAD berbantuan snowball drilling dan model Snowball Throwing terhadap Hasil Belajar Materi Segiempat* [The effectiveness of STAD model assisted by snowball drilling and Snowball Throwing models on quadrilateral material learning outcomes]. Semarang, Indonesia: Universitas Negeri Semarang.

Polly, D., McGee, J.R., Wang, C., Lambert, R.G., Pugalee, D.K. & Johnson, S. (2013). The Association between Teachers' Beliefs, Enacted Practices, and Student Learning in Mathematics. *Mathematics Educator, 22*(2), 11–30.

Purnamasari, N. (2014). Eksperimentasi model Jigsaw Snowball Drilling dan Peer Tutoring Snowball Drilling pada Materi Pokok Tabung, Kerucut, dan Bola Ditinjau dari Gaya Belajar Siswa [Experimentation of Snowball Drilling Jigsaw and Peer Tutoring Snowball Drilling models on the Main Material of Tubes, Cones, and Balls Judging from Student Learning Styles]. *Jurnal Elektronik Pembelajaran Matematika, 2*(1), 110–120.

Sumargiyani. (2014). *Peningkatan kemandirian belajar kalkulus lanjut menggunakan metode pengembangan pembelajaran kooperatif Snowball Drilling* [Improving student self-regulated learning in advanced calculus through the cooperative learning Snowball Drilling development method]. *Segitiga, 5*(1), 841–847.

Tirtarahardja, U. & La Sulo, S. L. (2005). *Pengantar Pendidikan* [Introduction to education]. Jakarta, Indonesia: Rineka Cipta.

Vijayakumari, S.N. & D'Souza, M.J.S. (2013). Metacognitive – Cooperative Learning Approach to Enhance Mathematics Achievement, *International Journal of Education and Psychological Research (IJEPR), 2*(2), 111–119.

Innovative Teaching and Learning Methods in Educational Systems – Retnowati et al. (Eds)
© *2020 Taylor & Francis Group, London, ISBN 978-1-03-224183-8*

What lecturers know about their role as an agent of learning: Levels of innovativeness in the learning process

A. Ghufron
Universitas Negeri Yogyakarta, Yogyakarta, Indonesia

ABSTRACT: Lecturers are definitively responsible for increasing their professional standing and role as the agents of learning, developing knowledge, technology, and arts, as well as public servants aiming to improve the quality of national education. The question then is whether lecturers have succeeded in their role as the agent of learning in the context of improving the quality of national education. One of the ways to answer that question is measuring the level of lecturer's innovativeness in the learning process through Loucks and Hall's version of the Concern-Based Adoption Model (CBAM). This model was used in the study for several reasons: (1) the concept provides a description of the lecturer's involvement in the learning policies; (2) the levels and application of the lecturer's innovativeness are described in detail; (3) it identifies the lecturer's difficulties in realizing innovation in the learning process; (4) it allows the development of strategies in realizing innovation in learning; and (5) it helps lecturers in describing the learning activities during the learning process.

1 INTRODUCTION

In the professional field, lecturers are definitively responsible for increasing their esteem and role as the agents of learning, developing knowledge, technology, and arts, as well as being public servants aiming to improve the quality of national education (RoI, 2005). Lecturers have a strategic position in the learning activities, which are centered in transmitting and transforming the learning experience to the students, and thus their role cannot simply be replaced by even the most advanced modern learning software. Collay (2011) states that teaching is leadership, and leadership begins in the classroom because teachers act with meaning, seeking greater learning for their students and colleagues. This notion is emphasized by a viewpoint in Indonesia that underlines the importance of national character values in each learning activity.

It is therefore absolutely relevant to argue that learning quality is highly influenced by the lecturer's competence and qualifications, as they are responsible for the learning activity, particularly in terms of their passion, creativity, innovation, and strategies in delivering the subject content. Darling-Hammond and Bransford (2005) state that teaching can be viewed as a field that sits at the intersection of other professional fields.

The question is then whether lecturers have succeeded in their role as the agent of learning in the context of improving the quality of national education. There have not been many studies or discussions addressing this issue appropriately, particularly as there is still a lack of understanding of the concept and implementation itself.

One of the problems is that there are different perceptions among lecturers on their role as the agent of learning. Most lecturers assume that teaching is enough. However, a professional lecturer should be able to manage the learning process according to the students' needs, while at the same guiding their students to be independent learners.

For this reason, there is a high demand for the Teacher Training Institution (LPTK) to study and develop the lecturers' professional competences, particularly in relation to learning innovation, in order to improve the quality of the students' learning in the future. This is imperative, as the success of the learning process is greatly related to the lecturer's innovation in the class. Leithwood stated that implementation is a process of behavioral change, in direction, driven by the innovation and occurring in stages and over time if obstacles to such growth are overcome (Miller & Seller, 1985).

The level of lecturer's innovativeness can be examined in the learning process in several ways. One of these ways is through a version of the Concern-Based Adoption Model (CBAM) by Loucks and Hall. This model was used in the study for several reasons. First, the concept provides a description of the lecturer's involvement in learning. Second, the levels and application of the lecturer's innovativeness are described in detail. Third, it identifies the lecturer's difficulties in realizing innovation in the learning process. Fourth, it allows the development of strategies in realizing innovation in learning. Fifth, it helps lecturers in describing the learning activities during the learning process (Vandenberghe, 1984).

The focus of this research is lecturers' work performance as the agent of learning. More specifically, this research aims to address the question of "what is the level of lecturers' innovativeness in the learning process?"

2 RESEARCH METHOD

This study is a part of research activities conducted by the Yogyakarta State University (YSU) lecturers, particularly in relation to the development of academic culture. YSU is making great strides in this field and has prioritized academic culture development (in many dimensions) in order to achieve world-class university-level teaching, whether in the form of theoretical studies (seminars, discussions, focus group discussions) or research activities.

Such activities and efforts have resulted in the development of academic culture in YSU in many dimensions. The findings of the research aim to contribute to the references and empirical studies on academic culture development at YSU.

This study employed the descriptive research method in order to present accurate data in a real context, while at the same time developing the appropriate research instruments to achieve the research objective. The final products of this study are the profile and measuring instruments of the level of lecturer's innovativeness in the learning process.

The research subjects were lecturers in all seven faculties of the YSU and they were randomly selected, taking into account the proportionate representation of the faculties and study programs. The number of subjects was 151 lecturers, which was 15% of the total population of 1047 lecturers.

The data collection technique was based on the CBAM inventory proposed by Hall & Loucks (1977), particularly in relation to the level of a lecturer's innovativeness in the learning process. The aspects of innovation can be grouped into eight levels: (1) Level 0: no implementation or non-use; (2) Level 1: orientation; (3) Level 2: preparation; (4) Level 3: mechanic; (5) Level 4a: routine; (6) Level 4b: refinement; (7) Level 5: integration; and (8) Level 6: renewal (Orr & Mrazek, 2009).

The data analysis technique used descriptive statistics (mean). The results of the calculation were then interpreted and analyzed in qualitatively. Ultimately, the research findings were expected to provide a profile of the level of lecturer's innovativeness in the learning process.

3 FINDINGS AND DISCUSSION

3.1 *Data description*

The findings of data selection through an inventory on the level of lecturer's innovativeness in the learning process are presented in Table 1.

The results of the lecturer's innovativeness level in conducting the learning process, as shown in Table, are described as follows.

Table1.   The level of lecturer's innovativeness in conducting the learning process.

Learning implementation aspects	Alternative answers (%)								
	1	2	3	4	5	6	7	8	Σ
a. Effective classroom management as required by the applied curriculum	0	3	1	7	17	66	39	18	151
b. Learning process that stimulates students	0	0	0	2	9	49	58	33	151
c. Learning tutorial for the students	0	0	0	7	18	63	43	20	151
d. Supporting material enrichment	0	1	1	5	18	59	46	21	151
e. Students feel that there needs to be more improvement in the subject matter	1	3	2	15	26	54	33	16	150
f. Educative communication between students and the lecturer	0	0	2	2	12	55	44	36	151
Total	1	7	6	38	100	346	263	144	905
Total (%)	0,.	0.77	0.66	4.2	11.05	38.23	29.06	15.91	100

1) Level 0: no implementation or non-use (the lecturers have no knowledge, no involvement, and nothing to do with activities related to learning process implementation) – 0.110%.
2) Level 1: orientation (the lecturers attempt to obtain information or learn new things that are relevant to the learning process implementation) – 0.773%.
3) Level 2: preparation (the lecturers prepare for the first time to understand new and relevant things regarding the learning process implementation) – 0.662%.
4) Level 3: mechanic (the lecturers apply innovation in the learning process implementation, but are still tied to the rigidly defined procedures) – 4.198%.
5) Level 4a: routine (the lecturers apply innovative activities in the learning process implementation as a habit without any variation) – 11.049%.
6) Level 4b: refinement (the lecturers look for variations in the learning process implementation to improve the quality of students' learning) – 38.232%.
7) Level 5: integration (the lecturers combines their efforts with others' in the learning process implementation to improve the quality of activities) – 29.060%.
8) Level 6: renewal (the lecturers assess and renew the quality of learning process implementation and modify the innovation activities to improve the quality of students' learning outcomes) – 15.911%.

4 DISCUSSION

The data given in Table 1 indicates that the majority of the lecturers (38.232%), in terms of their innovativeness in the learning implementation, are "looking for variations in implementing the learning process to improve the students' learning quality". Only 0.110%, is at Level 0, "having no knowledge, no involvement, and nothing to do with the related activities". It is important to note here that the data shown in Table 1 are nonlinear, which means that the average scores are not following the level of lecturer's innovativeness in the learning process.
    Here are some nonlinear findings that are worth examining further:

1) Instead of merely following the rigid policies set by the Ministry of Education and Culture, university, faculty, and department in carrying out the learning process, most of YSU's lecturers have tried to vary the learning activities.
2) However, some lecturers (0.11%) do not implement any innovation in conducting the learning activities, possibly due to their lack of concern over the rules applied in the field of learning. Perhaps, they assume that their teaching competence is already good enough, so there is no need for them to improve.

3) Another interesting phenomenon is that 29.060% of the lecturers are already at the "integration" level (combining their efforts with others to improve the quality of the activities). Furthermore, 15.911% are at the "renewal" level, meaning that they assess and renew the quality of the activities and modify their innovations to improve the quality of the students' learning outcomes.

The three findings above are in line with Spector, Ohrazda, Shaack, and Willey (2013), that "calling something an innovation implies that involves something novel or new, but often much more is implied". Most often, three additional aspects of innovations are implied: the existence of a prior state of affairs or a predecessor object; a process of change to previous states or objects or situations; and people.

5 CONCLUSIONS AND RECOMMENDATIONS

Drawing from the findings of this study, it can be concluded that the level of most lecturer's innovativeness in learning implementation is at the "refinement" level, which is realized in the form of "looking for variations in implementing the innovations to improve the quality of students' learning." Nevertheless, there are still some lecturers who do not implement or have not implemented, innovations in conducting the learning process.

Based on the conclusions summarized above, the following recommendations are proposed:

1. The level of lecturer's innovativeness in learning activities found in this study is worth further examination qualitatively, to investigate the data accuracy and its causes.
2. The instruments employed in measuring the level of lecturer's innovativeness should be tested in terms of validity, reliability, and discriminating power.
3. The instruments then should be developed sufficiently to measure and develop the lecturer's efforts in improving their innovativeness in the learning process.

REFERENCES

Collay, M. (2011). *Everyday teacher leadership: Taking action where you are*. San Francisco, CA: Jossey-Bass

Darling-Hammond, L. & Bransford, J. (2005). *Preparing teachers for a changing world: What Teachers should learn and be able to do*. San Francisco, CA: Jossey-Bass.

Hall, G. E., & Loucks, S. F. (1977). A Developmental Model for Determining Whether the Treatment is Actually Implemented. *American Educational Research Journal, 14*(3), 263–276. doi:10.3102/00028312014003263

Miller, J.P., & Seller, W. (1985). *Curriculum: Perspectives and practice*. New York, US: Longman Inc.

Orr, D., & Mrazek, R. (2009). Developing the level of adoption survey to inform collaborative discussion regarding educational innovation. *Canadian Journal of Learning and Technology, 35*(2), 1–19.

RoI. (2005). *Undang-undang No.14 Tahun 2005 tentang guru dan dosen* [Law number 14 Year 2005 concerning teachers and lecturers]. Jakarta, Indonesia: Ministry of Education, Republic of Indonesia.

Spector, J.M., Ohrazda, C., Schaack, A.V., & Willey D. A. (2013). *Innovations in instructional technology*. Abingdon, UK: Routledge.

Vandenberghe, R. (1984). Teacher's role in educational change. *Journal of In-Service Education, 11*(1), 14–25.

Development and quality analysis of a learning media electrical motor installation on the Android platform, for vocational students

B.N. Setyanto & H. Jati
Yogyakarta State University, Indonesia

ABSTRACT: The purpose of this study is to design a learning media Electrical Motor Installation (EMI) application with quality standardization. The research was conducted with the research and development approach. The product discussed is two applications that can be used on smartphones on the Android platform. The two applications have the same content but different interface designs, so that comparisons could be made to obtain quality products. Data was collected through a questionnaire, offline tests (direct testing), and online tests (cloud testing). Quality analysis was carried out with reference to ISO 25010. This study shows that the the first application (IML-WA) is higher quality than the second application (IML-BBM): an analysis of functional suitability and compatibility aspects scored the applications as equal in quality, while the usability and performance efficiency aspects show that IML-WA received a higher mean rank score than IML-BBM.

1 INTRODUCTION

1.1 *Research background*

Vocational High School is one of the highest secondary education levels for vocational education in Indonesia. According to the Law of Republic Indonesia No. 20 in 2003 Article 15, vocational schools aim to prepare students specifically to work in certain fields. Meanwhile, a decree from the former Minister of Education and Culture Republic Indonesia number 179342/MPK/KR/2014 stated that a teacher must develop methods of learning with creativity, dare to innovate and be able to carry out a learning process that is fun for students (Baswedan, 2014).

In the era of the education revolution 4.0, there are demands on teachers to develop technology-based learning media that meet quality standards, as the application of technology in education can be exciting opportunities that can potentially transform society for the better (Xing, 2017). Lecturers and students should apply education 4.0 principles in order to help the students to compete in the era of 4.0 (Candradewi, 2018).

Winarto and Winih Wicaksono, as the teachers responsible for electric motor installation (EMI) and the masters for the Expertise Program of the Electric Power Installation Engineering (TIPTL) at the Vocational High School in Yogyakarta described the problem as follows. (1) EMI are categorized as difficult lessons. (2) EMI requires the right media innovation for the learning process. (3) Around 80% of students have Android platform smartphones. (4) There is a need to create an Android learning media platform that can display short theories, and installation steps in the form of simulation images, exercises, and theoretical questions. (5) The learning process carried out is still conventional using the EMI practice module. (6) Can we make a learning media is an interesting and high quality? (7) Teaching materials of electronic school books (BSE) particularly in vocational high school.

International Data Corporation (IDC) stated that, in Q1 2017, worldwide, smartphone use had increased to 1.46 billion devices, and this was dominated by the Android platform (85.0%), iOS (14.7%), Windows Phone (0.1%), and (0.1%). Indonesian internet service provider association (APJII) showed that, in 2017, in Indonesia 143.26 million people were internet users: with a 50.08% owned smartphone or tablet devices, of which 70.96% was owned by urban dwellers, and of this student users age 13–18 years accounted for 75.70%.

Based on the Global State StatCounter for the period July 2017 to July 2018, in Indonesia the smartphone market was as follows: Android system 87.79%, IOs 5.65%, and Blackberry 0.37%. Android-based communication media can produce changes in scientific attitudes in the learning process and increase the completeness of learning outcomes (Juwarti, 2015). Similarly, Prasetyo (2015) stated that Android-based media can improve student motivation and cognitive learning outcomes. Sumari (2015) stated that Android-based m-learning learning media can improve critical thinking skills and students' learning independence.

It was noted by Vaugan (2011) that the combination of text, art, sound, animation, and video sent to users by computers or other manipulated electronic or digital means is considered as multimedia. UNESCO (2014) adds that m-learning as mobile learning involves the use of mobile technology either alone or in combination with information and other communication technologies, and thus allows learning anytime and anywhere.

Branch (2009) explained that the model Analyze, Design, Develop, Implement, and Evaluate (ADDIE) is a learning planning design model that emphasizes student-centered, innovative, authentic, and inspiring. In addition, the ADDIE model can be used to develop educational products and learning resources. In connection, Sumari (2015) found that the research process for the development of Android-based learning media can use the ADDIE model. Similarly, Prasetyo (2015) also stated that research on the development of Android-based learning media can use the ADDIE development model.

Software testing is a process of executing a software product on input data and output data analysis (Mili, 2015). Suman (2014) explained that the main purpose of software engineering is of high quality and sustainable for use. Furthermore, Thomas (2017) stated that in the design comparison of a product, consumers are asked to use two products and then determine which is better. Several years before, Panovski (2008) suggested that in the process of analyzing a software product there are significant differences between application software products and software development products.

Software quality standards, including ISO 9126, ISO 25010, McCall Model, DROMEY, and FURP, are international standard models used in measuring software quality. Hidayati, Sarwosri, and Ririd (2009) described ISO 9126 as a standard used as a quality model in developing learning media applications. However, in 2012 ISO 9126 evolved to ISO 25010, so the quality test was better when using ISO 25010 (ISO/IEC, 2011).

According to Wagner (2013), ISO 25010 establishes eight aspects of product quality standards: functional suitability, reliability, performance efficiency, usability, maintainability, security, compatibility, and portability. David (2011) proposed that the quality analysis used in software product research with the Android platform can use four of these eight aspects: functional suitability, compatibility, usability, and performance.

Therefore, a summary of the above description is that the Indonesian education minister made a statement concerning developments in the evolution of education 4.0, which might be associated with the increased use of Android smartphones, and the desire of subject teachers to have quality learning media to improve the quality of learning. So, research needs to be done to produce quality media. In developing a product, a comparison is needed to get a better product, so that at least two products have been developed.

1.2 Research focus

Based on this problem, this study focuses on obtaining quality learning media products based on ISO 25010 standards, by comparing the two products in term of quality, analyzed on aspects of functional suitability, compatibility, usability, and performance efficiency.

2   RESEARCH METHODS

This research used the research and Development (R&D) approach with the ADDIE learning design development model. The purpose of this development research is to produce a quality product in reference to SO 25010. The process of developing the learning media was done by using the world's most popular messenger application interface,s Whatsapp Messenger and BBM Messenger. On May 15, 2018, the Whatsapp Messenger application ranked first with score of 4.4 and BBM Messenger ranked second with score of 4.3 (Google Play Store, 2018). Two user interface models were used in order to determine the best choice of applications for students. The development followed the provisions of the design material concept by Google.

The steps taken in developing this learning media are described below.

2.1   *Analyze*

A pre-survey was conducted to establish the requirements for learning media in schools. The results were: (1) identifying the subjects that need learning media; (2) obtaining curriculum information used; (3) obtaining and identifying subject matter; (4) obtaining media information data used; (5) determining the application for development that is suitable for the media.

2.2   *Design*

The Android platform learning media IML was designed into two applications with the same learning matter, but with differences in terms of layout. Layout A was designed to resemble the Whatsapp Messenger application and was named IML-WA, and layout B was designed to resemble BBM Messenger and was named IML-BBM. These design stages included (1) a design a use case diagram, (2) design a flowchart, and (3) design the user interface.

2.3   *Develop*

The development stage was conducted by preparing the content of the learning material and determining the media developed. The media developed in this study is an application media platform for Android, that designs processes for an Adobe Flash CS6 application using adobe Integrated Runtime(AIR) version 30 (Brossier 2011). The product was tested to find out the validity based on judgments from experts on learning the material, the media, and the functional suitability. The process for the validity of the learning media was tested using the SPSS application (George &Mallery, 2003).

2.4   *Implement*

The implementation phase was conducted with cloud testing on compatibility and performance efficiency aspects. After cloud testing was completed and there were no more any revisions, the final stage of testing was conducted to find out the usability aspects of the two learning media for Vocational High School students in Indonesia.

2.5   *Evaluate*

After the implementation phase, the final development phase was the quality evaluation of the two media, developed with ISO 25010 analysis procedures on aspects of functional suitability, compatibility, performance efficiency, and usability. Usability testing was performed with a minimum of 20 users for better results of the study (Nielsen, 2012). After obtaining the expected results, the significant difference test was carried out between the two applications to find out which was higher quality (Niknejad, 2011).

3 RESULTS AND DISCUSSION

The final products of the design IML-WA and IML-BBM learning media applications are shown in Figure 1.

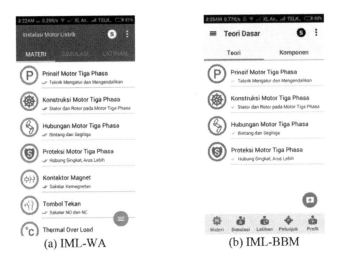

(a) IML-WA   (b) IML-BBM

Figure 1.   Main menu IML-WA and IML-BBM.

The validity of the two applications performed by media experts was "very feasible", with a mean score of 4.49. In addition, 100% of subject matter experts scored the validity as "valid".

3.1 *Functional testing*

Functional suitability testing was performed by two Android developers from PT. Putra Multi Cipta Teknikindo (PMCT) showed that the two applications functioned 100% successfully. Thus both applications functioned properly and satisfied the appropriate material criteria.

3.2 *Compatibility testing*

The result of compatibility online testing on Amazon apps & Games shows more than 208 devices supported. Compatibility of online testing on www.portal.kobiton.com and www.cloud.bitbar.com (Bitbar, 2018) is shown in Table 1. Direct testing is shown in Table 2.

The compatibility testing shows that the two applications were found to be 100% compatible and supported on any smartphones. So there are no obstacles to implementation.

Table 1.   Compatibility of online testing.

No	Smartphone	Version	Result
1	Acer Iconia 8	4.4.2	Running
2	Asus Fonepad	4.1.0	Running
3	Asus Google Nexus 7	4.1.2	Running
4	Lenovo S820	4.2.0	Running
5	Samsung J7 Prime	6.0.0	Running
6	Samsung J7 Pro	7.0.0	Running

Table 2. Compatibility direct testing.

No	Smartphone	Version	Result
1	Key touch S100	4.1.2	Running
2	XiaomiRedmi 2	4.4.4	Running
3	SmartfrenAndromaxU	4.0.4	Running
4	Oppo A37f	5.1.1	Running
5	Asus Z00	5.0.2	Running
6	Xiaomiredmi note4x	6.0.0	Running
7	Advan S5E 4GS	7.0.0	Running

3.3 Usability testing

Usability testing with 76 students shows that the IML-WA application has a reliable Cronbach's alpha score of 0.917 and IML-BBM score of 0.906. Thus, two applications have a score of more than 0.900 so they fall into the category of "excellent", Reliability IML-WA percentage of 83.50% and IML-BBM of 82.66% are to be in the category of "very good".

As shown in Table 3, nonparametric independent test results on usability aspects, using a Mann-Whitney U test, showed no significant differences between the two applications. However, the mean value showed that IML-WA was higher quality with a score of 79.90 compared to 73.10 for IML-BBM.

3.4 Performance efficiency testing

The performance efficiency results for the two applications are shown in Figures 2 and 3.

The data from Figures 2 and 3 were used to test any significant difference between the two applications on time usage. The result of a Mann-Whitney U test with SPSS is presented in Table 4.

The performance efficiency testing shows: (1) the IML-WA application has an average score of 4.92 seconds and the IML-BBM application has an average score of 5.53 seconds, thus it shows that both applications have the predicate "satisfied", as presented by Hoxmeier (2000); (2) there was no significant difference in the time used on the two applications, but the mean rank shows that IML-WA takes less time, with a score of 10.00, meanwhile IML-BBM had the score 15.00.

Table 3. Independent test usability aspects.

Ranks

	Group	N	Mean rank	Sum of ranks
Usability	IML-WA	76	79.90	6072.50
	IML-BBM	76	73.10	5555.50
	Total	152		

Test statistics	Usability
Mann-Whitney U	2629.500
Wilcoxon W	5555.500
Z	-.953
Asymp. Sig. (2-tailed)	.341

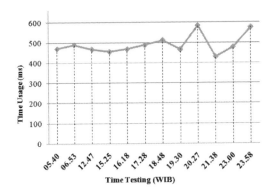

Figure 2.   Performance efficiency aspects IML-WA.

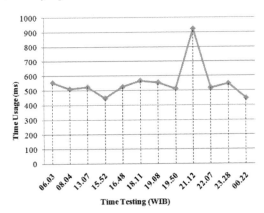

Figure 3.   Performance efficiency aspects IML-BBM.

Table 4.   Performance efficiency aspects (time usage).

Ranks

	Group	N	Mean rank	Sum of ranks
Usability	IML-WA	12	10.00	120.00
	IML-BBM	12	15.00	180.00
	Total	24		

Test statistics	Usability
Mann-Whitney U	42.000
Wilcoxon W	120.000
Z	-1.733
Asymp. Sig. (2-tailed)	.083
Exact Sig. [2*(1-tailed sig.)]	.089[b]

a. Grouping variable: Group
b. Not corrected ties

4 CONCLUSIONS

Based on the results of the research it can be concluded as follows:

1. Two learning media produce 100% functional suitability aspect quality, so the functions in the application work very well. The compatibility aspects can operate on more than 208 Android platform devices.
2. The level of user satisfaction based on usability aspect analysis shows that the IML-WA application is higher quality than IML-BBM.
3. The performance efficiency level of the two applications shows the IML-WA is higher quality than IML-BBM.
4. The results of the analysis and test differences indicate that the IML-WA application is higher quality and the best choice for the user.

REFERENCES

APJII. (2017). Hasil survey penetrasi dan perilaku pengguna internet Indonesia [Survey results on penetration and behavior of Indonesian internet user]. Retrieved from https://www.apjii.or.id

Baswedan, A. (2014). Surat menteri pendidikan dan kebudayaan [the letter of ministry of educational and cultural]. Retrieved from http://www.kemdiknas.go.id

Bitbar. (2018). Compatibility of online testing. Retrieved from www.cloud.bitbar.com

Branch, R.M. (2009). Instructional design: The ADDIE approach. New York, US: Springer.

Candradewi. (2018). Promoting education 4.0 in English for survival class: What are the challenges? *Journal of English Language, Literature, and Teaching, METATHESIS, 2*(1), 12.

David, A.B. (2011). Mobile application testing best practices to ensure quality - AMDOCS. Retrieved from http://www.globaltelecomsbusiness.com

George D. & Mallery P. (2003). *SPSS for Windows step by step: A simple reference 11.0 update (4th ed.).* Boston: Allyn & Bacon.

Google Play Store. (2018). *Messenger application detail.* Retrieved from https://play.google.com/store/apps/detail?id

Hidayati, A., Sarwosri, & Ririd, A.R.T.H. (2009). Analisa pengembangan model kualitas berstruktur hirarki dengan kustomasi ISO 9126 untuk evaluasi aplikasi perangkat lunak B2B [Analysis of the development of a hierarchical structured quality model with ISO 9126 customization for B2B software application evaluation]. *Presented in National Seminar of Electrical, Informatic, and It's Education* (pp. 1–8). Malang, Indonesia: Teknik Elektro Universitas Negeri Malang.

Hoxmeier, J.A. & Dicesare, C. (2000). System response Time and User Satisfaction: An Experimental Study of Browser Base Applications. AMCIS 2000 Proceedings (pp. 143). Retrieved from http://www.collector.org/achives/2000_april/03.pdf

IDC Team. (2018). *Smartphone market share.* Retrieved from https://www.idc.com

ISO/IEC. (2011). Systems and software engineering – systems and software quality requirements and evaluation (SQuaRE). Retrieved from https://www.iso.org.

Juwarti. (2015). Pengembangan media animasi berbasis Android materi sistem saraf untuk meningkatkansikap ilmiah dan hasil belajar kognitif siswa SMAkelasXI. [Development of android-based animation media nervous system material to improve scientific attitudes and cognitive learning outcomes for grade XI high school students]. Unpublished thesis, Program Pascasarjana Universitas Negeri Yogyakarta, Indonesia.

Kobiton. (2018). Compatibility of online testing. Retrieved from www.portal.kobiton.com

Mili, A. & Tchier, F. (2015). *Software testing – concept and operations.* USA: John Wiley & Sons, Inc.

Nielsen, J. (2012). How many test user in a usability study? Retrieved from http://www.nngroup.com/articles/how-many-test-users/

Niknejad, A. (2011). A quality evaluation of an android smartphone application. Retrieved from https://gupea.ub.gu.se/bitstream/2077/26728/1/gupea_2077_26728_1.pdf

Panovski, G. (2008). Product software quality. Unpublished thesis, department of mathematics and computing science. TechnischeUniversiteit Eindhoven, Germany.

Prasetyo, Y.D. (2015). Pengembangan Media Pembelajaran Kimia Berbasis Android padaMateriSistem-KoloiduntukmeningkatkanMotivasiBelajardanHasilBelajarKognitifPesertaDidik SMA [Development of Android-based Chemistry Learning Media on Colloidal System Material to improve Learning

Motivation and Cognitive Learning Outcomes of High School Students]. Unpublished thesis, Program Pascasarjana Universitas Negeri Yogyakarta, Indonesia.

StatCounter. (2018). Mobile & tablet android version market share worldwide. Retrieved from http://gs. statcounter.com.

Suman, M.W. (2014). A comparative study of software quality models. *International Journal of Computer Science and Information Technologies* (IJCSIT), 5(4), 5634.

Sumari, G.D. (2015). Pengembangan mobile learning berbasis Android materi sistem imun untuk meningkatkan kemampuan berpikir kritis dan kemandirian belajarsiswa kelas XI SMA [Developing Android Based Mobile Learning on Immune System to Improve the Critical Thinking and Learning Autonomy of Class XI Student of Senior High School]. Unpublished Thesis, Program Pascasarjana Universitas Negeri Yogyakarta, Indonesia.

Thomas, J.W. (2017). Product testing. Retrieved from https://decisionanalyst.com

UNESCO. (2014). Mobile learning. Retrieved from http://www.unesco.org

Vaugan, T. (2011). *Multimedia: Making it Work*. Eighth Edition. USA: McGraw-Hill.

Wagner, S. (2013). Software product quality control. New York, NY: Springer.

Xing B. & Marwala T. (2017). Implications of the fourth industrial age on higher education. Retrieved from https://www.researchgate.net

The readiness of vocational secondary schools on forming working characteristics for industry 4.0

H. Mulyani & I.W. Djatmiko
Graduate School, Yogyakarta State University

ABSTRACT: This study aims to find out (1) working characteristics required by industry, (2) the gap in the working characteristics between vocational secondary school graduates and the industrial world, and (3) the readiness of vocational secondary schools in facing Industry 4.0 in terms of practical learning. The method used in this paper was literature study, undertaken by searching for data in books, journals, government regulations, and other electronic sources. Results of this study are as follows. First, the working characteristics demanded by industries are communication, responsibility, cooperation, caring for the environment, safety and occupational health (K3) awareness, negotiation, problem-solving, analysis, life-long learning, management, adaptability, innovative, creativity, critical thinking, flexiblity, decision-making, entrepreneurship, leadership and discipline. Second, gaps in working characteristics were identified between vocational secondary school and industrial demand. Third, the practical learning syllabuses of vocational secondary schools are not ready yet to provide the working characteristics for graduates based on industry demand.

1   INTRODUCTION

The transformation of Industry 4.0 is a new challenge for educational institutions as labor providers and prospective employees in terms of gaps in competence. Industry 4.0 is otherwise known as the fourth and current era of industrial revolution. The world is strongly influenced by the advances in technology. According to Prifty et al. (2017), the 4.0 era is characterized by the development of technology that has changed several things rapidly, such as censorship, the physical system of cyberspace, the internet of things (IoT), and other intelligent networks that will surely affect every aspect of life. The rapid development of technology in this current era is considered to be one of the factors causing change in the industrial workplace. Furthermore, it involve changing processes in purchasing, producing, manufacturing, selling, or maintaining by involving concepts such as intelligent manufacturing, intelligent maintenance, and high-leveled automation and integration.

The current needs of industry not only rely on both technical and digital skills but also other prominent skills, often called soft skills. These soft skills are required by graduates to face competition in the labor market. In line with Walsh & Donaldson (2017), soft skills are required for readiness of learning, interdisciplinary exchanges, and innovation in Industry 4.0. Aime et al. (2017) stated that the learning process may form soft skills because the acquisition of hard skills and soft skills enable students to play an essential role in a complex industry, which becomes a new paradigm in Industry 4.0. In addition to this, Motyl et al. (2017) stated that in the context of eengineering education, soft skills are regarded as less real. However, soft skills are actually no less important and will be useful as Industry 4.0 persists. In accordance with Selamat, et al. (2017), the challenge of the 4.0 era is supported by character building, high order thinking, multiple intelligence, soft skills, life-long learning, and many other skills needed to deal with new challenges. Lorentz et al. (2015) stated that in this era, soft skills are more important because employees must be open-minded to change, have a greater flexibility to adapt to new roles and working environments, and should be accustomed to ongoing interdisciplinary learning to boost knowledge and achieve results in the working climate. Beer and Brooks (2011) stated that personality has been considered a prominent factor in personality studies, especially in predicting job performance. That is a different behavior, which varies from one person to another. Awadh (2012) also explained that personality traits of the employees have a positive relationship are organizational productivity. When personality traits are in accordance with organizational culture, the organizational productivity will improve.

The problem occurring at present is that learning in vocational secondary schools is not concerned with the soft skills that graduates must possess. Sailah (2007) stated that the ratio of soft and hard skills the business world needs is inversely proportional to the developments in the education system. Moreover, Suryanto (2013) said that in the current education system, soft skills only account for around 10% of the curriculum. In addition to this, Risma (2012) in Zahra et al. (2017) stated that learning in vocational secondary schools is 70% practice and only 30% theory because vocational graduates are required to have certain skills.

The soft skills included in Curriculum-2013 used by vocational secondary schools have not yet been well evaluated. The relevance of soft skills in the curriculum, especially the soft skills of the working characteristics of vocational students, is still irrelevant to the needs of Industry 4.0. Thus, there must be a further study related to the needs of the work characteristics in line with this latest era, and how to implement the correct way to develop the working characteristics of students in vocational secondary school.

This present study concerning soft skills in the form of working characteristics was based on research in vocational secondary schools. Working characteristics were chosen because learning in vocational secondary schools is dominated by practical learning. Another reason is that graduates from vocational secondary schools are prepared to begin work immediately. The second concern of this study was looking at the learning syllabus for the Electrical Power Installation subject to determine the working characteristics required. That subject was chosen due to the importance of electricity at present and in the future. Electricity is considered as a basic daily need for humans because it is used to supply electronic and digital equipment.

In brief, this study aimed to discover (1) the working characteristics required by the industry, (2) gap in the working characteristics between vocational secondary schools and the industrial world, and (3) the readiness of vocational secondary schools in facing Industry 4.0 in terms of practical learning.

2 MISMATCH OF SKILLS IN LEARNING OUTCOMES OF VOCATIONAL SECONDARY SCHOOLS

The compatibility between the competences of prospective employees and those desired by the industry has become a concern and a high priority in recruitment. Some companies have experienced problems after recruitment because the new employee did not have the skills required to do the job. ACT (2011) stated that a significant segment of the current workforce does not have the skills needed by industry and many employers find it difficult to find suitably skilled workers. Furthermore, Uzair-ul-Hasan& Noreen (2013) stated that the most easily recognizable factor is inadequate skills. Compatibility between the skills needed and provided play an important role in economic growth. Moreover, Allen &Velden (2001) identified four types of skill mismatch: the wrong skills, a lack of skills, a skills mismatch, and a skills surplus. Thus, the suitability of employees' skills in the industry remains essential.

Yang (2013) stated that skills that are only concentrated in institutions or schools have not significantly affected the supply and demand for highly skilled required. Furthermore, Bukamal & Mirza (2017), who conducted research in Bahrain, stated that although competent workers' output targets were clear in the national strategy, it was not clear how higher education institutions were expected to adhere to this strategy. As a result, this could be the main cause of poor results in accordance with the work produced by higher education institutions in Bahrain. Thus, it is necessary to have a match between education as a provider of human resources/labor and industry as service users.

3 DEMAND OF WORKING CHARACTERISTICS IN INDUSTRY 4.0

Many studies on the competency model of workers required to meet the challenges of Industry 4.0 have been carried out to obtain competency models that are suitable for industry needs. Prifti et al. (2017) provided a competency model required to meet the challenges of the current era with eight major competencies: (1) leadership and decision-making, (2) support and cooperation, (3) interaction and presentation, (4) analyzing and concluding, (5) creating and conceptualizing, (6) managing and executing, (7) adapting and overcoming, and (8) drawing conclusion and informing.

Furthermore, Prifti et al. (2017) described some basic characteristics required to form those eight competencies: (1) leadership and decision-making: making decisions, taking responsibility, and leadership skills; (2) support and cooperation: working together, collaborating with others, communication, respecting ethnicity, caring for the environment, being aware of ergonomics; (3) interaction and presentation: compromising, creating business networks, maintaining relationship with consumers, negotiating, emotional intelligence, presenting ability, and communicating; (4) analyzing and concluding: communication techniques, literacy, integrating information technology, economics, extracting business and social media values, problem-solving, optimization, analyzing ability, cognitive ability; (5) creating and conceptualizing: long-term learning, management ability, innovation, creativity, critical thinking, changing management, business strategy, abstracting ability, complexity management; (6) organizing and executing: project management, planning and managing work, management ability, customer oriented, customer relationship management, legislation awareness, safety awareness, individual responsibility; (7) adapting and overcoming: working in an interdisciplinary environment, inter-cultural competence, flexibility, adaptability and ability to change mindset, working and life balance; and (8) drawing conclusion and informing: self-management and regulation, understanding business models, entrepreneurship.

Another competency model was proposed by Fu & Yan (2017), who described competency needs in the digital era as three types of skill: technical and professional skills, ICT generic skills, and soft skills in ICT. Each of those three skills also requires certain characteristics to support the skills, as follows: technical and professional skills: installation and operation; ICT generic skills: understanding, using and adapting, ICT; life skills and soft skills: creativity, communication, critical and logical thinking, cooperation, and entrepreneurship.

Moreover, ADB (2014) stated that high-level technology companies need high skill levels. Several key characteristics include cooperation, communication, problem-solving, and management. In addition, workers also need literacy, numeracy, critical thinking, and digital skills. Motyl et al. (2017) stated that the skills an engineering graduate should have include numerical mathematical knowledge, problem-solving skills, creativity, designing, investigating and experimenting skills, information processing, computer programming, and knowledge of special software; and the characteristics needed are communication, teamwork, and leadership.

Furthermore, Berger (2016) classified the skills qualifications that must be provided by the education and vocational education in preparing the evolution of industrial technologies especially in the context of Industry 4.0. He also divided skills qualifications into four main categories: ICT knowledge, ability to work with data, technical capabilities and personal abilities. Each category is elaborated with the skills and characteristics that graduates must possess, as follows. (1) Knowledge of ICT: knowledge of basic information technology, ability to use and interact with computers and smart machines, such as robots, tablets, etc., understand machine to machine communication, IT security and data protection. (2) The ability to work with data: process and analyze data and information obtained from machines, understand visual data output and make decisions, knowledge of basic statistics. (3) Knowing technical capabilities: interdisciplinary and general knowledge about technology, specific knowledge about activities and manufacturing processes in place, technical knowledge of machines to carry out activities related to maintenance. (4) Personal ability: adaptation and ability to change, decision-making, teamwork, communication skills and mindset changes for long-term learning.

The skills qualifications that must be met by future workers are also described by Gehrke et al. (2015), divided into technical skills and personal skills. Technical skills include abilities and knowledge in the field of IT, data and information processing analysis, statistical knowledge, understanding organizations and processes, and the ability to interact with modern interfaces. Personal capabilities include self-management and time, the ability to adapt to change, the ability to cooperate, social ability, the ability to communicate. Furthermore, Bermúdez & Juárez (2017) found four characteristics of competence that must be met by workers in Industry 4.0: technology knowledge, innovative management, organizational learning, and the environment.

In line with those mentioned previously by experts, the skills qualifications needed in Industry 4.0 refers to working characteristics such as communication, responsibility, cooperation, caring for the environment, safety and occupational health (K3) awareness, negotiation, problem-solving, analysis, life-long learning, management, adaptation, innovation, creativity, critical thinking, flexibility, decision-making, entrepreneurship, leadership, and discipline.

4 VOCATIONAL SECONDARY SCHOOL AND WORKING CHARACTERISTICS

Guidelines for regulation of the education system are included in the education curriculum. The current curriculum used by vocational secondary schools is Curriculum-2013 in which characteristics are already included. According to Bialik et al. (2015), character describes how someone is involved and behaves and building students' characteristics at school is key to facing the challenges in the 21st century and the disruption era. In addition to this, Agboola & Tsai (2012) said that character is described as the way we express our inner being, the embedded value that is in us, and will make some of us go of our way to express compassion, caring, integrity, respect, and all other values that are in line with virtue.

Moreover, Batistich et al. (2000) said that character is the embodiment of a person's positive development intellectually, socially, emotionally, and ethically. A person with good character is thought to be the best type of person. Marshall et al. (2011) claimed that characteristics education is constantly trusted to be able to develop the good character of students. Students are nurtured toward seeing things from a different perspective, such as teaching them to be mature when facing challenges. Aristotle (2017) stated that character is a set of personal traits or dispositions that produce certain moral emotions, inform motivation, and guide behavior.

Building character values in Curriculum-2013 used by vocational secondary schools is written into the second core competency. The description of the working characteristics is as follows:

a. Core competencies: appreciating and practicing honesty, discipline, responsibility, care (mutual assistance, cooperative, tolerance, peace), politeness, responsive and proactive, and showing the attitude as a part of the solution to various problems in interacting effectively with the social and natural environment.
b. Basic competencies:

1) Implementing honest, disciplined, careful, critical, curious, innovative, and responsible behavior in carrying out work in the field of electric power installation.
2) Appreciating cooperation, tolerance, peace, politeness, democracy, in solving problems of different concepts of thinking in carrying out tasks in the field of electric power installation.
3) Demonstrating responsive, proactive, consistent, and interactive behavior effectively with the social environment as a part of the solution to various problems in carrying out work in the field of electric power installation.

Based on the syllabus of an electric power installation course in vocational secondary school, it can be known that the character values embedded in the content of the syllabus. But there are still many working characteristics that are not appropriate yet for the demands of Industry 4.0.

5 RESULT AND DISCUSSION

Based on the discussion about working characteristics between Industry 4.0 and vocational secondary school, there are still many gaps. There are 25 working characteristics obtained by discussions with experts in Industry 4.0 and the practical lesson syllabus of vocational secondary school. The 25 working characteristics are communication, responsibility, cooperation, caring for the environment, safety and occupational health (K3) awareness, negotiation, problem-solving, analysis, life-long learning, management, adaptation to change, innovation, creativity, critical thinking, flexibility, decision-making, entrepreneurship, leadership, honesty, disciplined, polite, responsive, proactive, conscientious, and curious. The gap of working characteristics between Industry 4.0 and vocational secondary school is shown in Table 1.

In summary, there are still many working character gaps between Industry 4.0 and vocational secondary school. Vocational secondary school is not yet ready to prepare the working characteristics for graduates based on industry demand.

Table 1. The gap of working characteristics between Industry 4.0 and vocational secondary schools.

	Working characteristics	Percentage
Working characteristics that have been fulfilled by the vocational secondary schools	Responsibility, cooperation, caring for the environment, problem-solving, innovation, critical thinking and discipline	28 %
Working characteristics that have not been fulfilled by the vocational secondary schools	Communication, safety and occupational health (K3) awareness, negotiation, analysis, life-long learning, management, adaptation to change, creativity, flexibility, decision-making, entrepreneurship and leadership	48%
Working characteristics that are not relevant to Industry 4.0 needs	Honest, polite, responsive, proactive, conscientious, and curious	24%

6 CONCLUSION

This paper aims to determine the readiness of vocational secondary schools for reforming characteristics for working qualifications in Industry 4.0. Results of the discussion are as follows. First, the working characteristics demanded by industries are communication, responsibility, cooperation, caring for the environment, safety and occupational health (K3) awareness, negotiation, problem-solving, analysis, life-long learning, management, adaptation, innovation, creativity, critical thinking, flexiblity, decision-making, entrepreneurship, leadership, and discipline. Secondly, there are still many gaps in working characteristics between vocational secondary school and industrial demand. Finally, the practical learning syllabus of vocational secondary schools is not yet ready to prepare the working characteristics for graduates based on industries demand.

ACKNOWLEDGMENTS

This study was done as part of a thesis project. The authors thank colleagues from graduate school Yogyakarta State University who provided insight that greatly assisted the study.

REFERENCES

ACT. (2011). *A better measure of skills gaps: Utilizing ACT skill profile and assessment data for strategic skill research*. USA: ACT.

ADB. (2014). *Sustainable vocational training toward industrial upgrading and economic transformation: A knowledge sharing experience*. Manila: Asian Development Bank.

Agboola, A. & Tsai, K.C. (2012). Bring character education into classroom. *European Journal of Educational Research, 1*, 163–170.

Aime S., Pironti, M., Quatraro, F., Zanetti, M., Forno, S., Natale, F., & Stecca, S. (2017). *UniTo & the challenges of industry 4.0*. Torino: University of Torino.

Allen, J. & Van der Valden, R. (2001). Educational mismatches versus skill mismatches: Effects on wages, job satisfaction, and on-the-job search. *Oxford Economic Papers, 53*, 434–452.

Aristotle. (2017). *A Framework for character education in schools*. UK, University Of Brimingham: The Jubilee Center for Character & Virtues.

Awadh, A.M. (2012). The impact of personality traits and employee work-related attitudes on employee performance with the moderating effect of organizational culture: The case of Saudi Arabia. *Asian Journal of Business and Management Sciences, 1*, 108–127.

Battistich, V. (2000). Character education, prevention, and positive youth development. *Journal of Primary Prevention, 21*, 75–99.

Beer, A., & Brooks C. (2011). Information quality in personality judgment: The value of personal disclosure. *Journal of Research in Personality, 45*, 175–185.

Berger, R. (2016). Skill development for industry 4.0. *Whitepaper Summary*, BRICS Skill Development Working Group.

Bermúdez, M.D., & Juárez, B.F. (2017). Competencies to adopt industry 4.0 for operations management personnel at automotive parts suppliers in Nuevo Leon. *Proceedings of the International Conference on Industrial Engineering and Operations Management Bogota, Colombia, October 25 –26,2017*.

Bialik, M., Bogan, M., Fadel, C. & Horvathova, M. (2015). *Character education for the 21st century: What should students learn*. Boston, Massachusetts: Center for Curriculum Redesign.

Bukamal, H., & Mirza, C. (2017). The mismatch between higher education and labor market needs: A Bahrain case study. *Gulf Affairs*, 14–16.

Fu, X. & Yan P. (2017). *Building digital competencies to benefit from existing and emerging technologies with special focus on gender and youth dimensions*. Report of the Secretary General, Commision on Science and Technology for Development, Geneva.

Gehrke, L., Kuhn, A. T., David, R., Paul, M., Bellman, C., Siemas, S., Dawood, D., Singh, L., Julie, K., Standley, M. (2015). *A discussion of qualifications and skills in the factory of the future: A German and American perspective*. Düsseldorf, Germany: VDI.

Lorentz, M., Rüßmann, M., Strack, R., Lueth, K.L., & Bolle, M. (2015). *Man and machine in industry 4.0: How will technology transform the industrial workforce through 2025?* Boston, MA: The Boston Consulting Group.

Marshall, J.C., Caldwell, S.D., & Foster, J. (2011). Moral education the character plus way. *Journal of Moral Education, 40*, 51–72.

Motyl, B., Baronio, G., Uberti, S., Speranza, D., & Filippi, S. (2017). How will change the future engineers' skills in the Industry 4.0 framework: A questionnaire survey. *Procedia Manufacturing, 11*, 1501–1509.

Prifti, L., Knigge, M., Kienegger, H., & Krcmar, H. (2017, February). A competency model for "industrie 4.0" employees. *Conference: Wirtschaftsinformatik (WI)*. St. Gallen, Switzerland.

Sailah, I. (2007, April). *Pengembangan soft skills dalam proses pembelajaran di perguruantinggi* [Soft skills development in the learning process in higher education]. Paper presented at development socialization soft skills, UNS, Solo.

Selamat, A., Alias, R.A., Hikmi, S.N., Puteh, M., & Tapsi, S. M. (2017). Higher education 4.0: Current status and readiness in meeting the fourth industrial revolution challenges. *Redesigning Higher Education towards Industry, 4*, 23–24.

Suryanto, D., Kamdi, W., & Sutrisno, S. (2014). Relevansi Soft Skill yang Dibutuhkan Dunia Usaha/ Industri dengan yang Dibelajarkan di Sekolah Menengah Kejuruan [The relevance of soft skils required by the business/industry world with those learned in vocational high schools]. *Teknologidan Kejuruan, 36*, 107–118.

Uzair-Ul-Hasan, M. & Noreen, Z. (2013). Educational mismatch between graduates possessed skills and market demands in Pakistan. *International Education Studies, 6*, 122–129.

Walsh, E. & Donaldson, B. (2017). *Relevant in the era of industry 4.0*. Paper presented at Canada's Growth Agenda and Industry 4.0 – A Perspective From BCG, Boston, MA.

Yang, J. (2013). Research on supply and demand of highly skilled in China. *Business and Management Research, 2*, 61–65.

Zahra, N., Suherman, A., & Permana, T. (2017). Pengaruh penerapan model pembelajaran kooperatif tipe student teams achievement division terhadap aktivitas, interaksi, danhasilbelajarsiswa SMK [The effect of the application of cooperative learning models is the type of student teams achievement division on the activities, interactions, and learning outcomes of SMK students]. *Journal of Mechanical Engineering Education, 4*, 213–218.

Innovative Teaching and Learning Methods in Educational Systems – Retnowati et al. (Eds)
© *2020 Taylor & Francis Group, London, ISBN 978-1-03-224183-8*

The assessment model for competency certification tests

W. Ramadani & D.L.B. Taruno
Yogyakarta State University, Special Region of Yogyakarta, Indonesia

ABSTRACT: Competency certification is one of the Indonesian Government's efforts to improve the quality of human resources. This a literature study focuses on authentic assessment used in competency certification tests in universities, especially at diploma levels. The implementation of the competency certification test requires an assessment model that can measure individual abilities in real terms. The results of this study recommend that assessment tools in competency tests use authentic assessment models. An authentic form of assessment recommended for competency certification test is performance assessment, written assessment, and interview. Authentic assessment is expected to be applied in competency certification tests and can measure the competencies of an individual in real-world tasks.

1 INTRODUCTION

Learning in higher education is a series activity that is carried out in a planned and patterned manner in order to create an atmosphere of learning effectively. The learning process cannot be separated from assessment as an evaluation of the learning process. Assessment is an integral part of the learning process, and the assessment of student's learning outcomes provides knowledge on their level of understanding and knowledge achieved (Boud, 2010; Brown & Race, 2013). The effectiveness and meaningfulness of the learning process require an appropriate and comprehensive assessment system (O'Neill, 2012), which includes an assessment of knowledge, attitudes, and skills (general skills and special skills). Instruments to be used in the assessment can be in the form of tests and non-tests. One type of test used in higher education to assess student learning outcomes is a competency certification test. In addition to assessing student learning outcomes, the competency certification test also aims to recognise the competence of students in carrying out specific tasks. Thus, the implementation of the certification test requires an assessment model capable of assessing the ability of students in real-world tasks.

Using authentic assessments in higher education is more realistic, because student's duties generally tend to lead to problem-solving in a real-world context. Students are not only introduced to theories/concepts in the scientific field but also encouraged to address relevant issues around them. Authentic assessment is favored by students because it emphasizes their subconscious abilities (Azim & Khan, 2012; Hidayat, 2017). In authentic assessment students are asked to perform real-world tasks that demonstrate their application of knowledge and skills (Mueller, 2005). Authentic assessment emphasizes the learner's ability to demonstrate real and meaningful knowledge (Mueller, 2005; Nurgiyantoro, 2011).

The purposes of this paper are to build and construct concepts more strongly based on existing empirical research. So, the results of this literature study should contribute to the understanding of the benefits, challenges, and steps involved in implementing authentic assessments, and how that can be applied in implementing competency certification tests. This paper should be useful for prospective professional instructors who are interested in exploring a more authentic assessment.

2 LITERATURE REVIEW

The literature review in this paper was carried out using two approaches. We used a systematic approach to learn more about the concept of assessment in learning. The assessment discussed in this paper focuses on the main topics of authentic assessment. The sources reviewed in this paper include (i) academic books from the library, (ii) scholarly websites such as Google experts, etc., (iii) online government curriculum guidelines and documents; and (iv) other online electronic resources. Priority was given to the latest, peer-reviewed literature based on authentic assessments in higher education for the competency certification test. The relevant literature was then classified into appropriate categories. The analysis and synthesis of the literature are guided by the preview, question, read and summary (PQRS) system. The second approach was a thematic one,with key themes arising from the review objectives and classification categories.

3 OVERVIEW OF LITERATURE STUDY

3.1 *Authentic assessment*

The purposes of assessment are to obtain various information about the extent of an individual's competencies. The assessment process cannot run well if there is no assessment tool, namely the instrument to access competencies. The instrument can be in the form of an observation demonstration checklist, product or service observation, portfolio, written question list, list of interview questions, etc. Authentic assessment aims to assess an individual's abilities in a real-world context, namely how individuals apply their knowledge, skills and are effective in authentic tasks. Inauthentic assessment students are asked to perform real-world tasks that demonstrate meaningful application of essential knowledge and skills. Thus, authentic assessment requires students to show performance in real-world tasks which shows theirimplementation of knowledge and skills. Authentic assessments include structured tasks, performance tasks, projects, portfolios, demonstrations, experiments, oral presentations, and simulations (Nitko & Brookhart, 2007).

Some authentic assessments that are often used include performance appraisals, criteria-based assessments, systematic observations by instructors or learners (peer and self-assessment), and portfolios (Wellington et al., 2002). According to Ormiston (2011) authentic assessment requires authentic learning c and mirrors the tasks and problem-solving that is required in real-life outside of school. Authentic assessment consists of three assessment techniques: direct measurement of student skills related to long-term educational outcomes such as success at work.: an assessment of tasks that require broad involvement and complex performance: process analysis that is used to produce student responses to the acquisition of existing attitudes, skills and knowledge.

Authentic assessment includes tasks that encourage students to use scientific knowledge in real contexts and demonstrate the skills and competencies needed to navigate critical situations in professional life (Ashford-Rowe, Herrington, & Brown, 2014; Larkin, 2014; Sridharan & Mustard, 2015). Authentic types of assessment include portfolios, research and reflective journals, presentations, discussions, and performance assessments (Sridharan & Mustard, 2015). Authentic assessment is based on various types of assignments assessing high order thinking with demonstrations of knowledge that can be applied to real-life contexts (Ashford-Rowe et al., 2014)

Some of the literature on authentic assessment e include assessment of performance (performance assessment), assessment of written tests, and interviews.

3.1.1 *Self-assessment*
Self-assessment is an assessment technique that asks students to express their strengths and weaknesses in the context of attaining competency attitudes, spiritual attitudes, and social attitudes. The instrument used is a self-assessment sheet (Supardi, 2015; Kunandar, 2013).

3.1.2 *Performance assessment*

In performance assessment the teacher observes and makes judgments about the how the students demonstrate their skills or competence in producing a product. Performance appraisal can take the form of (i) paper and pencil tests where target is for students to display their work, such as tool design, graphic design, etc.; (ii) identification tests, which aim to measure students' abilities in identifying something, such as finding damaged components of an object; (iii) simulation tests, which are carried out without the use of real tools to assess whether the student has mastered the skill of using imitation tools; and (iv) performance tests, which are done with a real tool to see whether the students have mastered or are skilled in using the tool. There are three tools or instruments to measure student performance according to Johnson, Penny, and Gordon (2009): (i) checklist, (ii) rating-scale, and (iii) rubric.

3.1.3 *Written assessment*

Written test assessment is carried out with a written test,where questions and answers are given to participants in writing. There are two forms of written test questions: (i) choose answers, divided into multiple choice, two choices (right-wrong, yes-no), matchmaking, and causation; and (ii) supply answers, divided into fill in or complete, short or long answers, and descriptions.

3.1.4 *Oral/interview assessment (Interview)*

Interview is one of the non-test assessment tools used to get certain information about the participants through conducting oral questions and answers unilaterally, face-to-face, and with the direction and goals that have been previouslyn determined. According to Arifin (2009), there are three objectives in conducting an interview: (i) to obtain information directly to explain a particular situation or condition; (ii) to complete a scientific investigation; and (iii) to obtain data in order to influence the situation or certain people.

3.2 *The role of authentic assessment in the competency certification test*

Competency certification is an important instrument in assessing the quality and quantity of labor. The benefits of certification are explained in the British Institute of Nondestructive Testing (2010), which provide broader recognition of workers who contribute to effective corporate management and condition-based and predictive management. It can be concluded that the certification of individual expertise in the process of granting personal certificates can be carried out systematically and objectively through competency tests. One way to bridge the gap between industry expectations and the ability of graduates is to assess student learning more authentically. In the competency certification test, students are required to carry out a real task to assess and determine whether the tested participants are competent or not yet only competent on the standards tested. The standards tested require relevance between learning in institutions and in world of work. One way to bridge the gap between industry expectations and the ability of graduates is to assess student learning more authentically.

Competency certification testing requires critical thinking skills, better known as high order thinking, which requires students to solve problems in the real world with the knowledge and experience they have. In accordance with Bloom's taxonomy, thinking skills are categorized into two groups. Low order thinking consists of understanding, knowledge, and application, whilet high order thinking requires evaluation, creativity, and analysis. Research has shown that someone with critical thinking skills performs better and provides more information by elaborating on the results of communication, observation, and experience (Paul and Elder, 2013; Puteh and Hamid, 2014). In addition, critical thinking and higher thinking control basic cognitive processes, namely meta-cognitive, and combining a comprehensive understanding of certain topics. By following basic problem-solving steps, students develop analytical skills from examining cases, improving assessment skills through evaluating different scenarios and acquiring synthesis skills through scenario reconstruction (McPeck, 2016).

Some studies show that there are benefits for both students and lecturers. The population of students in higher education varies (Higher Education Authority, 2016; Bohemia & Davison, 2012), bringing with them a wider range of backgrounds, hopes, skills, and experiences. Authentic assessment is believed to motivate and inspire students to explore the dimensions of themselves and the world they might ignore. (Lombardi, 2008; Bohemia & Davison, 2012). Authentic assessment has been linked to the development of many attributes by giving students real-world experience and overcoming competency gaps in institutions and the world of work (Sutherland & Markauskaite, 2012). Such attributes include critical thinking, teamwork, problem-solving, effective communication, and reflective practice (Keeling et al., 2013; Mueller, 2005). Professional motivation and learning and lecturer development can also be enriched, through networking and collaboration with students, institutions, and stakeholders.

3.3 Challenges of authentic assessment

The authentic assessment provides benefits for lecturers, assessors, and students, but in practice, authentic assessments experience several challenges. For example, some students might find authentic assessments challenging because of the increased emphasis on thinking skills and teamwork (Bohemia & Davison, 2012; Craddock & Mathias, 2009). In addition, Keeling et al. (2013) suggested that students also feel confused by complex authentic tasks regarding real-world problems and require high order thinking skills. Hart, Hammer, Collins, & Chardon (2011) state thate solution for dealing with the difficulties faced by students is providing clear guidelines and support. The development and provision of rubrics have been found to reduce student anxiety, increase student confidence (Litchfield & Dempsey, 2015) and help the sense of objectivity when using authentic assessments.

Other challenges may be experienced by students, assessors, and lecturers relate to resources, especially time and costs. Authentic tasks require more investment of time and resources than a more traditional assessment (Hart et al., 2011). These challenges can be overcome by choosing small-scale activities when time and financial support are relatively limited and provide detailed task in a distinct time (Litchfield & Dempsey, 2015).

3.4 Steps involved in implementing authentic assessment

A good assessment requires careful planning so that it can provide benefits for students, lecturers, and assessors, because this study focuses on the authentic assessment of competency certification tests. The following outlines some suggested steps in applying authentic judgments.

3.4.1 Identifyingdesired learning outcomes and alignment with tasks

Learning outcomes must be defined and constructively adjusted to the task (Biggs, 2014). This ensures the relevance between intended learning outcomes, teaching and learning activities, and assessment, so that it will increase students' understanding of what is needed to successfully complete the task (Mueller, 2005; Greenstein, 2012).

3.4.2 Development of rubrics and marking criteria

Effective rubric development must be in line with learning standards and outcomes (Stevens & Levi, 2013). This helps students to overcome obstacles and successfully complete the assessment (Keeling et al., 2013; Fook & Sidhu, 2010).

3.4.3 Implementation of assessment, scoring, and interpretation of results

In order to improve the effectiveness of the implementation of authentic assessments, it is important to see the details about students' experiences through feedback forms (Keeling et al., 2013; Fook & Sidhu, 2010).

3.4.4 *Evaluation and reflection*

Evaluation and reflection are important components of the assessment process of assessors, lecturers. and students (Keeling et al., 2013). The final evaluation and reflection aim to identify what attributes they have developed and how they can build this in the future real-world experience.

3.5 *Design authentic assessment*

A good assessment concept designmust cover three very important elements: student support, intellectual rigor, and significance (Gore et al., 2009). These three elements relate to one another and show that each assessment task must be designed to increase students' active involvement, help them to develop their higher thinking, and require them to clearly show what they have learned (Clare, 2007). In addition, Ashford-Rowe et al. (2014) suggested another eight important elements in determining authentic assessment:

a. The authentic assessment must challenge students to build and produce knowledge.
b. The skills and knowledge demonstrated by students through authentic assessment must enable them to successfully produce performance or products (results).
c. Authentic assessment designs must support the concept that skills, knowledge, and attitudes being assessed that can be transferred to other areas.
d. Critical reflection, self-evaluation, and self-development must be authentic components in a system.
e. There is a need for accuracy in assessment activities in developing learner intellectual input and evaluating how key skills and knowledge are relevant to work-related scenarios.
f. Languages, graphics, and topics should be used to provide assessment tasks that simulate a 'real-world' environment.
g. The essence of opportunities that formally combine to provide feedback.
h. The importance of creating opportunities for collaboration.

The most important aspect of authentic assessment lies in the design stage of authentic assessment tasks. The difficulty that occurs in the design stage is how to design authentic assessment tasks that mimic students' real-world experiences and provide meaningful learning experiences in the learning process (Rodríguez-Gómez & Ibarra-Sáiz, 2015). Peris-Ortiz & Lindahl (2015) proposed three principles that support the design of authentic assessments, which are challenges, reflections, and diversions.

a. The creation of challenging assessment tasks by designing assessment tasks that require various strategies in solving them.
b. An assessment assignment proposal that demands intelligence, which means it requires students to demonstrate learning and understanding and high order thinking through special assessment assignments.
c. Knowledge and skills learned in one area can be applied in other areas that are often not related (Peris-Ortiz & Lindahl, 2015).

4 CONCLUSION

This article has included concepts and definitions of authentic assessment in competency certification tests. Assessment is an integral part of the learning process and provides benefits for lecturers and students. The aim is to formulate an effective framework for the design of authentic assessment tasks, which is seful for those who are new to authentic assessments and those who have used authentic judgments that seek to redevelop their judgments.

Considering everything, assessment in competency certification test is a synergic blend of several forms of assessment that will help students develop their knowledge and understanding and acquire various skills. Assessment is an opportunity for learning so must provide students with the information needed that will improve their performance. As such, it must include a variety of ways that offer information about the outcomes of the education process and be a measurement that is appropriate to the curriculum objectives.

Synergic combination of several forms of assessment is known as authentic valuation. This assessment is an authentic assignment that is able to measure students' abilities in the world of work. There are many forms of authentic assessment that can be used in competency certification tests. But according to the certification competency test principle, based on the results of a literature study, the forms of assessment that are suitable for competency certification tests are self-assessment, interviews, written tests, and performance tests. The choice of the form of assessment in the competency certification test must reflect the type of competency that will be achieved in accordance with the assessment plan target previously prepared by the assessor team.

ACKNOWLEDGMENTS

The authors would like to thank the Post Graduate Program at Yogyakarta State University (YSU), the Special Region of Yogyakarta for supporting this work.

REFERENCES

Arifin, Z. (2009). *Evaluasipembelajaran* [Evaluation of learning]. Bandung, Indonesia: PT. RemajaRosdakarya.

Ashford-Rowe, K., Herrington, J., & Brown, C. (2014). Establishing the critical elements that determine authentic assessment. *Assessment & Evaluation in Higher Education, 39*, 205–222.

Azim, S., & Khan, M. (2012). Authentic assessment: An instructional tool to enhance student's learning. *Academic Research International, 2*, 314–320.

Biggs, J. (2014). Constructive alignment in university teaching. *HERDSA Review of higher education, 1*, 5–22.

Bohemia, E., & Davison, G. (2012). Authentic learning: The gift project. *Digital and Technology Education: An International Journal, 17*, 49–62.

Boud, D., & Dochy, F. (2010). *Assessment 2020: Seven propositions for assessment reform in higher education*. Sydney: Australian Learning and Teaching Council.

Brown, S., & Race, P. (2013). Using effective assessment to promote learning. In L. Hunt & D. Chalmers (Eds.), *University teaching in focus: A learning-centered approach*. London, England: Routledge.

Clare, B. (2007). Promoting deep learning: A teaching, learning and assessment endeavor. *Social Work Education, 26*, 433–446.

Craddock, D., & Mathias, H. (2009). Assessment options in higher education. *Assessment and Evaluation in Higher Education, 34*, 127–140.

Fook, C.Y., & Sidhu, G.K. (2010). Authentic assessment and pedagogical strategies in higher education. *Journal of Social Sciences, 6*, 153–161.

Gore, J., Ladwig, J., Elsworth, W., & Ellis, H. (2009). *Quality assessment framework: A guide for assessment practice in higher education*. Callaghan: The University of Newcastle.

Greenstein, L. (2012). Beyond the core: Assessing authentic assessment 21st century skills. *Principal Leadership, 13*, 36–42.

Hart, C., Hammer, S., Collins, P., & Chardon, T. (2011). The real deal: Using authentic assessment to promote student engagement in the first and second years of a regional law program. *Legal Education Review, 21*, 97–121.

Hidayat, T. (2017). Authentic assessment and its relevation with the quality of learning results (Lecturer nnd student perception of IKIP PGRI Bojonegoro). *Journal of Social Sciences Education, 27*, 92–103.

Higher Education Authority. (2016). *Higher education key facts and figures 2015/2016*. Dublin: Higher Education Authority.

Johnson, R.L., Penny, J.A., & Gordon, B. (2009). *Assessing performance: Designing, scoring, and validating performance tasks*. New York, NY: Guilford Press.

Keeling, S.M., Woodlee, K.M., & Maher, M.A. (2013). Assessment is not a spectator sport: Experiencing authentic assessment in the classroom. *Assessment Update, 25*(5), 12–13.

Kunandar. (2013). *Authentic assessment (Assessment of student learning outcomes based on the 2013 curriculum)*. Jakarta, Indonesia: Raja GrafindoPersada.

Larkin, T. L. (2014). The student conference: A model of authentic assessment. *International Journal of Engineering Pedagogy (IJEP)*, *4*, 36–46.

Litchfield, B.C., & Dempsey, J.V. (2015). Authentic assessment of knowledge, skills and attitudes. *New Directions for Teaching and Learning*, *2015*, 65–80.

Lombardi, M.M. (2008). *Making the grade: The role of assessment in authentic learning.*Educause Learning Initiative. *EDUCAUSE Learning Initiative.*

McPeck, J.E. (2016). *Critical thinking and education*. London: Routledge.

Mueller, J. (2005). The authentic assessment toolbox: Enhancing student learning through online faculty development. *Journal of Online Learning and Teaching*, *1*, 1–7.

Nitko, A.J., & Brookhart, S.M. (2007). *Educational assessment of students* (3rd ed.). New Jersey, NJ: Pearson Education.

Nurgiyantoro, B. (2011). *Authentic assessment in language learning*. Yogyakarta, Indonesia: Gajah Mada University Press.

O'Neill, G. (2012). *UCD assessment redesign project: The balance between assessment for and of learning. University College Dublin*. Retrieved from https://www.ucd.ie/t4cms/UCDTLA0044.pdf

Ormiston, M.J. (2011). *Creating a digital-rich classroom: Teaching & learning in a web 2.0 world*. Bloomington, USA: Solution Tree Press.

Paul, R., & Elder, L. (2013). *Critical thinking: Tools for taking charge of your professional and personal life*. Upper Saddle River, NJ: Pearson Education.

Peris-Ortiz, M., & Lindahl, J.M.M. (2015). *Sustainable learning in higher education: Developing competencies for the global marketplace*. Cham, Switzerland: Springer.

Puteh, M.S., & Hamid, F.A. (2014). A test on critical thinking level of graduating bachelor of accounting students: Malaysian evidence. *Procedia-Social and Behavioral Sciences*, *116*, 2794–2798.

Rodríguez-Gómez, G., & Ibarra-Sáiz, M.S. (2015). Assessment as learning and empowerment: Towards sustainable learning in higher education. In M. Peris-Ortiz & J.M. Lindahl (Eds.), *Sustainable Learning in Higher Education*. Cham, Switzerland: Springer.

Sridharan, B., & Mustard, J. (2015). *Authentic assessment methods: A practical handbook for teaching staff*. Victoria: Business and Law Learning innovations.

Stevens, D.D., & Levi, J. (2013). *Introduction to rubrics* (2nd ed.). Sterling, VA: Stylus.

Supardi. (2015). *Assessment authentic*. Jakarta, Indonesia: Raja GrafindoPersada.

Sutherland, L., & Markauskaite, L. (2012). Examining the role of authenticity in supporting the development of professional identity: An example from teacher education in Higher Education, *Higher Education*, *64*, 747–766.

Wellington, P., Thomas, I., Powell, I., & Clarke, B. (2002). Authentic assessment applied to engineering and business undergraduate consulting teams. *International Journal of Engineering Education*, *18*, 168–179.

Innovative Teaching and Learning Methods in Educational Systems – Retnowati et al. (Eds)
© 2020 Taylor & Francis Group, London, ISBN 978-1-03-224183-8

Contribution of a teacher competency test to identifying teacher performance in vocational high schools

U. Nursusanto & N. Yuniarti
Yogyakarta State University, Indonesia

ABSTRACT: Competence is the foundation of a teacher in carrying out learning. The expertise of a teacher needs to be developed out on an ongoing basis. Competency testing is one of the stages in developing teacher skills. This study used a literature study approach. The data are presented using tables, figures, and descriptive analysis. The first analysis found that mastery of the competency of the teacher was in "good" criteria at 71.45%. The second analysis found that in the implementation of the teacher competency test the average score was 86% with "very good" category. It can be concluded that overall mastery of competence and teacher competency testing played an important role in teacher performance in vocational secondary schools in improving the quality and learning process.

1 INTRODUCTION

Vocational high schools (SMK) have played an active role in producing skilled, creative, innovative, productive and independent quality human resources. Human resource development in vocational high schools will be maximized if supported by teachers or professional educatorss to always improve the quality and productivity of work. Professional skills of teachers are needed as educators, instructors, trainers, assessors, and evaluators of students in the learning process within the vocational high school.

Teachers as educators must develop human resources through vocational education, both at the vocational level and through training activities. Human resources competencies are developed by teachers in the learning process in vocational high schools in the form of improving the quality of skills and work productivity of students. According to the 2016 Human Development Index (HDI) report, Indonesia ranks 113th out of 187 countries in the world. This proves that Indonesia is still lacking in the development of quality human resources.

Teachers are one of the human resources who play an active role in the field of formal and non-formal education. Teachers in carrying out their duties as educators are required to be have relevant professional skills. This must be followed with up-to-date knowledge of technological developments in education and industry.

In Indonesia, teacher professionalism requirements have been documented since 2013, in the Regulation of The State Minister of Public Application and Reform Bureaucracy number 16 of 2009, which concerns teachers' functional positions and the credit numbers (RoI, 2009). The regulation states that teacher performance will be assessed every year and improved by sustainable professional development (PKB) for promotion to a higher level. PKB includes self-development, scientific publications in the form of classroom action research (CAR), and innovative work. The aim of the activity was to increase teacher knowledge and improve the quality of the learning process, which in turn has an impact on improving student learning outcomes.

The teacher has a very important task, function, and role in the learning process to improve students' learning motivation. The duties, functions and roles must be carried out professionally and continuously. The latest research conducted by the Indonesian Education Monitoring Network (JPPI), the Right to Education Index (RTEI) in 2016, stated that Indonesia was ranked 7th out of 14 countries, ranking below the Philippines and Ethiopia. Indonesia scored

77% from the five indicators measured in the RTEI study, including governance, availability, accessibility, acceptability, and adaptability. JPPI said that three of the five indicators need special attention: teacher quality (availability), a school environment that is not child friendly (acceptability), and discrimination against marginal groups (adaptability). Teachers, although central in the education system, are still considered very low in terms of possessing quality professional skills in Indonesia. This shows that professional teachers are urgently needed to improve the quality of education in Indonesia.

This article discusses a test of teacher competency in learning activities. The main discussion is presented using literature studies on the development of a teacher competency test model and how it is applied in the world of education. The results are presented in tables, graphs and descriptive analysis.

2 LITERATURE STUDY

2.1 *Teacher competence*

Teachers, as one of the main components in the world of education, play an important role in the learning process, which carried out in an effort to improve the competence of students. A good learning process must involve reciprocal communication between teachers as educators and students. Management of a good learning process must be balanced with the four competencies that must be mastered by a teacher: personality, professional, pedagogical, and social (RoI, 2007).

2.1.1 *Personal competence*
Personal competence is an ability that is directly related to a teacher's personal behavior in everyday life (Madjid, 2016). This was supported by Hakim (2015), who expressed "personality competencies that must be possessed by the teacher are stable, independent personality, having a work ethic, wise, positive thinking, noble character, and a role model for students and the surrounding community."

2.1.2 *Professional competence*
Professional competence includes actively creating challenges, improving skills and knowledge, and being an efficient facilitator for students (Bolitho & Padwad, 2013). This expression was supported by Hakim (2015), who stated that "The existence of a professional teacher in the field plays an important role in the learning process and the formation of mindset, attitudes and following the behavior of students,"

2.1.3 *Pedagogic competence*
Teachers as facilitators in the learning process have pedagogical competencies related to the level of understanding of students, instructional design, the implementation of learning diagnosis, and the evaluation of learning (Suarmika, 2017). According to Madjid (2016), pedagogic competence is knowledge about the learning and behavior or characteristics of students.

2.1.4 *Social competence*
Satori (2009) revealed that "social competence is the ability to understand themselves related to behavior, personality, soft skills, and hard skills as part of society and citizens in developing tasks". It is also supported by Hamidi and Indrastuti (2012), who stated that social competence is required by teachers to establish relationships with other people or the ability to interact socially and carry out social responsibility.

2.2 *Teacher professionalism*

The teacher as educators is one of the important elements in the education system (Madjid, 2016). According to government regulation number 19 of 2017, a teacher is a professional educator who has the main task of creating students' character and transferring knowledge in formal education, basic education, and secondary education (RoI, 2017). According to Yu,

Luo, Sun, and Strobel (2012), "professional" is an activity carried out to improve competency based on knowledge in accordance with the field of expertise mastered. Teacher professionalism is the ability and responsibility in carrying out duties as educators to always improve their competencies. Through these three definitions, it can be concluded that professional teachers have the abilities possessed by the teacher to convey and provide knowledge through structured learning activities in an effort to improve the ability of students.

2.3 Teacher competency test (TCT)

According to RoI (2015), a teacher competency test is an activity carried out as part of teacher performance evaluation in the framework of rank and position career coaching. According to Mulyawan (2012), teacher competency testing is carried out in the framework of mapping teacher competencies with the aim of improving the quality of education. Competencies tested in the teacher competency test include pedagogical competence and professional competence (Dharma, 2013). Thus, the teacher competency test is a set of activities carried out with teachers in the framework of competency mapping, as an assessment, with the aim of improving the quality and quality of education.

2.4 Teacher performance

Performance is work behavior that is carried out coherently in accordance with procedures to obtain results that meet the requirements of quality, speed, and quantity (Madjid, 2016). This is supported by Smith (200 stated 3), whoc that performance is the result or output of a process. The understanding of performance is described more clearly by Bernardin and Russel (1993): "Performance is the result obtained based on certain tasks or functions in a certain period." Based on the above definitions, it can be concluded that the teacher's performance is a result obtained by a teacher with ability and effort in a certain period.

3 RESEARCH METHOD

The research method used was literature study. The literature used for the discussion of teacher competency tests was the results of research, journals, and papers that have been implemented in the education system. The data obtained are presented in the form of graphs, tables and descriptive analysis.

4 RESULTS AND DISCUSSION

A teacher competency test is a step in the development of human resources, implemented by the government, in the context of competency mapping with the aim of improving the quality of education. Teachers in the education system are required to master four competencies: personal, professional, pedagogical, and social. Competence is the main requirement of a teacher in carrying out tasks, functions, roles, and responsibilities in the education system to get the desired results.

The results of research conducted by Madjid (2016) through a questionnaire given to 161 respondents are explained below.

4.1 Pedagogical competence

Questionnaire sheets related to pedagogical competencies were given to 161 respondents The six categories were: [1] mastering the characteristics of students; [2] mastering learning theory and learning principles; [3] curriculum related to the subject being taught; [4] carrying out educational learning; [5] utilizing information and communication technology for learning; [6] ability to communicate effectively, empathetically, and politely to students.

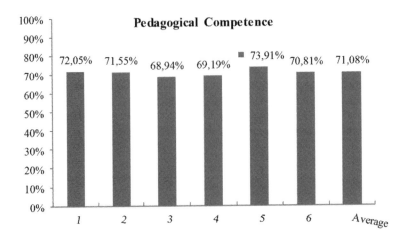

Figure 1. Percentage of pedagogical competence.

From the data, Madjid (2016) found that the score related to the use of information and communication technology for learning was the highest at 73.91%, followed by the mastery of the characteristics of students with a score of 72.05%. The lowest value was 68.94%, for curriculum development related to the subject being taught. Overall, the assessment of pedagogical competence obtained a value of 71.08%. This shows that the teacher has good pedagogical competence. Mastery of good pedagogical competencies will improve the success of good learning processes and outcomes.

4.2 Personal competence

Personal competence is directly related to the teachers' behavior. The results of the assessment through questionnaires related to personality competencies were assessed in four categories: [1] activities according to religious, legal and social norms; [2] attitude, showing themself to be honest, noble, and an example for students and the community; [3] a work ethic showing high responsibility; [4] upholding the teacher's code of ethics.

Figure 2 shows that the highest score is 73.04%, related to the ability to uphold the teacher's code of ethics and the lowest score is 68.32% for attitudes that are in accordance with religious, legal and social norms. Overall, the personality competency scored an overall 71.27%.

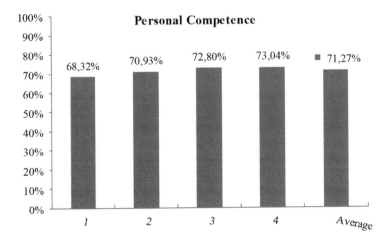

Figure 2. Percentage of personality competence.

Thus, the personality competence of teachers is in the "good" category. Teachers' personality competencies must emanate noble attitudes, behavior, and morality in daily life.

4.3 Social competence

Social competence is closely related to the life environment of a teacher, in school, the community, and as citizens. Social competence is the ability to interact to establish relationships between teachers and learners, fellow teachers, and other people around about. Assessment of social competence is carried out through three categories: [1] be inclusive and objective toward students; [2] communicate effectively, empathetically, and politely with fellow educators; [3] communicate with the professional community itself and other professions or in writing or other forms. The results of social competence research are shown in Figure 3.

The overall social competence score was 72.51%, in the "good" category. It can be concluded that the teacher is able to communicate with the community, the school environment and the environment where the teacher lives.

4.4 Professional competence

Professional competence is the ability of a teacher in mastering learning material widely and in depth, both subject matter curriculum and scientific structure and methodology. The questionnaire distributed to respondents in the assessment of professional competencies was carried out through four categories: [1] mastering the material, structure, concepts, and scientific mindset; [2] mastering competency standards and basic competencies of subjects; [3] developing professionalism in a sustainable manner; [4] utilizing information and communication technology to develop themselves.

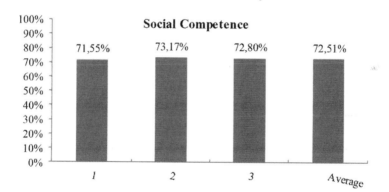

Figure 3. Percentage of social competence.

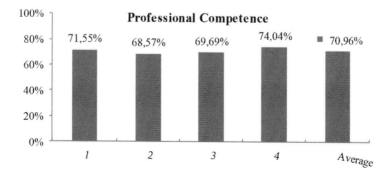

Figure 4. Percentage of professional competence.

227

The overall score for professional competence was 70.96%, in the "good" category. Teachers in the learning process are considered capable in carrying out their duties, functions and roles as educators. Teachers who are professional in carrying out their duties must meet the four criteria above.

The results of the overall teacher competency assessment obtained a mean of 71.45%, in the "good" category. Thus, teachers who achieve "good" criteria in the competency assessment are able to carry out tasks, functions and roles well.

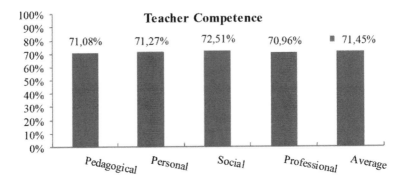

Figure 5. Results of the overall competency assessment.

The result of the second study was taken from a vocational and career education journal article with the title "Development of the catering system pedagogic competency test model on vocational teachers' professional education in culinary programs" by Handayani, et al. (2017). This discussed the development of a competency test model carried out in the Family Welfare Education Faculty of Engineering, Semarang State University. The subjects of this study were PKK students (in Culinary Art). The research procedure used by Plomp (1997) is phase, initial investigation, design phase, phase of realization, and phase of trial and revision (Figure 6).

The results and data obtained through the four procedures above are explained in the four models. The test model that was approved and developed was test model number 3, with a percentage approval of 80%.

Model 3 was carried out with teacher professional education program (PPG) participants graduated from education/non-education institutions. The development focus was the difference from the non-education participant who took the proficiency test and given the chance to

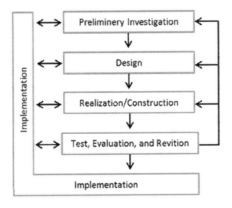

Figure 6. Research and development procedure scheme (Plomp, 1997).

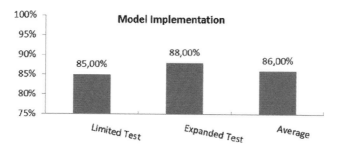

Figure 7. Results of the model implementation.

choose a concentration of subjects for pedagogical competence test. Other developments carried out on more specific assessment covered all the indicators of pedagogical competence. Activities during the development phase were assessment by experts for the guidebooks and competency testing tools carried out three times: limited trials (pre UT), after UT, and UP. After the assessment phase was completed, the next activities were observing the behavior of the training participants when learning, instructor activities, assessor activities, and evaluating the effectiveness of the model and assessing the practicality of the model.

The results of the observations observed by two observers in a limited trial obtained a scoree of 85% and the expanded trial the score was 88%. Therefore, the average level of implementation of the model is 86%, with very good criteria, as shown in Figure 7 below.

5   CONCLUSION

Competence is a single entity attached to a teacher. Competence is needed for a teacher to carry out tasks, functions, and roles as an educator. There are four competencies that a teacher must have: personality, professional, pedagogical, and social (PP No.19 of 2017). The results of the first study related to teacher competency obtained an overall assessment with "good" criteria: 71.27% for personality competency, 71.08% for pedagogical competence, 72.51% for social competence and 70.96% for professional competence.

As a foundation forcarrying out learning process, all four competencies should be mastered by teachers to improve their professionalism. Those competencies also help teachers to improve students' skills. Some factors that could possibly affect teacher competencies level are educational background, expertise of each teacher, health, economy, teacher income, discipline in work, and school supervision).

REFERENCES

Bernardin, H.J. & Russel, J.E. (1993). *Human Resource Management: An Experiential Approach.* New York, NY: McGraw-Hill.
Bolitho, R., & Padwad, A. (Eds.). (2013). *Continuing professional development lessons from India.* New Delhi: British Council.
Dharma, S. 2013. *Tantangan guru SMK abad 21* [*Challenges of 21st century vocational teachers*]. Jakarta: Direktorat Pembinaan Pendidik dan Tenaga Kependidikan Menengah Direktorat.
Hakim, A. (2015). Contribution of competence teacher (pedagogical, personality, professional competence and social) on the performance of learning. *The International Journal Of Engineering And Science (IJES),* 4(2), 1–12.
Hamidi & Indrastuti, S. (2012). *Analisis pengaruh kompetensi, latar belakang pendidikan terhadap kinerja dosen Politeknik Bengkalis dari perspektif pimpinan* [Analysis of the influence of competence, educational background on the performance of Bengkalis Polytechnic lecturers from the leadership perspective]. Politeknik Negeri Bengkalis.
Handayani, I., Basyirun, B., & Endroyo, B. (2017). Pengembangan model uji kompetensi pedagogik tata boga pada pendidikan profesi guru SMK program keahlian kuliner [Development of a cookery

pedagogical competency test model in vocational teacher professional education culinary expertise program]. *Journal of Vocational and Career Educational, 2*(2), 35–40.

Madjid, A. (2016). *Pengembangan kinerja guru melalui kompetensi, komitmen dan motivasi kerja* [Teacher performance development through competence, commitment and work motivation]. Yogyakarta, Indonesia: Penerbit Samudra Biru (Anggota IKAPI).

Mulyawan, B. (2012). Pengaruh pengalaman dalam pelatihan terhadap peningkatan kompetensi profesional guru [Effects of experience in training on improving teacher professional competence]. *Jurnal Ilmiah Ilmu Sosial, 11*(1), 45–65.

RoI. (2015). *Pedoman pelaksanaan uji kompetensi guru* [Guidelines for implementing teacher competency tests]. Jakarta, Indonesia: Ministry of Education and Culture, Republic of Indonesia.

RoI. (2007). *Peraturan Menteri Pendidikan Nasional Republik Indonesia Nomor 16 Tahun 2007 tentang Standar Kualifikasi Akademik dan Kompetensi Guru* [Regulation of Minister of National Education Number 16 of 2007 concerning Standard Academic Qualifications And Teacher Competency]. Jakarta, Indonesia: Ministry of National Education, Republic of Indonesia.

RoI. (2009). *Peraturan Menteri Negara Pendayagunaan Aparatur Negara dan Reformasi Birokrasi Number 16 tahun 2009, Tentang Jabatan Fungsional Guru dan Angka Kreditnya.* [RI government regulation no. 19 of 2009, concerning teacher functional positions and credit numbers]. Jakarta, Indonesia: Republic of Indonesia.

RoI. (2017). *Peraturan Pemerintah RI Nomor 19 Tahun 2017, tentang Perubahan Atas Peraturan Pemerintah Nomor 74 Tahun 2008 tentang Guru* [RI Government Regulation No. 19 of 2017, concerning Amendments to Government Regulation Number 74 of 2008 concerning Teachers]. Jakarta, Indonesia: Republic of Indonesia.

Said, A. (2018). Vocational teaching-learning through the eyes of undergraduate vocational students in Malta: A qualitative exploratory study. *International Journal for Research in Vocational Education and Training (IJRVET), 5*(1), 42–63.

Satori, D. (2009). *Profesi keguruan* [Teacher's profession]. Jakarta, Indonesia: Universitas Terbuka.

Smith, A.W (2003). *Management system analysis and applications.* New York, NY: Holt Saunders International Editions.

Suarmika, P.E. (2018, January). Teacher pedagogic competency and national examination result at elementary school. *SHS Web of Conferences 42,* 00030. doi: 10.1051/shsconf/20184200030.

Yu, J.H., Luo, Y., Sun, Y., & Strobel, J. (2012). A conceptual K-6 teacher competency model for teaching engineering. *Procedia-Social Behavioral Sciences, 56*(2012), 234–252. doi: 10.1016/j.sbspro.2012.09.651

Innovative Teaching and Learning Methods in Educational Systems – Retnowati et al. (Eds)
© 2020 Taylor & Francis Group, London, ISBN 978-1-03-224183-8

Performance in innovative Teacher Professional Development (TPD) in Indonesia: Does gender matter?

Wuryaningsih, M. Darwin & D.H. Susilastuti
Universitas Gadjah Mada, Yogyakarta, Indonesia

A.C. Pierewan
Universitas Negeri Yogyakarta, Yogyakarta, Indonesia

ABSTRACT: Innovative Teacher Professional Development (TPD) with web 2.0 technology is an appropriate learning model in the 21st century, due to its ability to overcome the digital divide in issues such as gender boundaries. This study used pre-test and post-test scores on teacher training programs in 2016. Participants were all teachers involved in the national TPD program totaling 427,189 teachers from all education levels in 34 provinces. This study examines the relationship between TPD's effectiveness and gender differences, and the test score increase generated for males versus females in TPD innovative learning versus traditional learning. Results show that there is a relationship between TPD effectiveness and gender difference. Innovative TPD performs better than traditional TPD, and females perform better than males.

1 INTRODUCTION

Information and communication (ICT) integration in teacher professional development (TPD) introduces new goals and structures (Twining, Raffaghelli, Albion, & Knezek, 2013) that support education reform in the 21st century and impact educational change (Bautista & Ortega-Ruiz, 2015). This innovation associated with technology (Rogers, 1995) has effected contemporary change from traditional patterns to online learning using web 2.0 technology (Brooks & Gibson, 2012).

In 2016 the Government of Indonesia implemented a new policy of TPD integrated with web 2.0 technology, which involved a total of 427,189 teachers at all levels in 34 provinces in Indonesia.

Innovation technology is an enabler for education transformation, but there are many failures due to complex interactions between users, technology and the environment (Huda & Hussin, 2016), such as teacher readiness (Kachelhoffer & Khine, 2009) and gender limitations. Women are seen as one of the inhibiting factors of online learning (The Australian Institute for Social Research, 2006) and men have a better attitude toward the use of technology (Cai, Fan, & Du, 2017), which will affect learning outcomes (Li & Tsai, 2017).

This study will determine the results of gender differences in innovative TPD by examining the relationship between TPD effectiveness and gender.

2 LITERATURE REVIEW

2.1 *Innovative TPD*

TPD can strengthens teaching and learning (Johnson, 2014), which influences teacher performance (Guskey, 2002; Borko, 2004; Lemoff, Bush, & Williams, 2005) and impacts student achievement (Heller, Daehler, Wong, Shinohara, & Miratrix, 2012; Kunter et al., 2013; Johnson, 2014).

Online TPD is an innovation (OECD, 2009) to improve teacher knowledge and ICT skills in pedagogical contexts related to computers and internet networks. The integration of ICT that is tailored to internal and external needs can benefit teachers (Albion, Tondeur, Forkosh-Baruch, & Peeraer, 2015). Teachers are expected to improve their competencies within their busy schedules, absorb a wide and unlimited source of information, available in real-time and on an ongoing basis (Dede, 2006) through the application of web 2.0 technology, which provides ease of learning preferred by teachers rather than just technology recognition material (Sánchez-García, Marcos, Guan Lin, & Escribano, 2013).

Online TPD integrated with the application of technology (Tondeur, van Braak, Ertmer, & Ottenbreit-Leftwich, 2016) brings positive changes to the teaching process (Toit, 2015). Levels of motivation (Keramati, Afshari-Mofrad, & Kamrani, 2011), self-efficacy (Kao & Tsai, 2009), intensity of the online learning (Chu & Chen, 2016), plus digital literacy (Mohammad yari & Singh, 2015) affect behavior and learning outcomes (Li & Tsai, 2017).

Innovative TPD has a positive impact on learning outcomes (Walczak & Taylor, 2018), increasing teacher beliefs (Fishman et al., 2013) and post-test scores (Fisher, Schumaker, Culbertson, & Deshler, 2010).

2.2 Gender difference

Gender differences in technology use have long been a concern in education (Cai et al., 2017). This is called the digital divide (OECD, 2001), where being female is seen as an inhibiting factor for online learning (The Australian Institute for Social Research, 2006). Meanwhile, online learning also has an important role in reducing the gap in relation to use and internet access (Olson et al., 2011).

Studies of the innovative influence of TPD on teacher and student performance provide an overview of demographic characteristics, such as gender, age, ethnicity, teaching experience, subject matter, etc. Many studies compare the differences and list them in results and findings sections. Some such studies comparing gender performance reveal that male teachers are more prepared to transfer self-efficacy to online learning (Hung, 2016), and appear more critical in online discussions, even though the frequency of participation and discussion is more prominent for females (Asterhan, Schwarz, & Gil, 2012) and women benefit more from virtual communities, especially those with less digital experience (Matzat, 2013).

Likewise, studies outside of TPD related to the use of computers in learning show that a positive relationship is shown by men with self-efficacy, whereas this relationship is negative in women (Deechuay, Koul, Maneewan, & Lerdpornkulrat, 2016) and men have higher competence than women (Vitanova, Atanasova-Pachemska, Iliev, & Pachemska, 2015). But other studies show that women have a greater chance of success than men in various learning models (Kazu & Demirkol, 2014).

Although there are studies that reveal differences in gender performance (Santos, Azevedo, & Pedro, 2013) toward acceptance of online learning (Tarhini, Elyas, Akour, & Al-Salti, 2016), others reveal no difference in gender variables (Smith, 2013) and no influence on online learning (Van Seters, Ossevoort, Tramper, & Goedhart, 2012).

2.3 Research question

a) How did TPD effectiveness relate to gender difference?
b) How was TPD effectiveness for males versus females in innovative TPD versus traditional TPD portrayed?

2.4 *Hypothesis*

a) H_0: $X = 0$; there is no correlation between TPD effectiveness and gender difference
b) H_a: $X \neq 0$; there is a correlation between TPD effectiveness and gender difference

3 METHOD

3.1 *Participants and research context*

This study aims to find out the effectiveness of TPD's policy innovation in Indonesia in terms of gender differences by examining correlations between score gain in males versus females, then comparing TPD in different models (the innovative TPD versus traditional TPD).

Participants were all teachers involved in the teacher training program in 2016, divided into three training models: traditional face-to-face, online, and blended (combination of both) with pre-test and post-test scores.

The total number of participants was 427,189 teachers from all levels (kindergarten, elementary school, junior high school, high school, vocational school) of 34 provinces in all districts of the country.

3.2 *Data analysis*

Data was analyzed using SPSS software with a Chi-Square test to see the relationship between score gain (before and after training) with gender of each TPD model. The best gender difference between men and women is shown by the highest percentage result between innovative TPD versus traditional TPD.

Score gain was obtained from difference between post-test and pre-test scores which is classified into three categories, "increase" when the result is positive, "decrease" when the result is negative, and "stable" when the result is 0. The effect size was seen through a Cramer's V and Phi test with the categories small (0.1), medium (0.3) and large (0.5). Given the amount of data, the test was then carried out on clustering in teaching levels.

4 RESULTS

This study aims to answer two research questions. The results were obtained by testing the relationship between increasing the TPD score gain and gender difference using a Chi-Square test. In general (N = 427,189) there is a relationship between the effectiveness of TPD as indicated by the increase of score with gender difference in each TPD model, i.e. online model (N = 63,986), blended model (N = 154,968) and traditional TPD (N = 208,235), with a significance value of $p = 0,000$ so that H_0 is rejected with a small effect size.

4.1 *The relationship between gain scores and gender difference based on teaching levels*

a) Kindergarten school: The Chi-Square test results show that there is no relationship between score gain and gender difference, with a significance value of $p = 0.489$ for the online model, $p = 0.719$ for the blended model, and $p = 0.127$ for the traditional face-to-face model, so H0 was accepted because $\alpha > 0.05$.
b) Elementary school: At the elementary level the test results show that there is a relationship between the increasing score with gender difference with a significance value of $p = 0.000$ in all learning models. This means that H0 is rejected because $\alpha < 0.05$. Size effect is shown by the Phi value = 0.034 and Cramer's V = 0.034 for the online model, Phi and Cramer's V = 0.027 for the blended model, and Phi and Cramer's V = 0.022 for the face-to-face model. This means that the effect is small (< 0.01).

c) Junior high school: Chi-Square test results for teachers in junior high school level show a relationship between increasing score and gender difference for the online model and blended model, with a significance value of p = 0.000, so that H0 was rejected. Whereas in the traditional model it was revealed that there was no relationship between the increase in value with gender difference, with a significance value of p = 1,000. The size effect shown is small in both models: online (0.036) and blended (0.22).

d) High school: For high school teachers, H0 was rejected. There is a relationship between the effectiveness of TPD and gender difference, with a significance value of p = 0.000 in the online model, p = 0.001 for the blended model, and p = 0.009 for the traditional model. The effect size shown by Phi at Cramer's V is small at 0.059 (online), 0.040 (blended), and 0.043 (face-to-face).

e) Vocational school: The results of teacher training at the vocational school level reveal no relationship between the effectiveness of TPD and gender difference for online and face-to-face models, with their respective significance values of p = 0.489 and p = 0.310, so that H0 was accepted (α > 0.05). The blended model H0 was rejected with a significance value of p = 0.002, with a small effect size of Phi and Cramer's = 0.036.

4.2   *TPD effectiveness in males versus females in the innovative TPD versus Traditional model*

Based on the results of the correlation between TPD effectiveness and gender difference, this describes the percentage of scores gain generated by males and females in the innovative TPD (online and blended model) versus traditional TPD face-to-face.

At the elementary level, the best performance was produced by the online model, with 77% for men versus 5.7% for women, while gender difference on other models tended to be non-existent.

At the junior high school level, results for innovative TPD in both models are similar, where women scored higher (80.9%) than men (79.1%).

At the high school level, the best performance for innovative TPD with an online model was by women (85%) compared to men (80.4). Likewise, at the vocational level the correlation was shown only by the blended model, and females scored higher (91.1%) than males (89.1%).

5   DISCUSSION

This study aims to determine the gender difference for innovative TPD by examining the relationship between TPD's effectiveness and gender difference, and seeing whether there are differences in the increase of training scores of men and women in innovative TPD versus traditional TPD.

The results show that there is a relationship between gain scores and gender, and the percentage of increase in women tends to be higher than men at all levels, except at the elementary school level, and innovative TPD fares better than traditional TPD.

These results are consistent with previous studies that innovative TPD increases post-test scores (Fisher et al., 2010) in women compared to men in all TPD models (Kazu & Demirkol, 2014), and provides a rebuttal to the potential for the digital inequality of woman (The Australian Institute for Social Research, 2006). This finding prompts further discussion on the digital divide.

6   CONCLUSION

The effectiveness of innovative TPD, demonstrated by the percentage increases scores before and after training, tends to be higher than traditional TPD. TPD effectiveness has a positive correlation with gender difference, where women tend to be better than men in all TPD models.

REFERENCES

Albion, P.R., Tondeur, J., Forkosh-Baruch, A., & Peeraer, J. (2015). Teachers' professional development for ICT integration: Towards a reciprocal relationship between research and practice. *Education and Information Technologies, 20*(4), 655–673. doi: 10.1007/s10639-015-9401-9

Asterhan, Christa S.C., Schwarz, B.B. and Gil, J. 2012. Small-Group, Computer-Mediated Argumentation in Middle-School Classrooms: The Effects of Gender and Different Types of Online Teacher Guidance. *British Journal of Educational Psychology 82*(3), 375–397.

Bautista, A., & Ortega-Ruiz, R. (2015). Teacher professional development: International perspectives and approaches. *Psychology, Society and Education, 7*(3), 240–251.

Borko, H. (2004). Professional development and teacher learning. *Educational Researcher, 8*(November2004), 3–15. doi: 10.3102/0013189x033008003

Brooks, C., & Gibson, S. (2012). Professional learning in a digital age. *Canadian Journal of Learning and Technology, 38*(2), 1–17.

Cai, Z., Fan, X., & Du, J. (2017). Gender and attitudes toward technology use: A meta-analysis. *Computers and Education, 105*, 1–13. doi: 10.1016/j.compedu.2016.11.003

Chu, T.-H., & Chen, Y.-Y. (2016). With good we become good: Understanding e-learning adoption by theory of planned behaviour and group influences. *Computers & Education, 92–93*, 37–52. doi: 10.1016/j.compedu.2015.09.013

Dede, C. (2006). A research agenda for online teacher professional development. *Journal of Technology and Teacher Education, 14*(4), 657–661. doi: 10.1177/0022487108327554

Deechuay, N., Koul, R., Maneewan, S., & Lerdpornkulrat, T. (2016). Relationship between gender identity, perceived social support for using computers, and computer self-efficacy and value beliefs of undergraduate students. *Education and Information Technologies, 21*(6), 1699–1713. doi: 10.1007/s10639-015-9410-8

Fisher, J.B., Schumaker, J.B., Culbertson, J., & Deshler, D.D. (2010). Effects of a computerized professional development program on teacher and student outcomes. *Journal of Teacher Education, 61*(4), 302–312. doi: 10.1177/0022487110369556

Fishman, B., Konstantopoulos, S., Kubitskey, B.W., Vath, R., Park, G., Johnson, H., & Edelson, D.C. (2013). Comparing the impact of online and face-to-face professional development in the context of curriculum implementation. *Journal of Teacher Education, 64*(5), 426–438. doi: 10.1177/0022487113494413

Guskey, T.R. (2002). Professional development and teacher change. *Teachers and Teaching: Theory and Practice.* doi: 10.1080/135406002100000512

Heller, J.I., Daehler, K.R., Wong, N., Shinohara, M., & Miratrix, L.W. (2012). Differential effects of three professional development models on teacher knowledge and student achievement in elementary science. *Journal of Research in Science Teaching, 49*(3), 333–362. doi: 10.1002/tea.21004

Huda, M.Q., & Hussin, H. (2016). Evaluation model of information technology innovation effectiveness case of higher education institutions in Indonesia. *2016 International Conference on Informatics and Computing (ICIC)*, 221–226. doi: 10.1109/IAC.2016.7905719

Hung, M.-L. 2016. "Teacher Readiness for Online Learning: Scale Development and Teacher Perceptions." *Computers & Education* 94:120–133.

Johnson, W.W. (2014). Why professional development matters. *Journal of Contemporary Criminal Justice, 30*(4), 360–361. doi: 10.1177/1043986214541602

Kachelhoffer, A., & Khine, M.S. (2009). Bridging the digital divide, aiming to become lifelong learners. *IFIP Advances in Information and Communication Technology, 302 AICT*, 229–237. doi: 10.1007/978-3-642-03115-1_24

Kao, C.P., & Tsai, C.C. (2009). Teachers' attitudes toward web-based professional development, with relation to internet self-efficacy and beliefs about web-based learning. *Computers and Education, 53*(1), 66–73. doi: 10.1016/j.compedu.2008.12.019

Kazu, I.Y., & Demirkol, M. (2014). Effect of blended learning environment model on high school students' academic achievement. *The Turkish Online Journal of Educational Technology, 13*(1), 78–87.

Keramati, A., Afshari-Mofrad, M., & Kamrani, A. (2011). The role of readiness factors in e-learning outcomes: An empirical study. *Computers and Education, 57*(3), 1919–1929. doi: 10.1016/j.compedu.2011.04.005

Kunter, M., Klusmann, U., Baumert, J., Richter, D., Voss, T., & Hachfeld, A. (2013). Professional competence of teachers: Effects on instructional quality and student development. *Journal of Educational Psychology, 105*(3), 805–820. doi: 10.1037/a0032583

Lemoff, A.S., Bush, M.F., & Williams, E.R. (2005). Structures of cationized proline analogues: Evidence for the zwitterionic form. *Journal of Physical Chemistry A, 109*(9), 1903–1910. doi: 10.1021/jp0466800

Li, L.-Y., & Tsai, C.-C. (2017). Accessing online learning material: Quantitative behavior patterns and their effects on motivation and learning performance. *Computers & Education, 114*(300), 286–297. doi: 10.1016/j.compedu.2017.07.007

Matzat, U. 2013. Do Blended Virtual Learning Communities Enhance Teachers' Professional Development More than Purely Virtual Ones? A Large-Scale Empirical Comparison. *Computers and Education* 60(1):40–51.

Mohammadyari, S., & Singh, H. (2015). Computers & education understanding the effect of e-learning on individual performance: The role of digital literacy. *Computers & Education, 82*, 11–25. doi: 10.1016/j.compedu.2014.10.025

OECD. (2001). Understanding the digital divide. *Industrial Law Journal, 6*(1), 52–54. doi: 10.1093/ilj/6.1.52

OECD. (2009). Measuring innovation in education and training. *OECD discussion paper.* Retrieved from https://www.oecd.org/edu/ceri/43787562.pdf

Olson, J., Codde, J., DeMaagd, K., Tarkleson, E., Sinclair, J., Yook, S., & Egidio, R. (2011). *An analysis of e-learning impacts & best practices in developing countries: With reference to secondary school education in Tanzania.* Michigan, US: Michigan State University Board of Trustees.

Rogers, E.M. (1995). *Diffusion of innovations. elements of diffusion.* New York, NY: Free Press.

Sánchez-García, A.-B., Marcos, J.-J.M., GuanLin, H., & Escribano, J.P. (2013). Teacher development and ICT: The effectiveness of a training program for in-service school teachers. *Procedia - Social and Behavioral Sciences, 92*(Lumen), 529–534. doi: 10.1016/j.sbspro.2013.08.713

Santos, R., Azevedo, J., & Pedro, L. (2013). Digital divide in higher education students' digital literacy. In S. Kurbanoğlu, E. Grassian, D. Mizrachi, R. Catts, & S. Špiranec (Eds.), *Worldwide Commonalities and Challenges in Information Literacy Research and Practice Vol 397* (pp. 178–183). Cham, Switzerland: Springer. doi: 10.1007/978-3-319-03919-0_22

Smith, N. V. (2013). Face-to-face vs. blended learning: Effects on secondary students 'perceptions and performance. *Procedia – Social and Behavioral Sciences, 89*(2013), 79–83. doi: 10.1016/j.sbspro.2013.08.813

Tarhini, A., Elyas, T., Akour, M.A., & Al-Salti, Z. (2016). Technology, demographic characteristics and e-learning acceptance: A conceptual model based on extended technology acceptance model. *Higher Education Studies, 6*(3), 72–89. doi: 10.5539/hes.v6n3p72

Toit, J. Du. (2015, September). *Background paper for ict in education statistics teacher training and usage of ict in education: New directions for the UIS global data collection in the post-2015 context.* Canada: Unesco Institute for Statistics, United Nations Educational, Scientific and Cultural Organization.

Tondeur, J., van Braak, J., Ertmer, P. A., & Ottenbreit-Leftwich, A. (2016). Understanding the relationship between teachers' pedagogical beliefs and technology use in education: a systematic review of qualitative evidence. *Educational Technology Research and Development, 65*(3), 555–575. doi: 10.1007/s11423-016-9481-2

Twining, P., Raffaghelli, J., Albion, P., & Knezek, D. (2013). Moving education into the digital age: The contribution of teachers' professional development. *Journal of Computer Assisted Learning, 29*(5), 426–437.doi: 10.1111/jcal.12031

The Australian Institute for Social Research. (2006). The digital divide – Barriers to e-learning. *Final Report, presented toDigital Bridge Unit, Science Technology and Innovation Directorate, DFEEST, The Australian Institute for Social Research.*

Van Seters, J.R., Ossevoort, M. A., Tramper, J., & Goedhart, M J. (2012). The influence of student characteristics on the use of adaptive e-learning material. *Computers and Education, 58*(3), 942–952. doi: 10.1016/j.compedu.2011.11.002

Vitanova, V., Atanasova-Pachemska, T., Iliev, D., & Pachemska, S. (2015). Factors affecting the development of ict competencies of teachers in primary schools. *Procedia – Social and Behavioral Sciences, 191*, 1087–1094. doi: 10.1016/j.sbspro.2015.04.344

Walczak, S., & Taylor, N.G. (2018). Geography learning in primary school: Comparing face-to-face versus tablet-based instruction methods. *Computers and Education, 117*, 188–198. doi: 10.1016/j.compedu.2017.11.001

Innovative Teaching and Learning Methods in Educational Systems – Retnowati et al. (Eds)
© 2020 Taylor & Francis Group, London, ISBN 978-1-03-224183-8

Needs identification of learning media for people with disabilities

Rizalulhaq & R. Asnawi
Yogyakarta State University, Yogyakarta, Indonesia

ABSTRACT: The paper aims to determine students' need toward the learning process with the use of media. Needs identification of learning media is a step taken in the research process. The availability of learning media, the use of media by educators, and the appropriate solutions to meet the needs of learning media, is an important factor that was examined at the Bantul Special School 1 in Yogyakarta. This research is a descriptive qualitative research using data collection techniques in the form of observation, interviews and literature studies. Based on the results of observations and interviews, the availability of a medium is minimal and there are still obstacles to its use by educators in the form of students who are lacking in the learning process. The learning process should be conducted in accordance with the plan. Therefore, it is necessary to analyze students' needs by adapting the steps from the Department of National Education (2004): studying the applicable curriculum; determining the competencies of students to be achieved; choosing and determining the material to be presented; choosing and determining the type of learning media.

1 INTRODUCTION

As the 21st century brings in revolutionary change in the way students study at schools and universities, technology continues to play a crucial role in helping students achieve more conceptual and practical knowledge of topics taught in classrooms. Students with special needs are also often able to study in a general classroom setting, access relevant technologies and use them for higher cognitive development, which helps them integrate with their surroundings. However, existing literature shows that although multiple learning tools exist that enhance learning in special needs students, they either cater to specific areas of development such as Mathematics and English, or that are targeted toward a specified category of students with special needs. (Khan, et al., 2016)

The success of an education is closely related to the abilities and learning achievements of students, and students'high learning abilities and achievements show the success of an educational institution in the learning process. The learning achievements of students is influenced by several factors. According to Slameto (2003), the factors that influence learning achievement are divided into internal factors that include physical factors, intelligence, motivation, attention, interests, talents, and readiness and external factors that consist of family, community, learning methods and curriculum factors. In the learning process, educators often have difficulty in explaining a subject to students, for example when comparing objects that are different in nature, type, shape, size, color or there are hidden parts of a tool. Therefore, it is necessary to improve teaching by educators so that in the future students taught can reap better results. Achievement can be obtained by someone when taking formal or informal education through a learning process.

Education is a complex process, along with human development. Various problems in the learning process must be harmonized so that learning conditions are created in accordance with the objectives and can be obtained optimally. But, in reality, the purpose of the learning program does not meet expectations either, because of the factors of educators, students, and infrastructure. In the learning process there are several factors that lead to the failure of the learning program, namely the factors of the educator (educator) and students and the availability of facilities and environmental factors. The first factor is the failure of educators to develop learning media so that it inhibits the delivery of teaching materials. The second factor is that students have different characteristics from one another so that it is necessary to know the needs of each individual. The lack of learning achievement process becomes an obstacle that can be overcome by developing learning media, one of which is learning/module application.

Learning media is a solution used by educators when teaching: educators can explain the material with a model, photo, or video about the object described. In line with the development of computer science and technology, learning media that can provide aspects of animation can improve students' understanding and interest in learning. Learning media can provide simple integration between text and graphics and also animation so that explanations can include videos, animations or text format.

Learning media is a tool or intermediary that is useful to facilitate the teaching and learning process, in order to streamline communication between educators and students. This is very helpful for educators in teaching and makes it easier for students to receive and understand lessons. This process requires educators who are able to harmonize between learning media and learning methods. The use of instructional media in the teaching and learning process can also generate new desires and interests for students, generate learning motivations, and even bring psychological benefits. In addition to being able to increase students' learning motivation, the media can also increase their understanding of the lesson. The media used is a tool for educators in teaching, for example, graphics, films, slides, photos, and learning using computers. The objective is to capture, process, and reconstitute visual and verbal information. As a tool for teaching, the media is expected to provide concrete experience, learning motivation, and enhance the absorption and retention of students' learning.

In the learning process in the classroom, the same method is used as for non-disabled students, so the learning is not maximized. During the learning process, children with special needs have difficulty receiving and understanding the material delivered by educators. It is also the case that not all educators can give special treatment to students with disabilities in overcoming learning difficulties. The lack of special learning media for students with disabilities is also one of the factors that mean students experience difficulties during the learning process.

2 MATERIALS AND METHODS

2.1 *Instructional media*

Materials in a Universal Design for Learning (UDL) classroom are different. These materials will be used to give students multiple means of representing concepts, engaging in learning the concepts, and demonstrating what they have learned. In a UDL classroom, instruction is more flexible and provides accessibility for all students. Teachers who use the principles of UDL in their classroom recognize that instruction does not come as a one-size-fits-all design; for example, digital content can be presented in different ways to meet the learning needs of each student. This content can include hyperlinks, glossaries, graphs, animation, and videos linked within the body of materials to aid understanding and expand content experience (Rose, Meyer, & Hitchcock, 2006). The UDL principles help teachers create classrooms where students can use technologies to move beyond being academic observers. These principles provide a model for self-actuated learning and universal access for all students. Regardless of students' disabilities or differentiated learning styles, every student needs and has the right to access the curriculum (Nelson, 2006). UDL should be part of the initial design of the curriculum, learning environments, and assessments. Pisha and Coyne (2001) call this approach "smart from the start." The study by Voltz, Sims, and Nelson (2010) includes several websites that will further your understanding of UDL.

Media is a communication channel conveyed from the Latin word and means "in-between"; the term refers to anything that carries information between the source and the recipient, such as video, television, diagrams, print materials, computers, and instructors. Media is considered a learning medium when they carry messages with instructional goals. The purpose of the media is to facilitate communication. Since the turn of the 21st century, teachers have used various types of audio and visual aids to help them teach. Recently, teachers have expanded their material reports and procedures to incorporate new technologies for learning, such as computers, compact discs, DVDs, satellite communications, and the internet.

One of the main functions of media is as a tool to facilitate teaching and learning. Arsyad (2011: 15-16), in his book *Learning Media* explained that the use of instructional media during the teaching orientation phase will help the learning process and the effectiveness of the delivery and content of subjects at the time, because it also evokes students'motivation, interest, and helps students to improve understanding, and to present data with interesting, reliable, easy data, and compact information.

Media can serve many roles in learning. Learning often depends on the presence of a teacher (instructor direction), and the media may be widely used by teachers. On the other hand, learning also often does not require a teacher. Student-directed learning is also called "self-learning", even though it is actually guided by whoever designs the media. The most common use of media in learning situations is for additional support from learning, to make the classroom "more alive". Of course, properly designed learning media can enhance and promote learning and support learning taught by the teacher, but their effectiveness depends on learning (Rabinowitz, Blumberg, and Everson, 2004).

In the teaching and learning process, the use of media has a very important function. In general, the function of the media is to channel learning messages/aspirations. It can also arouse curiosity and interest, generate motivation and stimulation and can affect student psychology. The use of media can also help students improve understanding, present material/ data with added interest, and make it easier to interpret data and compress information.

2.2   *Disability*

Disability refers to the loss or limitation of opportunities to take part in the normal life of the community on an equal level with others, due to physical, mental or social factors (GoURT, 2004). The International Classification of Functioning, Disability, and Health (ICF) argued that disability is an umbrella term, covering impairments, activity limitations, and participation restrictions. Impairment is a problem in body function or structure; an activity limitation is a difficulty encountered by an individual in executing a task or action; while a participation restriction is a problem experienced by an individual with involvement in life situations. Thus, disability is a complex phenomenon, reflecting an interaction between features of a person's body and features of the society in which he or she lives (WHO, 2002). (Ngoyani, 2010)

Disability is a complex and controversial problem. In considering the preferences of people with limitations in their lives, and remembering that acceptable terminology changes over time and from one culture to another, the two main terms "disruption" and "disability" are often used synonymously. However, the meaning is different, and it is important to make a difference between the two. "Disruption" is defined as "not having all or part of the body; have limbs, organs or mechanisms that are deformed." Some campaigners with mobility limitations question the use of this term, because of its negative implications; they prefer the more neutral term "conditions". A condition may or may not be considered a disorder and can or does not limit a person's ability to function. Conversely, the term "disability" as used by disabled people organizations (DPOs), emphasizes the public's rejection of the human rights of people who experience the disorder. In the words of the International Disabled Person, "Disability is the loss or limitation of activities caused by contemporary social organizations, which take little or no account of people with disabilities, and thus exclude them from the mainstream of social activities." The differences between the two terms are neatly summarized in the results of discussions issued by the UK Government's Department for International

Development: "Disabled people have long-term disruption that causes social and economic harm, denial of rights, and limited opportunities to play the same part in the life of their community" (Harris and Enfield, 2003).

U.S. federal law defines specific learning disabilities as psychological processing disorders that result in deficits in at least one of the academic skills (U.S. Office of Education, 1977). A child with this label does not have mental retardation, behavior disorders or other major disabilities. The child with LD has difficulty with processing skills such as memory, visual perception, auditory perception, or thinking; and, as a result, has trouble achieving in at least one subject such as reading, math, or writing (Lerner, 2003). Some of the typical characteristics associated with learning disabilities include problems in reading, mathematics, writing, and oral language; deficits in interpreting what is seen or heard; and difficulty with study skills, self-control, self-esteem, memory, and attention (Mercer, 1997; Steele, 2005).

3 METHODS

This research is descriptive qualitative research using data collection techniques and methods such as observation, interviews and literature studies from various supporting sources related to the identification of learning media for people with disabilities.

4 RESULTS AND CONCLUSION

The learning process at SMALB in Bantul Special School 1 in Yogyakarta uses direct speech and practicum methods. Educators first deliver the material using the direct speech method then use different methods for each disabled student to help them understand the learning. After the speech method, students with disabilities must do a practicum to clarify the material presented by educators, with different practicum methods for each student. In this case, the students need special attention during the implementation of this practicum to avoid later problems. Students who are blind, have speech impairments, are mentally disabled, and disabled at the time of practicum participate in the laboratory; but disabled students with visual impairments do not conduct experiments directly, while deaf, speech, grahita and daksa will participate to practice directly although they all need special attention. For visually impaired students, educators should only introduce practicum tools that require students to touch the tools or use braille. Whereas, deaf-speaking students, grahita, and daksa can participate directly with the style of each language.

As educators' awareness of their responsibilities toward accessibility of the learning environment for disabled students increases, significant debate surrounds the implications of accessibility requirements on educational multimedia. There would appear to be widespread concern that the fundamental principles of creating accessible web-based materials seem at odds with the creative and innovative use of multimedia to support learning and teaching, as well as concerns over the time and cost of providing accessibility features that can hold back resource development and application. Yet, the effective use of multimedia offers a way of enhancing the accessibility of the learning environment for many groups of disabled students. Using the development of 'Skills for Access' as an example, which is a web resource supporting the dual aims of creating optimally accessible multimedia for learning, the attitudinal, practical and technical challenges facing the effective use of multimedia as an accessibility aid in a learning environment will be explored. A holistic approach to accessibility may be the most effective in ensuring that multimedia reaches its full potential in enabling and supporting students in learning, regardless of any disability they may have, and this will be outlined and discussed.

Despite the many successes in educational practice for students with disabilities, there are many areas of special education where additional research is needed, especially as they relate to individuals with autism spectrum disorder and students with significant learning and behavioral disabilities. These students require more intensive interventions that are delivered in smaller groups over a longer period of time (D. Fuchs et al., 2014; Pyle & Vaughn, 2012). A systematic review of the research investigating the effects of intensive reading treatments for students with learning disabilities reveals not only progress in this area but also the urgent need for additional research into the effects of intensive treatments in all academic areas. (Vaughn and Swanson, 2015)

REFERENCES

Arsyad, A. (2011). *Media Pembelajaran [Learning Media]*. Jakarta: Raja Grafindo Persada.

Harris, A., & Enfield, S. (2003). *Disability, equality and human rights: A training manual for development and humanitarian organizations*. Oxfam GB.

Khan, Z.R., Ibrahim, Y.Y., Sadhwani, S.C., & Salum, S.T. (2016). *Educational Application for Special Needs is a Learning Tool the Way Forward?* UEA: University of Wollongong.

Ngonyani, M.S. (2010). Teachers' facilitation of learning for learners with disabilities in inclusive Classrooms in Tanzania: Teachers' use of interactive teaching methods in inclusive classrooms. *Thesis.* UNIVERSITETET I OSLO, *Africa*.

Slameto. (2003). *Belajar dan Faktor-faktor yang Mempengaruhinya [Learning and the Affecting Factors]*. Jakarta: Rineka Cipta.

Steele, M.M. (2005). Teaching students with learning disabilities: Constructivism or behaviorism?. *Current Issues in Education, 8*(10).

Rabinowitz, M., Blumberg, F.C., & Everson, H.T. (2004). *The design of instruction and evaluation: Affordances of using media and technology*. Routledge.

Vaughn, S., & Swanson, E.A. (2015). Special education research advances knowledge in education. *Exceptional Children, 82*(1), 11–24.

Voltz, D.L., Sims, M.J., & Nelson, B.P. (2010). *Connecting teachers, students, and standards: Strategies for success in diverse and inclusive classrooms*. USA: ASCD. Department of National Education. 2003. *Department of National Education Policy* 2004.

Implementing gamification to improve students' financial skills in business and management vocational schools

Sukirno, E.M. Sagoro, L.N. Hidayati, Purwanto & D.A.Y Wastari
Universitas Negeri Yogyakarta, Indonesia

ABSTRACT: The purpose of this research is to examine the effectiveness of implementing gamification to improve students' financial skills in business and management vocational schools. The design of this research is Classroom Action Research (CAR) with two cycles, consisting of four stages: planning, action, observation, and reflection. The research subjects were students of Vocational Business and Management in Yogyakarta, Indonesia. The data collection method used consists of tests, documentation, and field notes. The data analysis method is quantitative descriptive analysis. The results showed that the average value of students' financial skills increased from 56.96 at pre-test to 81.78 at post-test. The paired sample t-test test shows that there are differences in the average value of financial skills before and after the action. Thus, it can be concluded that the implementation of gamification can improve financial skills.

1 INTRODUCTION

Learning methods have a role as a tool for extrinsic motivation, as a teaching strategy and as a tool to achieve learning goals (Djamarah & Zain, 2013). The use of appropriate learning methods affects changes in student behavior during the learning process, in knowledge, attitudes, and skills. Business and Management Vocational Students face problems related to low financial skills. Financial skills in this study are those used in completing the accounting cycle.

Student motivation is one of the factors that influences student achievement (Lee, 2010; Al Othman & Shuqair, 2013; Long et al, 2013; Taurina, 2015; Chung & Chang, 2017; Bal Taştan et al. 2018), including skills (Bakar, 2014; Ahmed, 2016). Increasing student motivation is expected to improve students' skills. Learning methods can influence student motivation (Hanrahan, 1998; Harandi, 2015; Taheri et al, 2015). The use of appropriate learning methods can increase student motivation, which in turn will improve student skills.

Gamification can be used as a learning method to increase student motivation. Gamification is defined as a way of using gaming principles in non-game contexts (Attali & Arieli-Attali, 2014; Werbach & Hunter, 2012; Brull & Finlayson, 2016). Gamification can increase student motivation (Buckely & Doyle, 2014; Sagoro, 2016; Papp, 2017; Sailer et al, 2017; Alsawaier, 2018). Gamification is one of the learning methods that can be applied in accounting learning and the use of gamification can improve student skills (Su & Cheng, 2014; Lister, 2015; El Tantawi, Sadaf, & AlHumaid, 2016).

Implementation of gamification in accounting learning is still very limited, but could improve students' financial skills if applied to learning the accounting cycle. At every stage in the accounting cycle, which includes recording, classifying, summarizing, and reporting, students are given different challenges at different levels. Students are rewarded for each challenge resolved.

2 METHOD

This type of research is collaborative and participatory classroom action research (CAR), which means that this research was carried out by researchers who collaborate with accounting teachers. Classroom Action Research has four stages in its implementation: planning, implementing, observing, and reflecting. The research was carried out in two cycles and the four stages above, using the model proposed by Arikunto, Suhardjono & Supardi (2016). The subjects of this study were 34 accounting students of Vocational Business and Management in Yogyakarta, Indonesia.

2.1 Cycle 1

2.1.1 Planning
At this stage the activities to be carried out are agreed with the teacher: drafting the lesson plan using gamification, administering the required learning, preparing and preparing pre-test and post-test questions, preparing matching questions, preparing group lists, prizes and preparing field notes.

2.1.2 Implementation
At the implementation stage the teacher conducts learning activities in accordance with the lesson plan. The implementation activities began with pre-tests, teacher presentations, dividing the students into groups, and group discussions. In the group discussion, the students were given questions and answers to match up. The group would complete the game by pasting the results on the answer board according to the rules. Each question had a level of difficulty along with different points, where the points are accepted by the group if the answer is correct. Points collected were accumulated into the next cycle until there was a winner, who received a reward. After a break, they students were given a post-test question.

2.1.3 Observations
Observations were carried out in conjunction with the implementation of actions or during the learning process. The researchers were assisted by four observers who would record all the things needed during the action.

2.1.4 Reflection
This stage was carried out after the research results were processed and analyzed. This reflection was used to assess activities that had been carried out and evaluate the implementation of these learning activities to be improved in cycle II. The results of this reflection were taken into consideration in planning learning for the next cycle.

2.2 Cycle 2

In the second cycle, the activities were almost the same as in cycle 1, but in the implementation phase in cycle 2 there was discussion and a game playing "right or wrong". The actions in cycle 2 were improved based on the results of the reflection of cycle 1 to achieve indicators of success.

2.3 Data collection

Techniques of data collection in this study were as follows:

2.3.1 Test
The test in this study was used to measure students' financial skills for accounting cycles, in the form of a pre-test and post-test, before using gamification. This post-test is also used to determine the increase of student's financial skills in cycle 1 and cycle 2.

2.3.1 Field notes
This field notes were used to collect data of various aspects in the learning process, classroom atmosphere, teacher and student interaction, and student interaction with students during the research.

2.3.1 Documentation

This documentation is a data collection technique collected by researchers in the form of activity photos, the syllabus, lesson plans, student lists and student value lists, which are used as a basis for forming group during the research implementation.

2.4 Data analysis

Data was analysis was done using quantitative descriptive data, with percentages used to analyze the financial skills data. Data was processed based on the results of the pre-test and post-test in each cycle. The processed data is then analyzed to calculate the improvement of financial skills by using the following formula:

$$Me = \frac{\sum xi}{N} \tag{1}$$

where Me is the average (mean), $\sum xi$ is the sum of all values, and N is the number of individuals (Sugiyono, 2012).

After obtaining the average student score, the percentage of student learning completeness was calculating using the following formula:

$$KB = \frac{T}{Tt} \times 100\% \tag{2}$$

where KB is the completeness of learning, T is the nNumber of students who meet KKM (≥ 80), and Tt is the number of students taking the test (Trianto, 2009).

3 DISCUSSION

The implementation of the learning process by applying gamification in cycle 1 and cycle 2 shows an increase in students' financial skills. The data obtained in the first cycle and second cycle was used for processing, and only 28 students participated in both cycles. To answer the action hypothesis, a discussion was held on financial skills after the implementation of gamification. Table 1 and Figure 1 show the data obtained and the average value of students' financial skills in Business and Management Vocational in cycle 1 and cycle 2.

Table 2 shows the completeness of the financial skills in cycle 1 and cycle 2.

Table 1 shows that there has been an increase in the average learning outcomes in the first cycle and second cycle after the implementation of gamification. The average value at pre-test was 56.96, increasing by 18.57 or 32.60% to 75.53, then in the second cycle it increased by 6.26 or 8.28% to 81.79. Increases in students' financial skills can also be seen from cycle I and cycle II. These results show that the implementation of gamification can improve student s' financial skills.

Table 2 shows that there is an increase in the percentage of completeness of student financial skills from pre-test to the second post-test. The percentage of students who

Table 1. Average value of financial skills in cycle 1 and cycle 2.

Information	Pre-test	Post-test Cycle 1	Post-test Cycle 2
Value	1595	2115	2290
Average class value	56.96	75.53	81.79

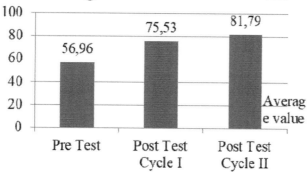

Figure 1.   Average value of financial skills in cycle 1 and cycle 2.

Table 2.   Completeness of financial skills in cycle 1 and cycle 2.

Value catagory	Complete learning outcomes		
	Pre-test	*Post-test* Cycle 1	*Post-test* Cycle 2
Value < 80	89.29%	57.14%	17.86%
Value ≥ 80	10.71%	42.86%	82.14%

reached the minimal criteria in the pre-test was 10.71%, which increased to 42.86%, then in the post-test cycle 2 grew to 82.14%. In the post-test cycle 2, 82.14% or 23 students reached the minimal criteria. Based on Table 1 and Table 2, it can be concluded that the implementation of gamification can improve the students' financial skills.

Another analysis was made using a paired sample t-test, which is based on the financial skills at pre-test and post-test 2. The results of the paired sample t-test calculations, made with the help of a computer program, show that the value of Sig. (2-tailed) is 0,000, because Sig. (2-tailed) <0.05, which means that there is a difference in students' financial skills between pre-test and post-test data. This means that there is an influence of gamification in improving the financial skills for accounting students.

4   CONCLUSION

The implementation of gamification can improve students' financial skills in a Vocational Business and Management course. The average value at pre-test was 56.96, which increased by 18.57 or 32.60% to 75.53 then increased by 6.26 or 8.28% to 81.79. Completeness of student learning outcomes in the pre-test was 10.71% then increased to 42.86% then increased again to 82.14%. In the post-test cycle 2, 82.14% or 23 students reached the minimal criteria. A paired sample t-test showed that the Sig. (2-tailed) value is 0,000, because Sig. (2-tailed) <0,005, which means that there are differences in students' financial skills between pre-test and post-test data, which means that gamification has an effect on improving students' financial skills in Vocational Business and Management at Yogyakarta, Indonesia.

Gamification should be used more often because it can improve financial skills or other learning goals, such as learning outcomes, motivation, engagement, etc. However, gamification should not be implemented in a short lesson, so that the learning process is more optimal. Th teacher should have time to the deliver material and students should have enough time for discussion, question and answer. Future research should examine not only financial skills nut also student learning activities or student motivation.

REFERENCES

Ahmed, Z.A.A. (2016). The Effect of Motivation on Developing EFL Learners' Reading Comprehension Skills. *International Journal of English Language Teaching, 4*(10), 1–9.

Alsawaier, R.S. (2018). The effect of gamification on motivation and engagement. *International Journal of Information and Learning Technology, 35*(1), 56–79.

Al Othman, F.H., & Shuqair, K.M. (2013). The impact of motivation on English language learning in the Gulf States. *International Journal of Higher Education, 2*(4), 123.

Arikunto, Suhardjono & Supardi. (2016). *Penelitian Tindakan Kelas [Classroom action research.].* Jakarta. PT Bumi Aksara.

Attali, Y., & Arieli-Attali, M. (2015). Gamification in assessment: Do points affect test performance? *Computers & Education, 83*, 57–63.

Bakar, R. (2014). The Effect of Learning Motivation on Student's Productive Competencies in Vocational High School, West Sumatra. *International Journal of Asian Social Science, 4*(6), 722–732.

Bal Taştan, S., Davoudi, S.M.M., Masalimova, A., Bersanov, A., Kurbanov, R., Boiarchuk, A., & Pavlushin, A. (2018). The Impacts of Teacher's Efficacy and Motivation on Student's Academic Achievement in Science Education among Secondary and High School Students. *Eurasia Journal of Mathematics, Science and Technology Education, 14*(6), 2353–2366.

Brull, S., & Finlayson, S. (2016). Importance of Gamification in Increasing Learning. The *Journal of Continuing Education in Nursing, 47*(8), 372–375.

Buckley, P., & Doyle, E. (2014). Gamification and student motivation. *Interactive Learning Environments, 24*(6), 1162–1175.

Chung, L.-Y., & Chang, R.-C. (2017). The Effect of Gender on Motivation and Student Achievement in Digital Game-based Learning: A Case Study of a Contented-Based Classroom. *Eurasia Journal of Mathematics, Science and Technology Education, 13*(6), 2309–2327.

Djamarah, S.B., & Zain, A. (2013). *Strategi Belajar Mengajar [Teaching and Learning Strategies].* Jakarta: PT Rineka Cipta.

El Tantawi, M., Sadaf, S., & AlHumaid, J. (2016). Using gamification to develop academic writing skills in dental undergraduate students. *European Journal of Dental Education, 22*(1), 15–22.

Hanrahan, M. (1998). The effect of learning environment factors on students' motivation and learning. *International Journal of Science Education 20*(6), 737–753.

Harandi, S.R. (2015). Effects of e-learning on students' motivation. *Procedia - Social and Behavioral Sciences 181* (2015), 423–430.

Lee, I.C. (2010). The Effect of Learning Motivation, Total Quality Teaching and Peer-Assisted Learning on Study Achievement: Empirical Analysis from Vocational Universities or Colleges' students in Taiwan. *The Journal of Human Resource and Adult Learning, 6*, (2).

Lister, M. (2015). Gamification: The effect on student motivation and performance at the post-secondary level. *Issues and Trends in Educational Technology, 3*(2).

Long, C., Ming, Z., & Chen, L. (2013). The Study of Student Motivation on English Learning in Junior Middle School – A Case Study of No.5 Middle School in Gejiu. *English Language Teaching, 6*(9).

Papp, T.A. (2017). Gamification Effects on Motivation and Learning: Application to Primary and College Students. *International Journal for Cross-Disciplinary Subjects in Education (IJCDSE), 8*(3).

Sagoro, E.M. (2016). Keefektifan Pembelajaran Kooperatif Berbasis Gamifikasi Akuntansi Pada Mahasiswa Non-Akuntansi [Effectiveness of Cooperative Learning Based on Accounting Gamification in Non-Accounting Students]. *Jurnal Pendidikan Akuntansi Indonesia;14* (2), 63–79.

Sailer, M., Hense, J.U., Mayr, S.K., & Mandl, H. (2017). How gamification motivates: An experimental study of the effects of specific game design elements on psychological need satisfaction. *Computers in Human Behavior, 69*, 371–380.

Su, C.-H., & Cheng, C.-H. (2014). A mobile gamification learning system for improving the learning motivation and achievements. *Journal of Computer Assisted Learning, 31*(3), 268–286.

Sugiyono. (2012). *Statistika untuk Penelitian [Statistics for Research].* Bandung: Alfabeta.

Taheri, M, Nasiri, E, Moaddab, F, Nayebi, N, Asadi Louyeh, A. (2015). Strategies to Improve Students' Educational Achievement Motivation at Guilan University of Medical Sciences. *Res Dev Med Educ;4* (2), 133–139.

Taurina, Z. (2015). Students' Motivation and Learning Outcomes: Significant Factors in Internal Study Quality Assurance System. *International Journal for Cross-Disciplinary Subjects in Education (IJCDSE), 5*(4), 2625–2630.

Trianto. (2009). *Mendesain Model Pembelajaran Inovatif-Progresif: Konsep, Landasan, dan Implementasi-nya pada Kurikulum Tingkat Satuan Pendidikan (KTSP) [Designing Innovative-Progressive Learning Models: Concepts, Platforms, and Implications in the Education Unit Level Curriculum (KTSP)]*. Jakarta: Kencana.

Werbach, K., & Hunter, D. (2012). *For the win: How game thinking can revolutionize your business*. Philadelphia, PA: Wharton Digital Press.

Author Index

Albana, L.F.A.N.F. 141
Ambarini, G. 37
Amiruddin, M.H.B. 3
Andrijati, N. 52
Ashadi 37
Asnawi, R. 82
Asnawi, R. 237

Bautista, A. 151
Budiningsih, C.A. 24

Cholifah, P.S. 184
Cholily, Y.M. 179
Chua, S-L. 151

Darmawanti, S. 12
Darwin, M. 231
Djatmiko, I.W. 209

Eriyanti, R.W. 174
Erviana, V.Y. 104

Fatmawati, L. 104

Ghufron, A. 129, 197
Guntoro, D.W. 159

Hadi, S. 135
Haryadi, S. 45
Hermawan, H.D. 32
Hermawati, D. 104
Hidayati, L.N. 242

Indartono, S. 159

Jati, H. 201

Keumalasari, R. 168
Kurniawan, H. 32
Kusnandi 97

Mahanani, P. 69
Mahendra, I.G.B. 59
Manalu, I.A. 82
Mardapi, D. 52
Marsudi, K.E.R. 113
Marsyaly, F.P. 135
Maryani, I. 104
Marzuki 77
Maulida, R.S. 18
Mulyani, H. 209
Mustadi, A. 104

Novamizanti, L. 190
Nursusanto, U. 223

Oksa, S. 122
Oktaviani, H.I. 184
Öztürk, T. 91

Pierewan, A.C. 231
Purwanto 242
Putri, R. 129

Ramadani, W. 216
Rauzah 97

Razali, N.B. 3
Razzaq, A.R.A. 3
Retnawati, H. 52
Rini, T.A. 69
Rizalulhaq 237
Rohiat, M.A. 3

Saputri Sukirno, A. 32
Sagoro, E.M. 242
Setyanto, B.N. 201
Sholekhah, I. 159
Soenarto, S. 122
Sugiman 24
Sujarwo 141
Sukirno 242
Sunarso, S. 113
Susilastuti, D.H. 231

Tan, C. 151
Taruno, D.L.B. 216
Tisngati, U. 24
Turmudi 168

Usman, H. 12, 18

Wangid, M.N. 104
Wastari, D.A.Y 242
Wellyana 77
Wiyono, G. 59
Wong, J. 151
Wuryaningsih, 231

Yuniarti, N. 223